AUTOMOTIVE FUEL ECONOMY Part 2

(Selected SAE Papers through 1979)

Prepared under the auspices of the
Fuel Economy Advisory Committee

Published by:
Society of Automotive Engineers, Inc.
400 Commonwealth Dr., Warrendale, PA 15096

The appearance of the code at the bottom of the first page of each article in this volume indicates SAE's consent that copies of an article may be made for personal or internal use, or for the personal or internal use of specific clients. This consent is given on the condition, however, that the copier pay the stated per article copy fee through the Copyright Clearance Center, Inc., Operations Center, P.O. Box 765, Schenectady, NY 12301 for copying beyond that permitted by Sections 107 or 108 of the U.S. Copyright Law. This consent does not extend to other kinds of copying such as copying for general distribution, for advertising or promotional purposes, for creating new collective works, or for resale.

Articles published prior to 1978 in similar SAE collective works may also be copied at a per article fee of $2.50 under the above stated conditions.

To obtain quantity reprint rates, permission to reprint an article, or permission to use copyrighted SAE publications in other works, contact the SAE Publications Division.

ISBN 0-89883-106-7
SAE/PT-79/18
Copyright © 1979 Society of Automotive Engineers, Inc.
Library of Congress Catalog Card Number: 76-25691
Printed in USA

PREFACE

SINCE 1973, FUEL ECONOMY has become a primary consideration in automotive design. Fuel shortages identified by long lines at the few open gas stations and a profusion of "out of gas" signs at the others brought home a message to the driving public that has radically shifted customer demand patterns for new cars. Federal regulation was also applied to force a pattern of fuel economy reduction on new model cars and Corporate Average Fuel Economy has become a critical target for all auto manufacturers.

Fuel economy improvements have been attained in a great number of innovative ways since the petroleum crisis first thrust its way to the attention of the American public. Engineers have documented these techniques in a large number of SAE papers, some of which were compiled in an SAE publication in 1976 titled "Automotive Fuel Economy." The book was Volume 15 in the Progress in Technology Series and it included 22 papers selected from all SAE fuel economy publications over the 10 year period from 1965 to 1975. The selected papers were subdivided into five categories of topics as follows:

1. Fuel Economy Test Procedures
2. Vehicle Usage Factors Affecting Fuel Economy
3. Vehicle Design Factors Affecting Fuel Economy
4. Fuel and Lubricant Effects on Fuel Economy
5. Analysis of Fuel Economy

Since that publication, in the short span of three years, over 80 additional papers have been presented on this critically important subject. This volume (PT-18) contains 17 of these papers in their entirety and abstracts of 25 others which the Fuel Economy Advisory Committee selected as being valuable references. Selections were based on the evaluation of each paper by qualified Readers Committees, as well as on the Fuel Economy Advisory Committee's judgement of how a paper rated in the five topic categories.

Ralph C. Stahman
A. P. Stanley Hyde
Richard F. Irwin
Fuel Economy Advisory Committee

Editorial Advisory Board Preparing PT-18

A. P. STANLEY HYDE
GENERAL MOTORS CORPORATION

RICHARD F. IRWIN
CHEVRON OIL COMPANY

RALPH C. STAHMAN
UNITED STATES ENVIRONMENTAL PROTECTION AGENCY

TABLE OF CONTENTS

Fuel Economy Test Procedures

Prediction of Dynamometer Power Absorption to Simulate Light Duty Truck
Road Load, Glenn D. Thompson 3

Vehicle Usage Factors Affecting Fuel Economy

Soak Time Effects on Car Emissions and Fuel Economy, Robert L. Srubar,
Karl J. Springer, and Martin E. Reineman 37
Ambient Temperature and Trip Length—Influence on Automotive Fuel
Economy and Emissions, B. H. Eccleston and R. W. Hurn 55
Urban Traffic Fuel Economy and Emissions—Consistency of Various
Measurements, Leonard Evans 65

Vehicle Design Factors Affecting Fuel Economy

Effects of the Degree of Fuel Atomization on Single-Cylinder Engine
Performance, William R. Matthes and Ralph N. McGill 79
Hybrid Vehicle for Fuel Economy, L. E. Unnewehr, J. E. Auiler,
L. R. Foote, D. F. Moyer, and H. L. Stadler 101
An Analytical Study of the Fuel Economy and Emissions of a Gas
Turbine-Electric Hybrid Vehicle, Sidney G. Liddle 119
The Optimization of Body Details—A Method for Reducing the Aerodynamic
Drag of Road Vehicles, W. H. Hucho, L. J. Janssen, and H. J. Emmelmann ... 133
An Analytical Study of Transmission Modifications as Related to Vehicle
Performance and Economy, Howard E. Chana, William L. Fedewa, and
John E. Mahoney 151
An Overall Design Approach to Improving Passenger Car Fuel Economy,
Edward K. Hanson 159
Engine Air Control—Basis of a Vehicular Systems Control Hierarchy,
Donald R. Stivender 181
Energy Conservation with Increased Compression Ratio and Electronic
Knock Control, James H. Currie, David S. Grossman, and James J. Gumbleton ... 213
A Fuel Economy Development Vehicle with Electronic Programmed Engine
Controls (EPEC), Bruce D. Lockhart 223
The Effects of Varying Combustion Rate in Spark Ignited Engines,
R. H. Thring 229

Fuel and Lubricant Effects on Fuel Economy

Fuel Economy Improvements in EPA and Road Tests with Engine Oil and
Rear Axle Lubricant Viscosity Reduction, Malcolm C. Goodwin and
Merrill L. Haviland 243

Analysis of Fuel Economy

Fuel Economy of Alternative Automotive Engines - Learning Curves and
Projections, Roy Renner and Harold M. Siegel 265
Light Duty Automotive Fuel Economy .... Trends Through 1979,
J. D. Murrell 279
Bibliography 309

Fuel Economy Test Procedures

Prediction of Dynamometer Power Absorption to Simulate Light Duty Truck Road Load*

Glenn D. Thompson
United States Environmental Protection Agency

WHEN VEHICLE EXHAUST EMISSION TESTS or vehicle fuel consumption measurements are performed on a chassis dynamometer, the dynamometer is usually adjusted to simulate the road experience of the vehicle. Specifically the dynamometer must simulate the road load of the vehicle.

The purpose of this study was to develop equations to predict the dynamometer adjustment forces appropriate to simulate the on road experiences of light duty trucks. To accomplish this, equations of road load versus speed were obtained from a diverse class of light duty trucks. Fifteen light duty trucks were chosen as the experimental sample. These trucks were chosen to approximately represent the sales weighting of light duty trucks. Each of the 15 trucks was track tested with varying payloads, such that the vehicle test weights ranged from the empty vehicle weight to the GVW. This resulted in a total of approximately 50 track tests. The test fleet is described in table 1.

The track measurements include the dissipative losses of the vehicle tires, wheel bearings and drive train. To determine a load value appropriate for adjusting a chassis dynamometer, the dissipative losses from the drive train and driving tires must be subtracted from the total system measurements. These dissipative losses were measured using a 48" diameter single roll electric dynamometer.

THE MEASUREMENTS

Commonly used methods for road-load determination are: the deceleration or coast down technique, drive line force or torque measurements, and manifold pressure measurements. The coast down method was selected as the approach best suited for this study since a method easily adaptable to a diverse class of vehicles was required. The concept of the coast down technique is to determine the rate of deceleration of a freely coasting vehicle; then, knowing the mass of the vehicle, the

---

*Paper 770844 presented at the Passenger Car Meeting, Detroit, September 1977.

ABSTRACT

When vehicle exhaust emission tests or vehicle fuel consumption measurements are performed on a chassis dynamometer, the dynamometer is usually adjusted to simulate the road experience of the vehicle. In this study, road load versus speed data were obtained from 15 light duty trucks. The road load of each truck was determined for different payloads, resulting in a total of approximately 50 road load measurements.

Dynamometer power absorption settings to simulate the measured road loads are computed. These dynamometer settings are regressed against vehicle frontal area and vehicle inertia weight. It is concluded that the dynamometer load settings are most accurately predicted on the basis of the vehicle frontal area.

Table 1 - Vehicle Identification

Identification Number	Vehicle Type	EPA Estimated Inertia Weight Class (lbs)	Frontal Area (ft^2)
5902	Ford F-100	4500	31.4
7001	Chevrolet Cheyenne 20	5000	32.0
7101	Chevrolet Scottsdale 10	4500	31.6
7201	Chevrolet Cheyenne 10	4500	31.6
7301	Chevrolet G-20 Van	4000	37.0
7402	Ford F-100 4X4	5000	31.4
7502	Ford F-250	4500	33.6
7602	Ford F-100 R XLT	5000	31.4
7701	Chevrolet G-10 Van	4500	37.0
7901	Chevrolet G-30 Van	6000	37.0
8002	Ford E-150 Van	4500	37.7
8202	Toyota Hilux SR-5	3000	21.2
8306	Toyota Hilux 2	3000	21.2
8507	Datsun	2500	21.2
9003	Dodge Tradesman 100 Van	4000	35.1

road-load force may be calculated by Newton's second law;

$$F = MA \quad (1)$$

TRACK MEASUREMENTS - Experimentally, it is not practical to measure the vehicle acceleration directly; however, the acceleration may be determined from the vehicle speed versus time record. Therefore the speed versus time data are the only measurements that are required on the test track. Ambient conditions were, however, also monitored to allow correction to a set of standard ambient conditions.

<u>Test Facility and Test Procedure</u> - All vehicle speed versus time data were collected on the skid pad of the Transportation Research Center of Ohio, in East Liberty, Ohio. This facility is a multilane, concrete, straight track with large turn around loops at each end. Approximately 1 kilometer of this straight track has a constant grade of 0.5% and this section was used for all measurements.

Prior to the coast down measurements, the vehicle tires were adjusted, when cold, to the manufacturers recommended pressures. After adjustment of the tire pressures, the vehicles were warmed up for approximately 30 minutes at about 50 mph.

Twenty coast downs were recorded for each vehicle at each test weight; ten in each direction of travel on the test track. Ten coast downs were conducted by accelerating the vehicle to approximately 65 mph, then shifting into neutral and recording speed versus time as the vehicle freely decelerated. The remaining ten coast downs were conducted in the same manner; however, the initial speed was approximately 40 mph. The two series of coast downs were necessary because the 1 km section of constant grade track was insufficient to coast most vehicles from 60 mph to a terminal speed near ten mph.

<u>Velocity Instrumentation</u> - The vehicle speed was measured by a police type Doppler radar. The instrumentation contained a noise discriminator system which rejected the Doppler pulse count any time the period between pulses differed significantly from the previous pulse separation.

Modifications were made to the standard configuration to increase the range. The length of the antenna horn was increased and aluminum corner reflectors, or strips of aluminum foil, were placed inside the target vehicle windows. These modifications increased the range from about 0.5 km to approximately 1.0 km. The Doppler frequency counter gate time was also increased from approximately 30 msec to 300 msec in an attempt to improve the system precision. This modification did increase the speed resolution; however, it also increased the total period the discriminator evaluated the Doppler signal for extraneous noise. The system noise is basically random; therefore, the probability the discriminator will reject a measurement of the Doppler frequency is linear with the counter gate time. The increase in the precision of each measurement was accompanied by a decrease in the number of speed versus time points measured during the coast down. Also, the range was greatly reduced since the probability of radar signal noise increases as the distance from the transmitter to the target increased. This modification was subsequently rejected and the final configuration of the system provided a range of about 1 km with a resolution of $\pm$ 1 mph ($\pm$ 1.6 km/hr).

A count of the Doppler frequency was recorded each second during the coast downs on a seven track digital magnetic tape recorder. This recorder and the support electronics were placed in a small van, parked on the track berm. Electric power was provided by an alternator, battery bank, and inverter on this van. An example of the speed versus time record of a light duty truck coast down is given in Figure 1.

<u>Ambient Conditions</u> - Coastdowns were conducted only when steady winds were less than 15 km/hr (9.3 mph) with peak wind speeds less than 20 km/hr (12.4 mph). Wind speed during the test period was measured with a photochopper type six-cup anemometer. The anemometer was located near one side of the test track, at one end of the 1 km test section. These data were recorded at one second intervals on the same magnetic tape as was used to record the vehicle speed. During test periods the ambient temperature was in the range of 5°C (41°F) to 35°C (95°F). The barometric pressure was between 102 kPa (30.2 in Hg) and 94 kPA (27.9 in Hg). The air moisture content ranged from 0.29 to 0.73 gm H_2O/kg dry air. These slowly varying ambient parameters were recorded by an observer, on a data sheet associated with each vehicle.

MASS MEASUREMENTS - In addition to the vehicle acceleration, the system mass must be measured in order to compute the road load force. The system mass must include the additional "effective equivalent mass" of the vehicle rotating components. The concept of effective equivalent mass is simply a convenient approach to include the dissipation of rotational kinetic energy from the tires, wheels and drive train. A precise statement of the equivalent effective mass is derived in section one of Appendix A.

The Gravitational Masses - The gravitational mass was measured by weighting each vehicle, with the driver, immediately after the coast downs. The vehicle scale of the TRC was used for all vehicle mass determinations. TRC personnel indicated calibration checks on this scale have repeatedly been within $\pm$ 10 pounds in the 0 to 10,000 pound range.

Equivalent Effective Mass of the Drive Wheels and Drive Train - The equivalent effective mass of the drive wheels, differential gears, drive shaft and transmission output shaft can be calculated if the dissipative forces acting on the drive train, and the time required for the drive train to freely coast through a speed interval are known. With this approach, the total effective equivalent mass of all drive train components is given by:

$$m_{Req} = \frac{\Delta t}{\int_{v_2}^{v_1} \frac{dv}{F}} \qquad (2)$$

where
m_{Req} = the effective equivalent mass of the drive train and driving tires
Δt is the time interval
F = the force at the tire roll interface
v_1 = the initial simulated vehicle speed
v_2 = the final simulated vehicle speed
The derivation of this equation is given in section 1 of Appendix A.

The force measurements were obtained by motoring the vehicle tires and drive train on an electric dynamometer. The dynamometer measurements are discussed in detail in the following section. The vehicle tire speed was monitored by chalking several white stripes on each vehicle driving tire and observing these chalk stripes with a stroboscopic tachometer. The appropriate frequencies corresponding to 60 mph and 10 mph were determined by observing the tachometer when motoring each vehicle with the dynamometer at these speeds. The vehicle drive train, with the transmission in neutral, was motored to approximately 65 mph by the dynamometer. The vehicle was lifted from the dynamometer and the time interval required for the vehicle to decelerate from 60 mph to 10 mph was timed using a stop watch. The initial and final speed points were determined by observing the frequency output of the strobocopic tachometer. Five time intervals were recorded for each drive wheel.

Equivalent Effective Mass of the Vehicle Non-Driving Wheels - The technique of motoring,

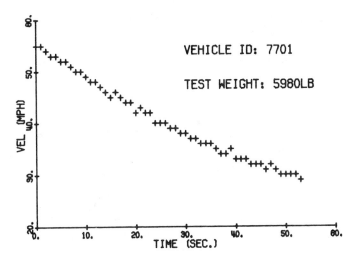

Fig. 1 - Example of typical coastdown speed versus time data

followed by coast down, was not used for the non-driving wheel effective mass determination because of the very low forces required to motor the front wheels. The equivalent effective masses of the front wheels were determined by the three wire torsional pendulum technique. With this method, the object is suspended by three wires, or placed on a platform suspended by three wires. The platform is rotated through a small angle, released and the period of the oscillations are timed. The tension in the supporting cables, induced by the mass of the object, causes a restoring torque on the oscillating system; the inertia of the object controls the acceleration of the system and, hence, the period of the oscillation. The total equivalent effective mass of the two non-driving tires is given by:

$$m_{feq} = \frac{g \; \ell^2}{2 \, \pi^2 \, LR^2} (m \tau_t^2 - m_p \tau_p^2) \qquad (3)$$

where
m_{feq} = the effective equivalent mass of both front wheels
g = the gravitional constant, 9.80 m/sec^2
ℓ = the length from the center of the pendulum to the point of attachment of the supporting wires, 0.532 m
L = the length of the supporting wires, 4.79 m
m = the total mass of the pendulum and tire wheel
τ_t = the time period of one oscillation of the pendulum with the tire and wheel
τ_p = the time period for one oscillation of the pendulum only
m_p = the mass of the pendulum only
R = the rolling radius of the tire

The derviation of equation 3 is given in section 2 of Appendix A.

For the determination of the rotational inertia of the non-driving wheel a vehicle tire and wheel, usually the spare, was placed on the platform. The platform was rotated approximately 0.5 radians and released. The time required for 10 oscillations was then timed with a stop watch. This measurement was repeated 4 times and then averaged.

The mass of each tire was then measured by weighing on a platform scale. This scale was a "shipping clerks" scale with a maximum capacity of 1000 lbs, and a resolution of $\pm$ 0.5 lb.

DYNAMOMETER MEASUREMENTS - The previously discussed measurements are sufficient to calculate the total vehicle road load. However, the total vehicle road load force includes the dissipation in the drive train from the rear wheels up to the point where the drive train is decoupled from the engine. When the vehicle is being tested on a dynamometer, the vehicle engine is required to overcome the drive train and driving tire losses prior to supplying power to the dynamometer. Consequently these losses should be subtracted from the total road load to determine the dynamometer adjustment.

The drive train and tire losses were measured by motoring the vehicle with a dynamometer, and recording the required dynamometer torques. The dynamometer used was one of the EPA light duty vehicle electric dynamometers. This dynamometer is a G.E. motor-generator type with a 48" diameter single roll. During these experiments the normal 0-1000 lb. load cell of the dynamometer was replaced with a more sensitive 0-300 lb. load cell.

Prior to all measurements the cold tire pressures were adjusted to the manufacturers recommended pressures. The vehicle weight was adjusted to be the same as the vehicle weight during the corresponding track measurement. The dynamometer force measurements were conducted on both the front and rear axles of the vehicle. During the rear axle measurements the transmission was shifted into neutral, as it was during the track coastdowns.

The vehicle was placed on the dynamometer, and then the vehicle and dynamometer were warmed up for 30 minutes at approximately 50 mph. After warm up, the torque necessary to motor the dynamometer and vehicle was measured at speeds from 60 to 10 mph in 5 mph decreasing speed intervals. For each measurement steady state dynamometer speed and torque signals were recorded on a strip chart for a period of approximately 100 seconds. The stabilized values were then read from the strip chart by the dynamometer operator.

After the measurements were completed with the full vehicle weight resting on the dynamometer rolls, the vehicle was then lifted until the vehicle tires were just contacting the dynamometer roll. The vehicle tires were considered to be just touching the dynamometer roll if a person could, with difficulty, manually cause the tire to slip on the roll when the roll was locked. With this test configuration the torque versus speed measurements were repeated as before. Finally, the torque required to motor only the dynamometer was recorded in the same manner.

The dynamometer speed data were converted to the units of m/sec. All torque data were converted to force in newtons at the tire-roll interface. A scatterplot of the data from one truck, after conversion to force at the tire-roll interface and subtraction of the force neccessary to motor the dynamometer, is given as an example in Figure 2. In addition, the difference between the force measurements when the full weight of the vehicle was on the dynamometer and the force measurements when the tire was just contacting the dyno roll, is also given in Figure 2.

DATA ANALYSIS

The speed versus time data must be analyzed to yield acceleration versus speed information. The various mass related measurements are reduced to a single total equivalent effective system mass. The product of the acceleration and the total equivalent effective mass then give the total vehicle road load force as a function of vehicle speed.

The dynamometer measurements are analyzed to yield drive train and tire dissipative forces as a function of speed. These forces are then be subtracted from the total road load forces to give the appropriate dynamometer force adjustment to simulate the vehicle road experience.

TRACK DATA ANALYSIS - The usual form of a vehicle deceleration curve is assumed to be a constant plus a term proportional to the velocity squared. However, the effect of a steady head-tail wind will appear as a linear-term. Figure 2 shows that the drive train losses appear to increase linearly with increasing speed. Also some published tire data (1)* have indicated the inclusion of a linear term may be desirable. For these reasons, a model equation was chosen of the form:

$$dv/dt = a_0 + a_1 v + a_2 v^2 \qquad (4)$$

Terms were added to equation 4 to account for effects of wind and track grade. The variables of the resulting equation can be separated and integrated to yield an expression for time as a function of velocity. Since these functions are inverse trigonometric or hyperbolic functions, their inverse may be taken to yield velocity as a function of time. The mathematics of the approach has been previously discussed in detail, (2) however, an extensive outline of the approach is presented in section 3 of Appendix A.

---

*Numbers in parentheses designate References at end of paper.

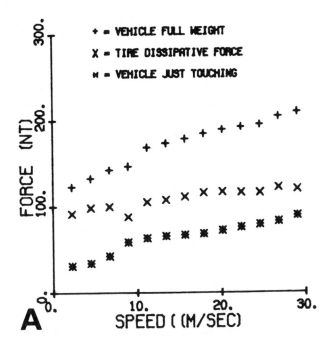

Fig. 2A - Examples of typical rear axel force measurements

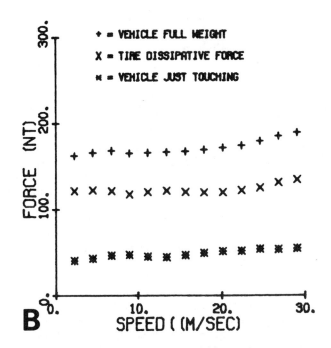

Fig. 2B - Examples of typical front axel force measurements

The velocity functions were fitted to the coast down data by the method of generalized least squares to determine the a_0, a_1, and a_2 of equation 4. By incorporating a directional variable to treat the effects of track grade and head-tail winds, all data sets were analyzed simultaneously. The fitting algorithm established values of parameters related to the initial conditions of each run and the three desired coefficients common to all runs.

Since the a_2 coefficient multiplies v^2, it is assumed to represent the aerodynamic drag of the vehicle. The aerodynamic drag is proportional to the air density; therefore all a_2 coefficients were corrected for differences between the ambient conditions during the test, and a set of standard ambient conditions chosen to be:

temperature	20°C (68°F)
barometric pressure	98 kPa (29.02 in Hg)
humidity	10 gm H_2O/kg dry air

The corrected acceleration coefficients for all vehicle tests are presented in table 1 of Appendix B.

THE TOTAL EFFECTIVE EQUIVALENT MASS OF THE VEHICLE - The total effective equivalent mass of the vehicle is the sum of the gravitational mass of the vehicle and the effective equivalent mass of the drive tires, drive train and non-driving wheels.

The Gravitational Mass - The gravitational mass was determined by the vehicle scale at the TRC. These data are presented in Table 2 of Appendix B.

The Drive Train Effective Equivalent Mass - The equivalent effective mass of the vehicle drive train and drive tire was determined by equation 2 from the drive train force and coast down measurements. The integral in the denominator of equation 2 was directly evaluated by numerical integration. Since F was measured at equally spaced speeds, and known to be quite nearly linear, the simple equally spaced trapezoidal integration algorithm was used. The results of this integration, the effective equivalent mass of the drive train is given in Table 2 of Appendix B.

The Front Wheel Effective Equivalent Mass - The front wheel effective equivalent mass was calculated by equation 3 using the platform the masses, and the timed periods of oscillations.

The rolling radius of the tire, required in equation 3, was assumed to be equal for all tires of the same nominal sizes. The values used in the calculations are given in Table 2.

These values are the average of measured rolling radii of 5 to 10 tires of each size. The equivalent effective masses of the vehicle

Table 2 - Rolling Radius versus Tire Size

Nominal Tire Size	Average Rolling Radius
14 inches	0.31 m
15 inches	0.34 m
16 inches	0.37 m
16.5 inches	0.35 m

non-driving wheels are presented in Table 2 of Appendix B.

The use of a standard rolling radius for each nominal tire size introduces some error, but this is slight compared to the total vehicle mass and it simplifies the calculation by reducing the number of measured parameters which need to be maintained. The use of the spare tire on the torsional pendulum also neglects the rotational inertia of the brake disk or drum. Several preliminary measurements which included brake disks and drums indicated the effective equivalent mass of the brake is only 10% of the effective equivalent mass of the wheel-tire combination. Since a single wheel-tire combination has a typical equivalent mass of 15 kg, neglecting the 10% effect of the brake introduces a probable error of only 3 kg in the total vehicle mass. This is less than the probable error in the measurement of the vehicle gravitational mass or the probable error in the determination of the equivalent effective mass of the drive train and rear tire.

The total equivalent effective mass of the vehicle system, the sum of the gravitational mass and the effective equivalent masses of the driving and non-driving wheels, is given in table 2 of Appendix B, along with each of the component masses.

DYNAMOMETER DATA ANALYSIS - The dynamometer measurements determine the dissipative losses of the driving tires and the drive train, and supply the necessary data to determine the rotational inertia of the rear wheels and drive train. The dynamometer measurements are conceptually simple since the dynamometer used, a 48" roll GE electric chassis dynamometer, measures the forces directly. The only arithmetic necessary is to convert from the force values at the dynamometer load cell to the force at the tire-roll interface. This conversion is simply the ratio of the length of the moment arms. In addition a conversion to MKS units of force was made at this time.

The data for the tire dissipative losses, the wheel bearing losses, and the drive train dissipative losses were all scatterplotted versus speed. An example of these scatterplots were given in figure 2 of the data collection section. These plots indicate the wheel bearing and drive train losses are linear with speed, while the tire losses are approximately constant with speed. Consequently a linear least squares regression was fitted to each data set of the drive train and rear tire losses, the rear tire losses, the drive train losses and the front tire losses. The coefficients from these regression analyses are given in Tables 3 through 6 respectively of Appendix B.

CALCULATION OF DYNAMOMETER ADJUSTMENT FORCES TO SIMULATE VEHICLE ROAD LOAD - The total vehicle road load is given by equation 1 as the product of the acceleration and the total system effective mass. The coefficients of this force were calculated and are presented in Table 1 of Appendix C. Also presented in Table 1 of Appendix C is the total road load force and power at 50 mph.

The total vehicle road load force is the sum of the tire rolling resistances; the dissipative losses of the drive train, wheel bearings, and brake drag; and the aerodynamic drag of the vehicle.

$$F_{TOT} = f_{tire} + f_{mech} + f_{aero} \qquad (5)$$

where

F_{TOT} = the total vehicle road load force
f_{tire} = the sum of the tire rolling resistances
f_{mech} = the mechanical dissipative losses
f_{aero} = the aerodynamic drag

The total vehicle road load force includes the dissipation in the drive train from the rear wheel up to the point where the drive train is decoupled from the engine. When the vehicle is being tested on a dynamometer, the vehicle engine is required to overcome the drive train and driving tire losses prior to supplying power to the dynamometer. Consequently these losses should not be included in the dynamometer adjustment force. The drive train losses are independent of the choice of a dynamometer, however, the tire rolling resistance will depend on the type of dynamometer. Therefore, to develop the appropriate dynamometer adjustment force, tire losses for that particular dynamometer must be subtracted from the total road measurements, in addition to the drive train losses.

Force Coefficients for Road Simulation on a Small Twin Roll Dynamometer - In order to calculate a force appropriate for adjusting a small twin roll dynamometer, two assumptions must be made about tire power dissipation on a small twin roll dynamometer.

Assumption 1: "Two on the rolls equals four on the road."

It is commonly stated that two tires dissipate as much energy on a small twin roll dynamometer as four tires dissipate on a flat surface. However, measurements on a sufficiently large sample of tires to definitively prove or disprove this concept have not yet been reported in the literature. There is some theoretical basis for this statement (3), and one study (4) has reported the power consumption of a bias ply tire on a small twin roll dynamometer to be very nearly twice the power consumption of the same tire, at the same inflation pressure, on a flat road. This same study, however, reported the power consumption of a radial tire on a small twin roll dynamometer to be significantly greater than the power consumption of the same tire or a flat road surface. The problem is further complicated since most discussions have been directed toward light duty vehicle tires. Truck tires are frequently constructed with a greater number of carcase plies and they are typically operated at higher inflation pressures.

Assumption 2: "Power dissipation on a large single roll is proportional to road power dissipation."

The assumption that tire power dissipation on a large single roll dynamometer is greater than, but proportional to, the power dissipation on a flat surface is much better documented. The relationship between tire losses on a large single roll and a flat surface, when determined by torque or power consumption measurements, has been shown theoretically (5) to be given by:

$$F_R = F_D / \sqrt{1 + \frac{r}{R_D}} \qquad (6)$$

where:
F_R = the rolling resistence of the tire on a flat road surface
F_D = the rolling resistance of the tire on a cylindrical dynamometer surface
r = the rolling radius of the tire
R_D = the radius of the dynamometer roll

The theoretical treatise used to develop equation 6 has also been used to predict the relationship between tire rolling resistances on a large single roll and on a flat surface when the measurements are obtained directly from spindle force transducers. This relationship has been experimentally tested (6) and appears valid.

To calculate a dynamometer power absorber setting for a twin roll dynamometer, the above two assumptions were used. The rolling radii given in Table 2 were inserted into equation 6. The correction factor, $\sqrt{1+r/R_D}$ ranged from .814 to .789. Since this value was very nearly constant, the value 0.8 was used to convert the rolling resistance measurements from all front and rear tires to estimates of the tire rolling resistance on a flat road.

To obtain the force coefficients appropriate for adjusting a small twin roll dynamometer, the estimate of the flat surface tire rolling resistances for both front and rear tires were subtracted from the total road forces, as required by assumption 1. In addition, the drive train losses were also subtracted. The resulting coefficients are given in Table 2 of Appendix C, as are the force and horsepower at 50 mph.

Force Coefficients for Road Simulation on a Single Large Roll Dynamometer - The appropriate adjustment force for a large roll dynamometer can be obtained directly since the tire and drive train dissipative losses were measured on this dynamometer. To obtain the force coefficients appropriate for adjusting a 48" roll dynamometer, the coefficients of the tire and drive train losses, given in Table 3 of Appendix B, were subtracted from the total force coefficients, given in Table 1 of Appendix C. The resulting net force coefficients, representing the sum of the non-driving tire and wheel bearing losses plus the vehicle aerodynamic drag, are presented in Table 3 of Appenidx C. The forces at 50 mph and the appropriate power setting for a large single roll dynamometer to simulate the vehicle road load at 50 mph are also presented in Table 3.

PREDICTION OF THE DYNAMOMETER POWER ABSORPTION TO SIMULATE THE VEHICLE ROAD LOAD

Emission certification and fuel economy measurements are performed primarily on twin small roll dynamometers. Consequently, the prediction equations are developed using the calculated dynamometer power absorption for this type of dynamometer. After the form of the equation has been chosen for the twin small roll dynamometer, a similar equation is presented for the single large roll dynamometer.

TWIN SMALL ROLL DYNAMOMETER - The ability to predict the small twin roll dynamometer power absorber setting as a function of the vehicle test mass, the EPA inertia weight category and the vehicle frontal area, will be discussed in the following sections.

Vehicle Test Mass as a Predictor of the Dynamometer Power Absorber Setting - The vehicle test mass was defined as the total mass of the vehicle including the driver and any payload. Figure 3, the plot of dynamometer power absorber setting at 50 mph versus vehicle test mass, indicates the power absorber setting increases with increasing test mass until about 2000 kg. Above 2000 kg the power absorber setting is approximately constant. This occurs because the measurements at the higher test masses resulted from increasing the vehicle payload with sandbags. Since sand is relatively dense, the height of the sandbags never exceeded the height of the sides of the pick up truck bed. Consequently the test payloads did not directly affect the aerodynamic drag significantly in the case of pick-up trucks and had no direct effect on the aerodynamic drag of the vans. Any aerodynamic effect which might occur from increasing the vehicle payload under these conditions would be an indirect result from changes in the vehicle ground clearance or the aerodynamic angle of attack.

Increasing the vehicle payload does increase the tire rolling resistance. However, the assumptions about the rolling resistance of tires on a twin roll dynamometer, and the resulting corrections to the total road load force attempt to remove the tire rolling resistance from the dynamometer adjustment. Investigating the calculated twin roll dynamometer power absorber settings, given in Appendix C, shows that the calculated dynamometer power absorber settings for different test masses of each vehicle are generally within ± 1 horsepower of the mean power absorber setting for that vehicle. Furthermore, the value of the power absorber settings do not systematically increase or decrease with changes in the vehicle test mass. This demonstrates that the data analysis has quite successfully removed the tire rolling resistance.

Since the test mass of any vehicle can vary significantly without systematic effect on the dynamometer power absorber setting, the test mass is not a logical parameter to use to predict the dynamometer power absorber setting. In addition this data analysis indicates increasing the payload of a vehicle does not adequately simulate vehicles of larger mass. The simulation is inadequate because an increase in payload may not affect the aerodynamic drag, while vehicles which are heavier when empty tend to be physically larger and hence have larger aerodynamic drag forces.

<u>Inertia Weight Category as a Predictor of the Dynamometer Power Absorber Setting</u> - The scatterplot of the dynamometer power absorber setting at 50 mph versus the estimated EPA inertia weight category of the vehicle, Figure 4, places all calculated power absorber settings for each vehicle at a single abscissa position. This position is approximately the curb weight of the vehicle plus 300 pounds. This approach is consistent with the previously demonstrated test mass independence of the calculated twin roll dynamometer power absorber setting.

The data plotted in Figure 4 appear approximately linear in the estimated EPA inertia category, except for the notable exception of the heaviest inertia category vehicle. This vehicle is a GM van. Because of this exception, a linear regression would under estimate the dynamometer power absorber settings for the majority of vehicles which are in the 4000 to 5000 pound categories, while still over estimating the dynamometer power absorber setting for the heaviest vehicle. Furthermore, a linear model is not theoretically logical since the twin roll dynamometer power absorber setting represents the aerodynamic drag of the vehicle, and there is no theoretical reason to anticipate the aerodynamic drag would increase linearly with inertia weight.

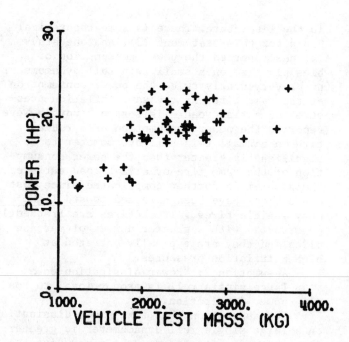

Fig. 3 - Twin roll dynamometer power at 50 mph versus total vehicle test mass

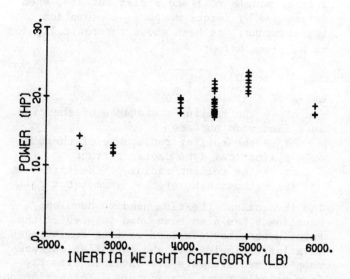

Fig. 4 - Twin roll dynamometer power at 50 mph versus inertia weight category

A theoretically based model can be developed based on several logical assumptions (7). The first assumption is that, because of similarities in manufacturing technology, the density of light duty trucks is approximately constant. Stated as an equation, the assumption is:

$$W \sim V \qquad (7)$$

where

W = the inertia weight category of the vehicle (i.e., the empty weight plus a standard small payload)
V = the volume of the vehicle

The vehicle volume is approximately equal to the product of the three major dimensions. The second assumption is that each of the major vehicle dimensions may be expected to increase approximately equally with an increase in weight. Consequently each major dimension is proportional to the cube root of the vehicle weight. That is:

$$D \sim W^{1/3} \qquad (8)$$

where
D = any of the major vehicle dimensions of the height width and length.

The twin roll dynamometer power absorber setting is primarily the aerodynamic drag of the vehicle. The aerodynamic drag is proportional to the frontal area which is approximately equal to the product of the vehicle height and width. Consequently the twin roll dynamometer power absorber setting should be proportional to the weight of the vehicle to the two-thirds power.

$$H_p \sim W^{2/3} \qquad (9)$$

The previous arguments are hardly rigorous, therefore a model of the form:

$$H_p = cW^x \qquad (10)$$

was chosen which allowed the exponent to vary. This model will predict a dynamometer power absorber setting of zero horsepower for a vehicle of zero mass, which is theoretically appropriate. Also, if x is less than 1, the model predicts the slope of the horsepower versus weight curve will decrease as the weight increases. This is also theoretically logical; and consistent with the observed data.

The model, equation 10, unfortunately cannot be conveniently fitted to the data by least squares process. The fitting process is difficult since the normal equations resulting from the least squares criterion are non-linear. These equations can be solved simultaneously by numerical methods, however a simpler approach is to "linerize" equation 10 by the following logarithimic transformation.

$$\begin{aligned} \ln H_p &= \ln c\,W^x \\ &= \ln c + \ln W^x \\ &= \ln c + x \ln W \end{aligned} \qquad (11)$$

Identifying $\ln H_p$ as the dependent variable and $\ln W$ as the independent variable, equation 11 can now be fitted by a simpler linear regression. The results of this regression are:

REGRESSION OF TWIN ROLL DYNAMOMETER POWER AT 50 MPH VERSUS VEHICLE INERTIA WEIGHT CATEGORY

Regression Model $\ln H_p = \ln c + x \ln W$
$\ln H_p$ = the natural logarithm of the dynamometer power absorber setting in horsepower
$\ln W$ = the natural logarithm of the vehicle inertia weight category in pounds
$\ln c$ = -2.531
x = 0.6509
Sample Size = 54
Multiple Correlation Coefficient .767
R^2 = 589
Standard Error of the Regression .1137
Converting to the form of the original model, the prediction equation is:

$$H_p = .0796\, W^{0.651} \qquad (12)$$

The correlation coefficient of the regression may indicate this regression does not fit the data very well, however the statistics of this regression cannot be readily interpreted since they are the statistics of the regression performed on the transformed parameters.

<u>Frontal Area as a Predictor of Dynamometer Power Absorber Setting</u> -The assumptions regarding tire rolling resistance on a small twin roll dynamometer, and the resulting corrections to the total road load force, remove the tire dissipation from the small twin roll dynamometer adjustment force. In addition, the drive train losses are subtracted, therefore equation 12 shows only a portion of the mechanical losses, the non-driving wheel bearing losses, and the non-driving wheel brake drag remain in addition to the aerodynamic drag. These remaining mechanical losses are probably weakly dependent on the vehicle mass since the vehicle bearing and brake size depend on vehicle mass. However, the random nature of these forces, especially the brake drag, will predominate over any systematic effects for a sample size of 15 vehicles. Consequently the remaining mechanical losses can be expected to appear as random "noise" superimposed on the primary force of the aerodynamic drag.

Since the aerodynamic forces predominate in the small twin roll dynamometer adjustment, an aerodynamic model is the logical choice for predicting this dynamometer adjustment force. The aerodynamic drag of the vehicle is theoretically given by:

$$F_{aero} = \frac{1}{2} \rho C_D A v^2 \qquad (13)$$

where
F_{aero} = the aerodynamic drag force
ρ = the air density
C_D = the drag coefficient of the vehicle
v = the vehicle velocity

If the drag coefficients of the vehicles are approximately constant, then the drag force, and hence the power should be linear in frontal area. In Figure 5, the plot of twin roll dynamometer adjustment power versus frontal area, the vehicles with frontal areas below 34 ft^2 are pick-up trucks while the remaining vehicles are vans. For the pick-up trucks the dynamometer power absorber setting at 50 mph does appear to be linear with frontal area, however the data occur in two major groups with no intermediate points. The power absorber setting for vans may also be linear with frontal area, however since no vans with small frontal areas were included in the test fleet this cannot be empirically determined. Small vans were not included since few, if any, such vehicles are currently being sold.

A simple regression line could be fitted to the dynamometer power absorber setting versus frontal area, however if a linear model were chosen this would under estimate the power absorber setting for the large pick-up trucks and over estimate the setting for vans. If a non-linear model were chosen this could adapt to the reduced road load of the large frontal area vans, but would be inappropriate for any possible pick-up trucks with large frontal areas. To avoid this dilemma the frontal area variable was "factored" into two variables one giving the frontal areas of the pick-up trucks only, and a second variable having the values of the frontal areas of the vans only. This is approximately equivalent to separating the data into two groups, vans and pick-ups. A general linear regression of these independent variables was first performed. As theoretically predicted, the intercept was nearly zero and there was relatively little statistical confidence that the intercept was not zero. Consequently a linear regression was performed forcing the regression line through zero. The results of this regression are:

REGRESSION OF TWIN ROLL DYNAMOMETER ADJUSTMENT POWER AT 50 MPH VERSUS VEHICLE FRONTAL AREA

Regression Model: $H_p = a A_{pu} + b A_{van}$

H_p = the dynamometer power absorber setting (horsepower)
A_{pu} = the frontal area of the pick-up trucks (ft^2)
A_{van} = the frontal area of the vans (ft^2)
a = .633 hp/ft^2
b = .511 hp/ft^2
Sample Size = 54
Multiple Correlation Coefficient .856
R^2 = .733
Standard Error of the Regression 1.59

In order to evaluate the area based prediction model versus the model using the vehicle inertia weight category, the prediction equations are plotted in figures 6 and 7. Also plotted in these figures are mean values of the calculated power absorber settings for each

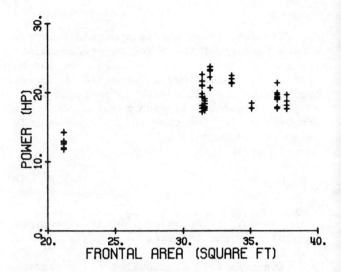

Fig. 5 - Twin roll dynamometer power at 50 mph versus vehicle reference frontal area

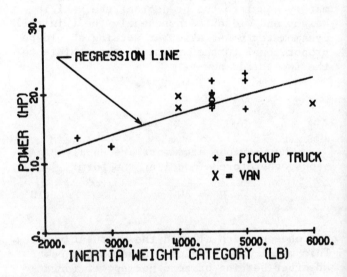

Fig. 6 - Mean twin roll dynamometer power at 50 mph versus vehicle inertia weight category

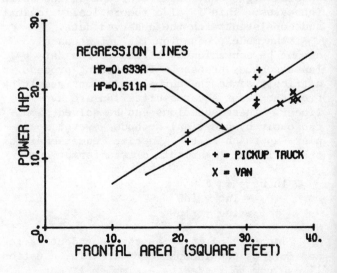

Fig. 7 - Mean twin roll dynamometer power at 50 mph versus frontal reference area

vehicle versus the predictor variable. Plotting only the mean values reduces the number of points plotted so that the vehicle types may be identified. These mean values are given in Table 1 of Appendix D.

Figure 7, the plot of the power absorber setting versus inertia weight category shows the fitted model to be a reasonable appearing choice for these data. However the worse case error of the fitted line from the data is about 4.5 horsepower. There are two additional cases of errors of about three horsepower and 4 errors of approximately two horsepower. The average dynamometer setting is about 18 horsepower, therefore a two horsepower error is an error slightly greater than 10%.

Figure 7, the plot of power absorber setting versus frontal area, with the vehicles divided into categories, shows the worse case error in this approach is only approximately 2.5 horsepower. In addition there is only one other data point which is farther than two horsepower from the regression lines.

An additional problem in the treatment of vans when using inertia weight as the predictor of the dynamometer power absorber setting is also apparent from Figures 6 and 7. The inertia weights of the vans tested varied by 2000 pounds or about 40% of the mean inertia weight of the vans. The dynamometer power absorber setting only varied about 1.5 horsepower or less than 8%. Consequently the inertia weight regression equation predicts significantly different dynamometer power absorber settings for different inertia weight vans, while little difference was observed. Figure 7 indicates no differences should be expected in the power absorber setting for the vans since the frontal areas, and hence the aerodynamic drag of the vehicles were approximately equal. The prediction model based on vehicle frontal area correctly predicts nearly equal dynamometer power absorber settings for all of the vans in the test fleet.

It is concluded that the prediction system based on the vehicle frontal area, when the vehicles are divided into the categories of vans and pick-ups, is significantly better than a prediction system based on the vehicle inertia weight categories. Both models had two parameters which were fitted to the data, therefore the advantage of frontal area as a predictor of the dynamometer power absorber setting is not merely the result of a more flexible model. Only prediction systems based on vehicle frontal area will be considered in the remainder of this paper.

LARGE ROLL DYNAMOMETER ADJUSTMENT FORCE - Equations to predict the power absorber settings for large single roll dynamometer are developed since these equations may be useful at the present, or in future work. Since the majority of EPA testing is conducted on small twin roll dynamometers it is assumed that the form of the prediction system which is chosen for the twin roll dynamometer will also be used for large single roll dynamometers. Therefore only the necessary modifications to the small twin roll dynamometer prediction equations, will be developed.

The appropriate adjustment force for a large roll dynamometer can be obtained directly since the tire and drive train dissipative losses were measured on this dynamometer. To obtain the force coefficients appropriate for adjusting a 48" roll dynamometer, the coefficients of the tire and drive train losses, given in Table 1 of Appendix B, were subtracted from the total force coefficients, given in Table 1 of Appendix C. The resulting net force coefficients, representing the sum of the non-driving tire and wheel bearing losses plus the vehicle aerodynamic drag, are presented in Table 3 of Appendix C. The forces at 50 mph and the appropriate power setting for a large single roll dynamometer to simulate the vehicle road load at 50 mph are also presented in Table 3.

The differences between the small twin roll dynamometer power absorber loads and the power absorber adjustment for a large single roll is primarly the front tire rolling resistances. Since tire rolling resistances are approximately proportional to the vertical load on the tire, the differences in the power absorber loads should be linear in the vehicle test mass. The power abosrber setting for a small twin roll dynamometer, given in Table 2 of Appendix C was subtracted from the power absorber setting appropriate for a large single roll dynamometer, given in Table 3 of Appendix C. The difference is scatterplotted in Figure 8 versus the vehicle test mass.

The scatterplot of the differences in the dynamometer power absorber settings indicate these differences are approximately linear in the vehicle test mass, although a significant amount of data scatter is present. A regression, using the model that the tire rolling resistance is proportional to the vehicle test mass was performed. The results of this regression are:

REGRESSION OF THE DIFFERENCES IN DYNAMOMETER POWER ABSORBER SETTINGS VERSUS VEHICLE TEST

MASS -

Regression Model: $H_p = aM$
H_p = the difference in dynamometer power absorber settings (horsepower)
M = the vehicle test mass (kg)
a = 0.000887
Sample Size = 54
Standard Error of the Regression 0.702

The results of this regression are appended to the prediction equations. The dynamometer power absorber setting for a single large roll dynamometer, predicated by the vehicle frontal area and weight is given by:

$$H_p = .63 A + .0004W \text{ for pick-up trucks}$$
$$H_p = .51 A + .0004W \text{ for vans} \quad (14)$$

where
A = the vehicle frontal area (ft^2)
W = the vehicle weight (lb)

A factor of 1/2.2 has been introduced in the weight coefficient since the independent variable is vehicle weight in pounds while the regression used the vehicle mass in kg.

COMPARISON WITH DATA FROM VEHICLE MANUFACTURERS - In order to compare the results of this study with other data sources, light duty truck dynamometer power absorption data were requested from the major light duty truck manufacturers. Response to this request was very cooperative and data were submitted for 20 vehicles. These data were generated by the particular recommended practices for road load determination adopted by the submitting corporations. In general these recommended practices are minor variations of coast down techniques where the time required to coast down through a known speed range centered about 50 mph is recorded on the road. The coast down times are then corrected to a standard set of ambient conditions, and corrected for the difference between the total equivalent effective mass of the vehicle and the nearest inertia step which can be simulated on a test dynamometer. The coast down times are then duplicated on the dynamometer, and the power absorption at 50 mph is determined from the dynamometer. The submitted data are given in Table 1 of Appendix D.

The submitted frontal area data included the frontal area of the vehicle protuberances such as mirrors; while the previously reported basic frontal areas excluded these protuberances. In order to have an equivalence of the data for comparison, the frontal areas of the protuberances of the EPA test vehicles were estimated and added to the basic vehicle frontal area. In addition, the means of the EPA dynamometer power absorption measurements at 50 mph were computed for each vehicle. These data are also presented in Table 1 of Appendix D.

In order to best compare the submitted data with the data of this report, the means of the power at 50 mph for those instances where multiple identical vehicles were tested by different sources were computed. Two Chevrolet C10 pick-ups with estimated frontal areas of 31.9 ft^2, were tested by EPA. The mean of the EPA measurements was 18.2 horsepower. The mean of the GM measurements on three similarly identified vehicles, but with reported frontal areas of 32.7 ft^2, was 18.1. Similarly the mean of the EPA measurements on three GM G series vans was 19.2 horsepower while the mean of the GM measurements for four of these vehicles was 19.0 horsepower. The differences between these

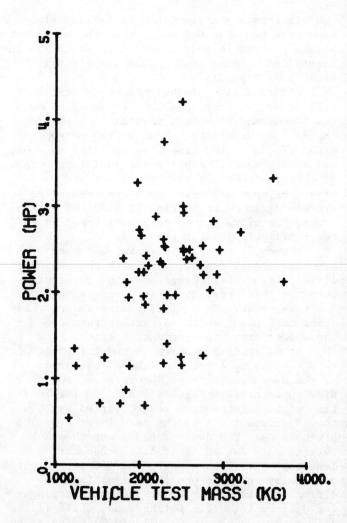

Fig. 8 - Differences between large single roll dynamometer and small twin roll dynamometer powers at 50 mph versus vehicle mass

means, approximately one-half of one percent of the measured values in the case of the pick-ups, and about one percent of the measured values in the case of the vans was not deemed statistically significant. It is therefore concluded that little if any systematic difference exists between the measurement sources.

A final prediction equation was developed by incorporating the submitted data. The representative data for each vehicle, presented in Table 1 of Appendix D, were regressed against the reported frontal areas in the same manner as the previous frontal area based regression. The results of this regression are:

REGRESSION OF TWIN ROLL DYNAMOMETER ADJUSTMENT POWER AT 50 MPH VERSUS VEHICLE FRONTAL AREA INCLUDING SUBMITTED DATA

Regression Model: $H_p = a A_{pu} + b A_{van}$
H_p = dynamometer power absorption (horsepower)
A_{pu} = the frontal area of the pick-up trucks (ft^2)

A_{van} = The frontal area of the vans (ft^2)
a^{van} = 0.58
b = 0.50
Sample Size = 35
Standard Error of the regression 1.5

In the case of light duty vans the coefficient of the vehicle frontal area changed very little. The slight reduction probably occured because inclusion of the vehicle protruberance areas increased the regressed frontal areas. Some reduction may have resulted from differences in test conditions, vehicle types, or standard ambient conditions.

The coefficient of the frontal area of pick-up trucks changed significantly from 0.63 to 0.58. This change primarily resulted from a change in the nature of the test fleet. The EPA fleet included a step side truck with an externally mounted spare tire, and a vehicle with large mirrors and cab lights. These configurations would have higher aerodynamic drag than the more basic vehicles represented by the submitted data. In addition, for the classification of vehicles, a van was defined as: a light duty truck having an integral enclosure, fully enclosing the driver compartment and load carrying device, and having no body sections protruding more than 30 inches ahead of the leading edge of the windshield. Because of the hood length restriction, some closed bed vehicles were included in the pick-up truck category. Aerodynamically, the drag coefficient of these vehicles are probably lower than the traditional open bed pick-up. The presence of these vehicles, such as a GM Blazer, in the regression sample balanced the effect of the high drag vehicles and reduced the slope of the pick-up truck regression line.

CONCLUSIONS

It is concluded that an aerodynamic related system is the preferred method of predicting the dynamometer power absorption to simulate light duty truck road load. The vehicle frontal area is the simplest and most significant aerodynamic parameter. A prediction system based on the vehicle frontal area is more logical and is more accuracte than prediction systems based on the vehicle weight. The area based prediction system can be improved by separating the vehicles into categories of vans and pick-up trucks. This considers the differences between the average drag coefficients of vehicles in these classes.

Little, if any differences were observed between the means of measurements on similar vehicles from different sources. Therefore, it is concluded that the results of the measurement techniques employed are equivalent.

Differences were observed between the prediction equations based on the data of this study only, and after inclusion of data contributed by light duty truck manufacturers. These differences are attributed to changes in the sample population resulting from the incorporation of the additional data.

REFERENCES

1. J.D. Walter and F.S. Conant, "Energy Losses in Tires", Tire and Science and Technology, TSTCA, Vol. 2, No. 4, November 1974.
2. G.D. Thompson, "The Vehicle Road Load Problem - Approach by Non-Linear Modeling." ISETA Fourth International Symposium on Engine Testing Automation, Vol. II. Published by Automotive Automation, Croydon, England.
3. S. Clark, University of Michigan, unpublished discussions.
4. W.B. Crum, "Road and Dynamometer Tire Power Dissipation." Society of Automotive Engineers, 750955.
5. S.K. Clark, "Rolling Resistance Forces in Pneumatic Tires." University of Michigan Report DOT-TSC-76-1, prepared for Department of Transportation, Transportation Systems Center, Cambridge, Mass., January 1976.
6. D.J. Schuring, "Rolling Resistance of Tires Measured Under Transient and Equilibrium Conditions on Calspan's Tire Research Facility." DOT-TSC-OST-76-9, March 1976.
7. C.W. LaPointe, unpublished discussion.

ACKNOWLEDGEMENT

The author wishes to acknowledge with gratitude the editorial assistance of Nancy Henderson, and also her help in preparation of the manuscript.

APPENDIX A

A1 - DERIVATION OF THE CONCEPT OF EQUIVALENT EFFECTIVE MASS

The total energy of the decelerating vehicle system is the sum of the translational kinetic energy of the vehicle and the rotational kinetic energy of any vehicle components in rotational motion. For all mechanical components of the wheels and drive train, the rotational velocity is proportional to the vehicle velocity; therefore, the energy of the system may be written as:

$$E = 1/2 \, mv^2 + 1/2 \, (\sum_i I_i \alpha_i^2) v^2 \qquad \text{(A1-1)}$$

where:

E = the total system energy
m = the vehicle mass
v = the vehicle speed
I_i = the rotatational inertia of the ith rotation component
α_i = the proportionality constant between the rotational velocity of the ith rotating component and the vehicle speed

Differentiating equation (1) with respect to time, and comparing the resulting expression for power with the similar time derivative of a purely translational system, the generalized force on the system may be expressed as:

$$F = (m + \sum_i I_i \alpha_i^2) A \qquad \text{(A1-2)}$$

where

F = the generalized system force
A = the translational acceleration of the system

Defining M as the "total effective mass of the system", where:

$$M = m + \sum_i I_i \alpha_i^2 \qquad \text{(A1-3)}$$

Equation (2) now has the familiar form

$$F = MA \qquad \text{(A1-4)}$$

The $\sum_i I_i \alpha_i^2$ term is identified as the "equivalent effective mass" of the rotating components and may be designated by:

$$m_{eq} = \sum_i I_i \alpha_i^2 \qquad \text{(A1-5)}$$

The equivalent effective mass, defined by equations (3) and (5), is simply one approach to include the effect of the rotational kinetic energy of the system.

<u>Derivation of the Expression for the Effective Equivalent Mass of the Drive Train and Drive Wheels</u> - The effective equivalent mass of the drive wheels and drive train was calculated using a coast down method to determine the drive wheel and drive train inertia. This technique was chosen because it gives a measure of rotational inertia of all drive train components. The rotational inertia of the system is given by the familar equation:

$$\tau = I \frac{d\omega}{dt} \qquad \text{(A1-6)}$$

where

- I = the rotational inertia
- τ = the torque necessary to motor the system
- ω = the angular velocity of the system

Assuming that the torque is a function of velocity, the variables of equation (6) may be separated and the integrals formed:

$$\int_{t_2}^{t_1} dt = I \int_{\omega_2}^{\omega_1} \frac{d\omega}{\tau} \tag{A1-7}$$

where

- ω_1 = the initial angular velocity
- ω_2 = the final angular velocity
- t_1 = the initial time
- t_2 = the final time

Integrating the left hand side and solving for I yields:

$$I = \frac{(t_1 - t_2)}{\int_{\omega_2}^{\omega_1} \frac{d\omega}{\tau}} \tag{A1-8}$$

Equation (8) gives the inertia as a function of the torque required to motor the tire and wheel. However the effective equivalent mass is actually the desired quantity. Substituting the definition of effective equivalent mass and

$$\tau = FR$$
$$\omega = v/R \tag{A1-9}$$

where

- R = the rolling radius of the tire
- F = the force at the tire roll interface
- v = the simulated translational velocity of the vehicle

into equation (8) yields:

$$m_{Req} = \frac{\Delta t}{\int_{v_2}^{v_1} \frac{dv}{F}} \tag{A1-10}$$

where

- m_{Req} = the effective equivalent mass of the drive train and driving tires
- Δt is the time interval
- v_1 = the initial simulated vehicle speed
- v_2 = the final simulated vehicle speed

It should be noted that both F and Δt of equation (10) are negative since the drive train is decelerating. Equation (10) indicates that the dissipative forces of the drive train and the time required for the drive train to coast over a known speed range are the parameters necessary to calculate the drive wheel and drive train effective equivalent mass.

A2 - THREE WIRE TORSIONAL PENDULUM

The total system energy of the three wire pendulum is the sum of the rotational kinetic energy, the vertical translational kinetic energy, and the potential energy from the gravitational field of the earth.

$$E = \frac{1}{2} I \dot{\theta}^2 + \frac{1}{2} M \dot{h}^2 + Mgh \tag{A2-1}$$

where:
- E = the total system energy
- I = rotational inertia of the pendulum
- M = the pendulum mass
- g = the acceleration due to gravity
- $\dot{\theta}$ = the time derivative of the angular displacement of the pendulum from equilibrium
- h = the vertical displacement of the pendulum
- $\dot{h}$ = the time derivative of h

Because of the physical constraints imposed by the construction of the system, h and θ are not independent variables. The variable h can be expressed as a function of the physical constants of the system and the variable θ. That is:

$$h = h(\theta) \tag{A2-2}$$

Consequently,

$$\dot{h} = h'\dot{\theta} \tag{A2-3}$$

where

$$h' = \partial h/\partial \theta \tag{A2-4}$$

Substituting (3) into (1) yields:

$$E = \frac{1}{2} I \dot{\theta}^2 + \frac{1}{2} M h'^2 \dot{\theta}^2 + Mgh \tag{A2-5}$$

The system energy is a constant of the motion. This is a reasonable assumption for any time interval during which viscous damping effect of the air is small. In general, since the motion of the pendulum is slow the air damping effect is quite small. If necessary, the effects of air damping may be treated within this framework by classical perturbation theory techniques.

Considering the system energy as a constant of the motion and solving for $\dot{\theta}^2$ yields

$$\dot{\theta}^2 = 2(E - Mgh)/(I + Mh'^2) \tag{A2-6}$$

Separating the variables of equation (6) yields:

$$dt = \sqrt{\frac{I + Mh'^2}{2(E - Mgh)}} \, d\theta \tag{A2-7}$$

The motion of the pendulum is cyclic; and approximately sinesodial in nature. The time required for the pendulum to move from θ_{max}, the maximum displacement, to the maximum displacement in the opposite directions, $-\theta_{max}$, is one half the period of the motion.

Therefore the period of the motion is

$$\tau = 2\int_{-\theta_{max}}^{\theta_{max}} \sqrt{\frac{I + Mh'^2}{2(E - Mgh)}} \, d\theta \tag{A2-8}$$

where

τ = the period of the pendulum motion.

Equation (8) is very similar in form to the familiar case of the simple pendulum. However, a term containing h' is present in (8) which is not present in the case of the simple pendulum. Also h is a more complex function of θ in the case of the three wire torsional pendulum than it is in the case of the simple pendulum.

The function of $h(\theta)$ is now constructed from the geometry of the pendulum. Figure 1 shows the pendulum in the equilibrium position.

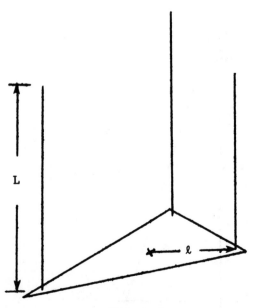

Figure 1

L= the length of the supporting cables
ℓ= radius of gyration from the center of the platform to the attachment point of any supporting cable.

Considering a top view of the pendulum system when an image of the equilibrium position and a displaced position are superimposed.

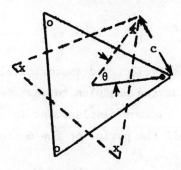

Figure 2

- o   denotes the supporting cable attachment point in the equilibrium posistion

- x   denotes the supporting cable attachment point after the platform is displaced through the angle θ

- θ = the angular displacement

The distance c is a chord of the circle formed by the loci of the supporting cable attachment points as the platform is rotated. The length of c is given by:

$$c = 2 \ell \sin \frac{\theta}{2} \qquad (A2-9)$$

A side view of the plane formed by a supporting cable in both the equilibrium position and in the displaced position is given in Figure 3.

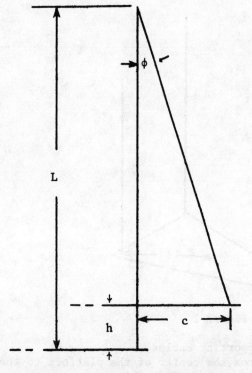

Figure 3

From Figure 3

$$h = L - L \cos \phi \quad (A2-10)$$

and

$$\sin \phi = c/L \quad (A2-11)$$

where:

ϕ = the angle the displaced supporting cable makes with the vertical

h = the vertical displacement of the platform.

Equations (9), (10) and (11) may be used to express h as a function of ℓ, L, and θ. First, eliminating c from equation (9) and (11)

$$\sin \phi = \frac{2\ell}{L} \sin \frac{\theta}{2} \quad (A2-12)$$

Using the trigonometric identity:

$$\sin^2 x + \cos^2 x = 1 \quad (A2-13)$$

equation (10) may be written

$$h = L - L\sqrt{1 - \sin^2 \phi} \quad (A2-14)$$

Substituting equation (12) into equation (14) yields

$$h = L - L\sqrt{1 - \frac{4\ell^2}{L^2} \sin^2 \frac{\theta}{2}} \quad (A2-15)$$

Using the identity:

$$\cos x = 1 - 2\sin^2(x/2) \quad (A2-16)$$

Equation (15) becomes:

$$h = L - \sqrt{L^2 - 2\ell^2 + 2\ell^2 \cos \theta} \quad (A2-17)$$

Differentiating equation (17) with respect to θ to form h' yields:

$$h' = \frac{\ell^2 \sin \theta}{\sqrt{L^2 - 2\ell^2 + 2\ell^2 \cos \theta}} \quad (A2-18)$$

The only remaining quantity to be constructed is an expression for E, the system energy. Since the motion is oscillatory, the system must stop for an instance when it reverses direction at $\theta = \theta_{max}$. At this instance, since $\dot{h}$ and $\dot{\theta}$ vanish, equation (1) becomes:

$$E = Mgh(\theta_{max}) \qquad (A2\text{-}19)$$
$$= Mg[L - \sqrt{L^2 - 2\ell^2 + 2\ell^2 \cos\theta_{max}}]$$

Since the system energy is a constant of the motion, E may always be expressed as a function of θ_{max}.

Equations (8), (17), (18), and (19) represent the mathematical system which must be integrated. This system is repeated here for clarity:

$$\tau = 2 \int_{\theta_{max}}^{\theta_{max}} \sqrt{\frac{I + Mh'^2}{2(E - Mgh)}} \, d\theta \qquad (A2\text{-}20)$$

where:

$$h = L - \sqrt{L^2 - 2\ell^2 + 2\ell^2 \cos\theta}$$

$$h' = \frac{\ell^2 \sin\theta}{\sqrt{L^2 - 2\ell^2 + 2\ell^2 \cos\theta}} \qquad (A2\text{-}21)$$

$$E = Mgh(\theta_{max}) \qquad (A2\text{-}22)$$
$$= Mg[L - \sqrt{L^2 - 2\ell^2 + 2\ell^2 \cos\theta_{max}}]$$

A simple approximation can be developed which can be integrated in closed form. This approximation yields a simple algebraic expression for I as a function of τ. This enables easy approximate calculation of the inertia of the system from the period.

In practical applications, when a three wire torsional pendulum is used to determine the inertia of an object, the maximum displacement angle, θ_{max}, must remain small to prevent crossing or twisting of the supporting cables. The maximum displacement angle would typically be in the range $1/2 \leq \theta_{max} \leq 1$ radians.

Since θ_{max} is small, expansions of L, h and h' in powers of θ will converge rapidly. Therefore it is appropriate to expand h in a MacLaurian series. That is:

$$h(\theta) \simeq h(0) + h'(0)\theta + 1/2 h''(0)\theta^2 + \ldots \qquad (A2\text{-}23)$$

Considering the terms in the above equation:

$$h(0) = L - \sqrt{L^2 - 2\ell^2 + 2\ell^2 \cos\theta} \Big/ \theta = 0$$
$$= L - \sqrt{L^2 - 2\ell^2 + 2\ell^2} \qquad (A2-24)$$
$$= L - L$$
$$= 0$$

Likewise:

$$h'(0) = \frac{\ell^2 \sin\theta}{\sqrt{L^2 - 2\ell^2 + 2\ell^2 \cos\theta}} \Big/ \theta = 0 \qquad (A2-25)$$
$$= 0$$

and:

$$h''(0) = \frac{\ell^2 \cos\theta}{\sqrt{L^2 - 2\ell^2 + 2\ell^2 \cos\theta}} + \frac{\ell^4 \sin^2\theta}{\sqrt{L^2 - 2\ell^2 + 2\ell^2 \cos\theta}^3} \Big/ \theta = 0 \qquad (A2-26)$$
$$= \ell^2/L$$

Therefore the expansion for $h(\theta)$, accurate to order θ^2 is,

$$h(\theta) \simeq \ell^2 \theta^2 / 2L \qquad (A2-27)$$

By inspection of equations (23) through (26) it is apparent that the expansion for h' accurate to order θ^2 is:

$$h'(\theta) \simeq \ell^2 \theta / L \qquad (A2-28)$$

Equation (28) is actually accurate to order θ^3, since $h'''(0)$ vanishes.

Substituting equations (27) and (28), the expansions for both $h(\theta)$ and $h(\theta_{max})$ into equation (20):

$$\tau \simeq 2 \int_{\theta_{max}}^{\theta_{max}} \sqrt{\frac{I + M(\ell^2\theta/L)^2}{2(Mg\ell^2\theta_{max}^2/2L - Mg\ell^2\theta^2/2L)}} \, d\theta \qquad (A2-29)$$

$$= \frac{2}{\sqrt{Mg\ell^2 L}} \int_{\theta_{max}}^{\theta_{max}} \sqrt{\frac{IL^2 + M\ell^4 \theta^2}{\theta_{max}^2 - \theta^2}} \, d\theta \qquad (A2-30)$$

The integral in equation (30) is a complete elliptic integral, and may be evaluated from tables. However, in a typical torsional pendulum system $L \gg \ell$ and $\theta < 1$ so that the first term of the numerator will dominate. Specifically, with the EPA three wire torsional pendulum used in the tire experiments:

$$I \simeq 2\text{kg m}^2$$
$$L \simeq 5\text{m}$$
$$\ell \simeq 0.5\text{m} \quad (A2\text{-}31)$$
$$M \simeq 30\text{kg}$$
$$\theta_{max} \simeq 0.5 \text{ rad}$$

Therefore, in this case:

$$IL^2 \simeq 50\text{kg m}^4 \quad (A2\text{-}32)$$

and

$$M\ell^4\theta_{max}^2 \simeq 0.469\text{kg m}^4 \quad (A2\text{-}33)$$

Neglecting the second term introduces an error of approximately 1%. This is probably less error than was introduced by the expansions; therefore it is consistant with the accuracy of this approximation to neglect this term. Consequently, neglecting this term:

$$\tau \simeq \frac{2}{\sqrt{Mg\ell^2 L}} \int_{-\theta_{max}}^{\theta_{max}} \sqrt{\frac{IL^2}{\theta_{max}^2 - \theta^2}} \, d\theta \quad (A2\text{-}34)$$

Equation (34) can be integrated in closed form. This is apparent by making the change of variables:

$$\theta = x\theta_{max} \quad (A2\text{-}35)$$

With this change of variables equation (34) becomes:

$$\tau \simeq 2\sqrt{\frac{IL}{Mg\ell^2}} \int_{-1}^{1} \frac{dx}{\sqrt{1-x^2}} \quad (A2\text{-}36)$$

or

$$\tau \simeq 2\sqrt{IL/Mg\ell^2} \, [\sin^{-1}x \rfloor_{-1}^{1}] \quad (A2\text{-}37)$$

$$\tau \simeq 2\pi\sqrt{IL/Mg\ell^2} \quad (A2\text{-}38)$$

Solving equation (38) for I:

$$I \simeq Mg\ell^2\tau^2/4\pi^2 L \quad (A2\text{-}39)$$

The final result of all of the approximations, equation (39), can be checked by direct numerical integration. Values for wheel inertia have been numerically computed by the method of Gauss-Chebyshev quadrature, and these values have been in very good agreement with those derived from equation (39).

Equation (39) gives the rotational inertial of the platform, either with or without any object on the pendulum. The inertia of the platform only may be expressed as;

$$I_p = \frac{g\ell^2 m_p \tau_p^2}{4\pi^2 L} \qquad (A2-40)$$

where

τ_p = the period of oscillation for the pendulum only

m_p = the mass of the pendulum

The rotational inertia of the pendulum platform with the vehicle tire is given as

$$I_{pt} = \frac{g\ell^2 m\tau_t^2}{4\pi^2 L} \qquad (A2-41)$$

where

m = the total mass of the pendulum with the tire

τ_t = the period of oscillation of the pendulum with the tire

The rotational inertia of the tire-wheel is given from equations (40) and (41) as:

$$I_t = \frac{g\ell^2}{4\pi^2 L} (m\tau_t^2 - m_p\tau_p^2) \qquad (A2-42)$$

The definition of the equivalent effective mass gives the relationship between the effective equivalent mass and the rotational inertia as:

$$m_{feq} = I/R^2 \qquad (A2-43)$$

where

m_{feq} is the effective equivalent mass of the tire-wheel combination.

Therefore,

$$m_{feq} = \frac{g\ell^2}{4\pi^2 LR^2} (m\tau_t^2 - m_p\tau_p^2) \qquad (A2-44)$$

A3 - DERIVATION OF THE COASTDOWN EQUATION

The acceleration of a freely coasting vehicle is assumed to be a polynomial function of the velocity that is:

$$\frac{dv}{dt} = A(v) = -(a_o + a_1 v + a_2 v^2) \qquad (A3-1)$$

The a_o term is usually identified as representing tire losses, however, both tire losses and mechanical drive train losses probably appear in both the a_o and a_1v term. The a_2v^2 term of (1) is identified as the aerodynamic term and the velocity in this term must be the vehicle air speed. Therefore, if any ambient wind exists, equation (1) must be written as:

$$A = [a_o + a_1v + a_2(v - ws)^2] \quad (A3-2)$$

where:

    s = a unit vector in the direction of vehicle travel
    w = the wind velocity vector.
    NOTE: scaler multiplication of w and s is always to be understood.

This analysis only considers the effect of the component of wind in the direction of vehicle travel. Also equation (2) is only valid when (v - ws) is greater than or equal to zero. If the component of wind speed in the direction of vehicle travel exceeds the vehicle speed, then the aerodynamic drag term must be replaced by an aerodynamic driving term by changing the sign of a_2. The effect of cross winds can be treated in this analysis if sufficient aerodynamic information is available about the vehicle to allow the cross wind effect to be approximated as a simple polynomial expression of the vehicle speed.

A term must be added to equation (2) to describe the effects of grade. The grade term is equal to g sin γ where γ is the angle between the test surface and the horizontal and g is the gravitational acceleration. Since γ is a very small angle the approximation of using the grade is quite accurate. Inserting this grade term, equation (2) becomes:

$$A = -[a_o + gs + a_1v - a_2(v - ws)^2] \quad (A3-3)$$

where:

    g = a vector in the track direction of increasing grade with a magnitude equal to the product of the test surface grade and the gravitational acceleration.
    NOTE: scaler multiplication of g and s is always understood.

Expanding and regrouping, equation (3) becomes:

$$A = -\{[a_o + a_2(ws)^2 + gs] + [a_1 - 2a_a(ws)]v + a_2v^2\} \quad (A3-4)$$

Integration and Inversion

Using A = dv/dt and separating variables of equation (4):

$$-dt = \frac{dv}{[a_o + a_2(ws)^2 + gs] + [a_1 - 2a_2(ws)]v + a_2v^2} \quad (A3-5)$$

Equation (5) can be integrated in closed form.

The integrals of equation (2) will depend on the relative magnitude and sign of a_o, a_1, a_2, ws and gs. The appropriate choice for the form of the integrals is more easily seen if the variable transformation:

$$v = \frac{2u - [a_1 - 2a_2(ws)]}{2a_2} \quad (A3-6)$$

and

$$B^2 = \frac{4a_2[a_0 + a_2(ws)^2 + (gs)] - [a_1 - 2a_2(ws)]^2}{4} \quad (A3-7)$$

are applied to equation (5). This yields:

$$dt = -du/(B^2 + u^2) \quad (A3-8)$$

Because of the constraint $[v - (ws)] \geq 0$, u will always be non-negative. The B is not so strongly restrained, B^2 may be either negative, positive or zero.

Case A; $B^2 < 0$

If B^2 is negative the transformation:

$$B^2 = -D^2 \quad (A3-9)$$

yields:

$$dt = -du/(u^2 - D^2) \quad (A3-10)$$

Now both u^2 and D^2 are positive, hence both u and D are real.

The integral of equation (10) can take either of two forms:

$$t = \frac{1}{D}[\tanh^{-1}(u/D)] + C_1 \quad (A3-11)$$

or

$$t = \frac{1}{D}[\coth^{-1}(u/D)] + C_2 \quad (A3-12)$$

The terms C_1 and C_2 are constants of integration, determined by the initial conditions.

The choice of equation (11) or (12) to represent the coast down data must depend on the physical possibilities of the motion of the system. If equation (11) is used, the magnitude of u cannot exceed the magnitude of D since the argument of the $\tanh^{-1}$ function is bounded between +1 and -1. But, if u is less than D, then equation (10) shows dt/du is positive. Hence, considering the transformations (6) and (7), dv/dt is positive and the vehicle must be accelerating. Consequently equation (12) describes a "coast-up" which can only occur if there is a grade or wind driving term of sufficient magnitude that the vehicle freely accelerates from its initial speed to some higher steady state speed.

If equation (12) is to be used to describe the coast down data, u must be greater than D since the magnitude of the argument of the $\coth^{-1}$ function must exceed 1. The physical interpretation of this can be seen by rewriting equation (10) as an expression for du/dt.

$$du/dt = -(u^2 - D^2) \quad (A3-13)$$

If the vehicle is initially at some velocity where u > D, the rate of change of u is negative hence the vehicle will decelerate until u = 0 or u = D. If the u = D condition occurs, du/dt vanishes and the vehicle will remain at the velocity where u = D. Physically this means there must be a grade or wind effect acting as a driving term. The vehicle will decelerate until the drag forces are balanced by this driving term, and the vehicle will remain in motion at a constant speed.

Case B; B = 0

The case B = 0 is a special case that can theroretically occur, however in practice it will probably never be observed because of experimental error. It is presented since it is the transition between the two classes of motion, and numerical analysis difficulties may occur in this area.

If: B = 0

$$dt = -du/u^2 \qquad (A3-14)$$

and

$$t = 1/u + C_3 \qquad (A3-15)$$

where C_3 is the constant of integration to be determined by the initial conditions of the system.

Case C; $B^2 < 0$

In the case $B^2 < 0$ equation (8) may be integrated in the form:

$$-t = (1/B)[\tan^{-1}(u/B)] + C_4 \qquad (A3-16)$$

or:

$$t = (1/B)[\cot^{-1}(u/B)] + C_5 \qquad (A3-17)$$

Again C_4 and C_5 are constants of integration.

Equations (16) and (17) are equivalent. This may be shown by inverting equation (16) to yield:

$$u = B \tan[-B(C_4 + t)] \qquad (A3-18)$$

Using the trignometric identity relating the tangent and cotangent functions, equation (18) may be written as:

$$u = B \cot(Bt + \pi/2 + BC_4) \qquad (A3-19)$$

Inverting equation (17) yields:

$$u = B \cot[B(t - C_5)] \qquad (A3-20)$$

Comparing equations (19) and (20) these equations differ only by the sign of the constants of integration, and by the phase angle of $\pi/2$. Either equation may be used to describe the vehicle coast down. However, since the $B^2 < 0$ case requires the $\coth^{-1}$ solution, equation (20) will be used for the case $B^2 > 0$ because of the convenient parallel nature of the cot and coth functions.

Applying the original transformations (6) to (20) and solving for v yields:

$$v = \frac{1}{a_2}\{B \cot[B(t - C_5)] - a_1/2 + a_2(ws)\} \qquad (A3-21)$$

where:

$$C_5 = -\cot^{-1}\{[2a_2 v_o + a_1 - 2a_2(ws)]/(2B)\} \qquad (A3-22)$$

The velocity v_o is the initial vehicle speed at t = 0.

Applying the same operations to equation (11) yields:

$$v = \frac{1}{a_2}\{D \coth[D(t - C_2)] - a_1/2 + a_2(ws)\} \qquad (A3-23)$$

and

$$C_2 = -\coth^{-1}\{[2a_2v_o + a_1 - 2a_2(ws)]/2D\} \qquad (A3-24)$$

Either equation (21) or (23) must be fitted to the speed time data by the method of least squares. The values of the coefficients a_o, a_1, and a_2 will determine which expression, (21) or (23) is the appropriate form to describe the data.

APPENDIX B - TABLE 1
AMBIENT CORRECTED ACCELERATION COEFFICIENTS

ID	WT (LB)	A0 (M/SEC**2)	A1 (1/SEC)	A2 (1/M)
7101	4110	0.1364E+00	0.3978E-02	0.4116E-03
7001	6050	0.1199E+00	0.8064E-02	0.1288E-03
7701	4500	0.1117E+00	-0.5550E-02	0.9136E-03
5902	4270	0.1464E+00	0.7684E-02	0.2577E-03
8206	2750	0.1210E+00	0.6234E-02	0.3516E-03
5902	4330	0.7434E-01	0.1432E-01	0.1712E-03
7001	4930	0.1879E+00	0.3511E-02	0.3653E-03
7502	4540	0.1444E+00	0.2891E-02	0.4456E-03
7402	5540	0.1803E+00	0.1476E-02	0.3774E-03
7001	5530	0.2035E+00	-0.3568E-03	0.4210E-03
7001	4620	0.1919E+00	0.7623E-02	0.2025E-03
7901	5690	0.1109E+00	0.1009E-02	0.4044E-03
7001	5500	0.1833E+00	0.2383E-02	0.3268E-03
7301	4560	0.1225E+00	0.2950E-02	0.4384E-03
7402	5760	0.1425E+00	0.6534E-02	0.2282E-03
7201	5110	0.7001E-01	0.7501E-02	0.2279E-03
7701	4990	0.1015E+00	0.4450E-02	0.3541E-03
5902	5020	0.1575E+00	0.1879E-02	0.4400E-03
8002	4370	0.6343E-01	0.8668E-02	0.3015E-03
7201	5480	0.1091E+00	0.1483E-02	0.3848E-03
7301	5050	0.8896E-01	0.7113E-02	0.2559E-03
8002	5530	0.1101E+00	0.2612E-02	0.3664E-03
8206	4050	0.1339E+00	0.3006E-02	0.2814E-03
8507	3900	0.1540E+00	0.5791E-02	0.1975E-03
8002	6320	0.8744E-01	0.5471E-02	0.2508E-03
7301	6420	0.1276E+00	0.2068E-02	0.3076E-03
7502	5500	0.1580E+00	0.4098E-02	0.2756E-03
7901	7900	0.8297E-01	0.4818E-02	0.1645E-03
7502	6240	0.1114E+00	0.8449E-02	0.1471E-03
7001	8190	0.1188E+00	0.5222E-02	0.1503E-03
8306	3490	0.1408E+00	0.5705E-02	0.2666E-03
8507	3370	0.1574E+00	0.6569E-02	0.2244E-03
9003	3970	0.1493E+00	0.3367E-02	0.4221E-03
9003	4380	0.8530E-01	0.1036E-01	0.1798E-03
9003	4810	0.1482E+00	-0.5306E-03	0.5140E-03
7701	5620	0.8704E-01	0.4485E-02	0.3136E-03
7201	6060	0.6875E-01	0.6287E-02	0.1920E-03
7101	5020	0.9678E-01	0.7696E-02	0.2047E-03
7901	7040	0.1083E+00	0.2564E-02	0.2221E-03
8002	5010	0.1228E+00	0.5966E-04	0.4946E-03
7701	5980	0.9093E-01	0.3926E-02	0.2649E-03
7502	5020	0.1120E+00	0.8399E-02	0.2351E-03
8206	3550	0.1387E+00	0.3739E-02	0.3136E-03
7201	4540	0.6536E-01	0.8017E-02	0.2499E-03
7301	4060	0.8686E-01	0.6857E-02	0.3856E-03
7101	4500	0.7461E-01	0.9948E-02	0.2024E-03
7402	5110	0.1519E+00	0.3603E-02	0.3729E-03
7602	5030	0.6956E-01	0.9136E-02	0.1548E-03
7101	5330	0.1080E+00	0.4081E-02	0.3279E-03
7901	6060	0.6192E-01	0.6517E-02	0.1909E-03
7001	6490	0.1244E+00	0.6162E-02	0.1420E-03
8306	4140	0.1251E+00	0.4669E-02	0.2295E-03
8306	2700	0.9182E-01	0.1190E-01	0.1881E-03
8507	2560	0.1643E+00	0.8565E-02	0.2637E-03

APPENDIX B - TABLE 2
VEHICLE MASSES

ID	GRAV MASS (KG)	DTT EFF MASS (KG)	FT EFF MASS (KG)	TOTAL VEH MASS (KG)
7101	1868.2	33.811	26.04	1928.0
7101	2045.5	33.811	26.04	2105.3
7101	2281.8	33.811	26.04	2341.7
7101	2422.7	33.811	26.04	2482.6
7201	2490.9	64.657	31.75	2587.3
7201	2063.6	64.657	31.75	2160.0
7201	2322.7	64.657	31.75	2419.1
7201	2754.5	64.657	31.75	2851.0
7502	2063.6	42.007	44.30	2149.9
7502	2281.8	42.007	44.30	2368.1
7502	2500.0	42.007	44.30	2586.3
7502	2836.4	42.007	44.30	2922.7
7602	2286.4	43.537	27.34	2357.2
7602	2013.6	43.537	27.34	2084.5
7602	2518.2	43.537	27.34	2589.1
8002	1986.4	30.971	31.92	2049.3
8002	2277.3	30.971	31.92	2340.2
8002	2513.6	30.971	31.92	2576.5
8002	2872.7	30.971	31.92	2935.6
8206	1250.0	31.074	20.75	1301.8
8206	1840.9	31.074	20.75	1892.7
8306	1227.3	20.605	22.91	1270.8
8306	1586.4	20.605	22.91	1629.9
8306	1881.8	20.605	22.91	1925.3
8507	1163.6	30.183	22.80	1216.6
8507	1531.8	30.183	22.80	1584.8
8507	1772.7	30.183	22.80	1825.7
9003	1804.5	42.142	19.64	1866.3
9003	1990.9	42.142	19.64	2052.7
9003	2186.4	42.142	19.64	2248.1
5902	1940.9	46.852	26.90	2014.7
5902	1968.2	46.852	26.90	2041.9
5902	2281.8	46.852	26.90	2355.6
5902	2500.0	46.852	26.90	2573.8
7001	2100.0	55.724	46.67	2202.4
7001	2240.9	55.724	46.67	2343.3
7001	2500.0	55.724	46.67	2602.4
7001	2513.6	55.724	46.67	2616.0
7001	2750.0	55.724	46.67	2852.4
7001	2950.0	55.724	46.67	3052.4
7001	3722.7	55.724	46.67	3825.1
7301	1845.5	50.140	28.90	1924.5
7301	2072.7	50.140	28.90	2151.8
7301	2295.5	50.140	28.90	2374.5
7301	2918.2	50.140	28.90	2997.2
7402	2322.7	43.537	28.81	2395.0
7402	2518.2	43.537	28.81	2590.5
7402	2618.2	43.537	28.81	2690.5
7701	2045.5	41.779	27.16	2114.4
7701	2268.2	41.779	27.16	2337.1
7701	2554.5	41.779	27.16	2623.5
7701	2718.2	41.779	27.16	2787.1
7901	2586.4	56.206	38.04	2680.6
7901	2754.5	56.206	38.04	2848.8
7901	2977.3	56.206	38.04	3071.5
7901	3200.0	56.206	38.04	3294.2
7901	3590.9	56.206	38.04	3685.2

APPENDIX B - TABLE 3
DRIVE TRAIN + REAR TIRE
FORCE COEFFICIENTS

ID	WT (LB)	A (NT)	B (KG/SEC)
7101	4110	125.089	1.907
7101	4500	142.537	2.218
7101	5020	163.138	2.473
7101	5330	176.577	2.081
7201	5480	147.625	1.704
7201	4540	94.002	2.297
7201	5110	120.298	2.601
7201	6060	170.826	2.007
7502	4540	126.448	2.814
7502	5020	149.018	2.951
7502	5500	170.187	3.685
7502	6240	212.167	2.824
7602	5040	132.137	1.656
7602	4430	115.029	1.510
7602	5530	156.920	1.471
8002	4370	79.075	3.480
8002	5010	104.432	3.995
8002	5530	83.967	4.311
8002	6320	121.000	4.973
8206	2750	81.356	2.150
8206	4050	135.113	2.290
8306	2700	85.214	2.253
8306	3490	128.722	2.343
8306	4140	152.310	2.400
8507	2560	100.226	2.213
8507	3370	113.064	2.145
8507	3900	143.116	2.404
9003	3970	96.395	2.009
9003	4380	110.992	1.914
9003	4810	122.549	2.225
5902	4310	105.970	2.939
5902	5020	135.467	2.922
5902	5500	163.721	2.760
7001	4620	127.895	1.958
7001	4930	139.843	1.841
7001	5530	160.010	1.823
7001	6050	146.015	2.995
7001	6490	172.954	3.111
7001	8190	251.360	3.301
7301	4060	83.631	1.685
7301	4560	95.228	1.584
7301	5050	114.625	1.665
7301	6420	195.448	2.476
7402	5110	157.907	3.445
7402	5540	163.957	3.215
7402	5760	172.719	3.372
7701	4500	109.004	1.846
7701	4990	106.903	2.312
7701	5620	116.962	2.690
7701	5980	130.141	2.818
7901	5690	121.809	2.681
7901	6060	133.591	2.955
7901	7040	160.318	2.992
7901	7900	200.388	2.728

APPENDIX B - TABLE 4
REAR TIRE
FORCE COEFFICIENTS

ID	WT (LB)	A (NT)	B (KG/SEC)
7101	4110	84.692	.239
7101	4500	102.140	.549
7101	5020	122.740	.804
7101	5330	136.180	.412
7201	5480	116.237	-.236
7201	4540	62.614	.357
7201	5110	88.909	.661
7201	6060	139.438	.067
7502	4540	88.802	.229
7502	5020	111.373	.366
7502	5500	132.542	1.099
7502	6240	174.522	.239
7602	5040	101.569	.100
7602	4430	84.462	-.045
7602	5530	126.352	-.084
8002	4370	70.134	.349
8002	5010	95.492	.864
8002	5530	75.027	1.180
8002	6320	112.060	1.842
8206	2750	56.494	.200
8206	4050	110.251	.340
8306	2700	61.014	-.020
8306	3490	104.522	.070
8306	4140	128.109	.127
8507	2560	77.627	-.324
8507	3370	90.466	-.392
8507	3900	120.517	-.133
9003	3970	83.001	.325
9003	4380	97.598	.230
9003	4810	109.156	.542
5902	4310	65.601	-.633
5902	5020	95.098	-.650
5902	5500	123.353	-.812
7001	4620	76.228	-.480
7001	4930	88.175	-.597
7001	5530	108.342	-.614
7001	6050	94.346	.558
7001	6490	121.286	.673
7001	8190	199.693	.863
7301	4060	64.873	-.087
7301	4560	76.469	-.187
7301	5050	95.867	-.107
7301	6420	176.690	.705
7402	5110	135.962	2.803
7402	5540	142.012	2.573
7402	5760	150.774	2.730
7701	4500	68.629	.214
7701	4990	66.529	.680
7701	5620	76.588	1.058
7701	5980	89.767	1.186
7901	5690	88.516	.353
7901	6060	100.298	.627
7901	7040	127.025	.663
7901	7900	167.096	.400

APPENDIX B - TABLE 5
DRIVE TRAIN FORCE COEFFICIENTS

ID	WT (LB)	A (NT)	B (KG/SEC)
7101	4110	40.398	1.668
7101	4500	40.398	1.668
7101	5020	40.398	1.668
7101	5330	40.398	1.668
7201	5480	31.388	1.940
7201	4540	31.388	1.940
7201	5110	31.388	1.940
7201	6060	31.388	1.940
7502	4540	37.645	2.586
7502	5020	37.645	2.586
7502	5500	37.645	2.586
7502	6240	37.645	2.586
7602	5040	30.568	1.555
7602	4430	30.568	1.555
7602	5530	30.568	1.555
8002	4370	8.940	3.131
8002	5010	8.940	3.131
8002	5530	8.940	3.131
8002	6320	8.940	3.131
8206	2750	24.862	1.950
8206	4050	24.862	1.950
8306	2700	24.200	2.273
8306	3490	24.200	2.273
8306	4140	24.200	2.273
8507	2560	22.598	2.537
8507	3370	22.598	2.537
8507	3900	22.598	2.537
9003	3970	13.394	1.684
9003	4380	13.394	1.684
9003	4810	13.394	1.684
5902	4310	40.369	3.572
5902	5020	40.369	3.572
5902	5500	40.369	3.572
7001	4620	51.668	2.437
7001	4930	51.668	2.437
7001	5530	51.668	2.437
7001	6050	51.668	2.437
7001	6490	51.668	2.437
7001	8190	51.668	2.437
7301	4060	18.758	1.772
7301	4560	18.758	1.772
7301	5050	18.758	1.772
7301	6420	18.758	1.772
7402	5110	21.946	.642
7402	5540	21.946	.642
7402	5760	21.946	.642
7701	4500	40.374	1.632
7701	4990	40.374	1.632
7701	5620	40.374	1.632
7701	5980	40.374	1.632
7901	5690	33.293	2.328
7901	6060	33.293	2.328
7901	7040	33.293	2.328
7901	7900	33.293	2.328

APPENDIX B - TABLE 6
FRONT TIRE FORCE COEFFICIENTS

ID	WT (LB)	A (NT)	B (KG/SEC)
7101	4110	100.119	.147
7101	4500	101.676	.376
7101	5020	98.805	.544
7101	5330	102.698	.705
7201	5480	73.717	.284
7201	4540	74.043	.941
7201	5110	71.504	.587
7201	6060	75.010	.592
7502	4540	53.455	-.047
7502	5020	78.666	.021
7502	5500	72.612	.671
7502	6240	138.780	-.413
7602	5040	116.928	.666
7602	4430	111.475	.910
7602	5530	128.301	1.112
8002	4370	125.442	-.570
8002	5010	123.561	.636
8002	5530	133.580	.760
8002	6320	151.844	.204
8206	2750	54.797	.367
8206	4050	61.980	.171
8306	2700	68.994	.107
8306	3490	75.065	.146
8306	4140	75.603	.222
8507	2560	53.466	-.591
8507	3370	56.870	-.294
8507	3900	70.427	-.503
9003	3970	136.528	-.634
9003	4380	148.811	-.418
9003	4810	159.544	-.405
5902	4310	132.806	.737
5902	5020	132.485	1.966
5902	5500	145.088	2.543
7001	4620	110.863	.088
7001	4930	108.425	.391
7001	5530	116.727	.514
7001	6050	118.585	.634
7001	6490	129.009	.416
7001	8190	132.334	.513
7301	4060	94.076	.447
7301	4560	107.055	.543
7301	5050	109.559	.858
7301	6420	117.703	1.015
7402	5110	******	******
7402	5540	******	******
7402	5760	******	******
7701	4500	99.552	.535
7701	4990	109.240	.372
7701	5620	111.976	.563
7701	5980	113.440	.552
7901	5690	132.923	-.204
7901	6060	121.567	-.043
7901	7040	145.946	.103
7901	7900	173.943	.405

APPENDIX C - TABLE 1
TOTAL VEHICLE ROAD LOAD
FORCE COEFFICIENTS

ID	WT (LB)	F0 (NT)	F1 (KG/SEC)	F2 (KG/M)	F50 (NT)	HP50 (HP)
5902	4330	0.1518E+03	0.2924E+02	0.3496E+00	0.9799E+03	29.369
5902	5020	0.3710E+03	0.4426E+01	0.1036E+01	0.9877E+03	29.603
5902	5500	0.2811E+03	0.1957E+02	0.5878E+00	0.1012E+04	30.331
7001	4620	0.4226E+03	0.1679E+02	0.4459E+00	0.1021E+04	30.601
7001	4930	0.4404E+03	0.8227E+01	0.8560E+00	0.1052E+04	31.530
7001	5530	0.5325E+03	-0.9334E+00	0.1101E+01	0.1062E+04	31.830
7001	6050	0.3419E+03	0.2300E+02	0.3673E+00	0.1039E+04	31.141
7001	6490	0.3796E+03	0.1881E+02	0.4333E+00	0.1016E+04	30.451
7001	8190	0.4543E+03	0.1997E+02	0.5747E+00	0.1188E+04	35.607
7101	4110	0.2630E+03	0.7670E+01	0.7936E+00	0.8309E+03	24.904
7101	4500	0.1571E+03	0.2094E+02	0.4260E+00	0.8380E+03	25.116
7101	5020	0.2266E+03	0.1802E+02	0.4793E+00	0.8688E+03	26.040
7101	5330	0.2682E+03	0.1013E+02	0.8141E+00	0.9014E+03	27.017
7201	4540	0.1412E+03	0.1732E+02	0.5397E+00	0.7978E+03	23.912
7201	5110	0.1694E+03	0.1815E+02	0.5514E+00	0.8504E+03	25.488
7201	5480	0.2823E+03	0.3837E+01	0.9955E+00	0.8653E+03	25.935
7201	6060	0.1960E+03	0.1792E+02	0.5473E+00	0.8700E+03	26.075
7301	4060	0.1672E+03	0.1320E+02	0.7422E+00	0.8328E+03	24.961
7301	4560	0.2636E+03	0.6348E+01	0.9434E+00	0.8767E+03	26.276
7301	5050	0.2112E+03	0.1689E+02	0.6076E+00	0.8923E+03	26.744
7301	6420	0.3825E+03	0.6198E+01	0.9219E+00	0.9816E+03	29.420
7402	5110	0.3637E+03	0.8629E+01	0.8931E+00	0.1003E+04	30.062
7402	5540	0.4672E+03	0.3824E+01	0.9776E+00	0.1041E+04	31.201
7402	5760	0.3834E+03	0.1758E+02	0.6139E+00	0.1083E+04	32.459
7502	4540	0.3105E+03	0.6215E+01	0.9580E+00	0.9280E+03	27.814
7502	5020	0.2653E+03	0.1989E+02	0.5567E+00	0.9879E+03	29.609
7502	5500	0.4087E+03	0.1060E+02	0.7128E+00	0.1002E+04	30.032
7502	6240	0.3256E+03	0.2469E+02	0.4298E+00	0.1092E+04	32.729
7602	4430	0.9496E+02	0.2743E+02	0.2393E+00	0.8276E+03	24.805
7602	5030	0.1640E+03	0.2154E+02	0.3649E+00	0.8275E+03	24.802
7602	5540	0.2093E+03	0.2173E+02	0.3916E+00	0.8905E+03	26.690
7701	4500	0.2361E+03	-0.1173E+02	0.1932E+01	0.9387E+03	28.135
7701	4990	0.2373E+03	0.1040E+02	0.8276E+00	0.8831E+03	26.468
7701	5620	0.2283E+03	0.1177E+02	0.8228E+00	0.9023E+03	27.044
7701	5980	0.2534E+03	0.1094E+02	0.7382E+00	0.8667E+03	25.977
7901	5690	0.2973E+03	0.2705E+01	0.1084E+01	0.8993E+03	26.954
7901	6060	0.1764E+03	0.1857E+02	0.5439E+00	0.8630E+03	25.866
7901	7040	0.3567E+03	0.8446E+01	0.7315E+00	0.9109E+03	27.301
7901	7900	0.3058E+03	0.1776E+02	0.6061E+00	0.1005E+04	30.122
8002	4370	0.1300E+03	0.1776E+02	0.6179E+00	0.8357E+03	25.047
8002	5010	0.2874E+03	0.1396E+00	0.1157E+01	0.8687E+03	26.037
8002	5530	0.2837E+03	0.6730E+01	0.9441E+00	0.9057E+03	27.145
8002	6320	0.2567E+03	0.1606E+02	0.7361E+00	0.9834E+03	29.474
8206	2750	0.1575E+03	0.8115E+01	0.4578E+00	0.5675E+03	17.009
8206	4050	0.2535E+03	0.5689E+01	0.5326E+00	0.6467E+03	19.383
8306	2700	0.1167E+03	0.1512E+02	0.2390E+00	0.5741E+03	17.207
8306	3490	0.2295E+03	0.9299E+01	0.4346E+00	0.6544E+03	19.614
8306	4140	0.2408E+03	0.8989E+01	0.4419E+00	0.6625E+03	19.856
8507	2560	0.1999E+03	0.1042E+02	0.3209E+00	0.5931E+03	17.776
8507	3370	0.2495E+03	0.1041E+02	0.3556E+00	0.6597E+03	19.772
8507	3900	0.2812E+03	0.1057E+02	0.3606E+00	0.6976E+03	20.908
9003	3970	0.2787E+03	0.6284E+01	0.7878E+00	0.8126E+03	24.355
9003	4380	0.1751E+03	0.2127E+02	0.3691E+00	0.8348E+03	25.020
9003	4810	0.3332E+03	-0.1193E+01	0.1156E+01	0.8838E+03	26.489

APPENDIX C - TABLE 2
SMALL TWIN ROLL
DYNAMOMETER FORCE COEFFICIENTS

ID	WT (LB)	F0 (NT)	F1 (KG/SEC)	F2 (KG/M)	F50 (NT)	HP50 (HP)
5902	4330	-0.4729E+02	0.2558E+02	0.3496E+00	0.6991E+03	20.954
5902	5020	0.1486E+03	-0.1990E+00	0.1036E+01	0.6616E+03	19.830
5902	5500	0.2598E+02	0.1461E+02	0.5878E+00	0.6462E+03	19.368
7001	4620	0.2213E+03	0.1467E+02	0.4459E+00	0.7718E+03	23.132
7001	4930	0.2315E+03	0.5955E+01	0.8560E+00	0.7921E+03	23.742
7001	5530	0.3008E+03	-0.3290E+01	0.1101E+01	0.7772E+03	23.294
7001	6050	0.1199E+03	0.1961E+02	0.3673E+00	0.7416E+03	22.228
7001	6490	0.1277E+03	0.1550E+02	0.4333E+00	0.6906E+03	20.699
7001	8190	0.1370E+03	0.1643E+02	0.5747E+00	0.7914E+03	23.719
7101	4110	0.7475E+02	0.5693E+01	0.7936E+00	0.5984E+03	17.935
7101	4500	-0.4635E+02	0.1853E+02	0.4260E+00	0.5806E+03	17.403
7101	5020	0.8966E+01	0.1527E+02	0.4793E+00	0.5898E+03	17.676
7101	5330	0.3670E+02	0.7568E+01	0.8141E+00	0.6125E+03	18.358
7201	4540	0.4870E+00	0.1434E+02	0.5397E+00	0.5906E+03	17.701
7201	5110	0.9682E+01	0.1521E+02	0.5514E+00	0.6251E+03	18.735
7201	5480	0.9895E+02	0.1859E+01	0.9955E+00	0.6378E+03	19.115
7201	6060	-0.6946E+01	0.1545E+02	0.5473E+00	0.6118E+03	18.337
7301	4060	0.2128E+02	0.1114E+02	0.7422E+00	0.6410E+03	19.212
7301	4560	0.9802E+02	0.4292E+01	0.9434E+00	0.6652E+03	19.937
7301	5050	0.2810E+02	0.1452E+02	0.6076E+00	0.6561E+03	19.664
7301	6420	0.1282E+03	0.3050E+01	0.9219E+00	0.6569E+03	19.689
7402	5110	0.1394E+03	0.5212E+01	0.8931E+00	0.7021E+03	21.042
7402	5540	0.2290E+03	0.2340E+00	0.9776E+00	0.7226E+03	21.657
7402	5760	0.1382E+03	0.1386E+02	0.6139E+00	0.7547E+03	22.620
7502	4540	0.1590E+03	0.3484E+01	0.9580E+00	0.7155E+03	21.444
7502	5020	0.7562E+02	0.1699E+02	0.5567E+00	0.7335E+03	21.985
7502	5500	0.2069E+03	0.6598E+01	0.7128E+00	0.7105E+03	21.294
7502	6240	0.3731E+02	0.2224E+02	0.4298E+00	0.7491E+03	22.453
7602	4430	-0.9236E+02	0.2518E+02	0.2393E+00	0.5900E+03	17.684
7602	5030	-0.4137E+02	0.1937E+02	0.3649E+00	0.5739E+03	17.200
7602	5540	-0.2499E+02	0.1935E+02	0.3916E+00	0.6031E+03	18.077
7701	4500	0.6118E+02	-0.1396E+02	0.1932E+01	0.7142E+03	21.407
7701	4990	0.5631E+02	0.7926E+01	0.8276E+00	0.6469E+03	19.388
7701	5620	0.3708E+02	0.8842E+01	0.8228E+00	0.6457E+03	19.353
7701	5980	0.5046E+02	0.7917E+01	0.7382E+00	0.5962E+03	17.868
7901	5690	0.8686E+02	0.2580E+00	0.1084E+01	0.6341E+03	19.005
7901	6060	-0.3439E+02	0.1577E+02	0.5439E+00	0.5899E+03	17.679
7901	7040	0.1050E+03	0.5506E+01	0.7315E+00	0.5935E+03	17.788
7901	7900	-0.3242E+00	0.1479E+02	0.6061E+00	0.6329E+03	18.971
8002	4370	-0.3540E+02	0.1481E+02	0.6179E+00	0.6042E+03	18.108
8002	5010	0.1032E+03	-0.4191E+00	0.1157E+01	0.5875E+03	17.608
8002	5530	0.1079E+03	0.2047E+01	0.9441E+00	0.6252E+03	18.739
8002	6320	0.3664E+02	0.1129E+02	0.7361E+00	0.6567E+03	19.683
8206	2750	0.4360E+02	0.5711E+01	0.4578E+00	0.3999E+03	11.987
8206	4050	0.9085E+02	0.3330E+01	0.5326E+00	0.4313E+03	12.928
8306	2700	-0.1151E+02	0.1278E+02	0.2390E+00	0.3934E+03	11.792
8306	3490	0.6163E+02	0.6853E+01	0.4346E+00	0.4319E+03	12.944
8306	4140	0.5363E+02	0.6436E+01	0.4419E+00	0.4182E+03	12.535
8507	2560	0.7243E+02	0.8615E+01	0.3209E+00	0.4253E+03	12.746
8507	3370	0.1090E+03	0.8422E+01	0.3556E+00	0.4749E+03	14.233
8507	3900	0.1058E+03	0.8541E+01	0.3606E+00	0.4769E+03	14.292
9003	3970	0.8968E+02	0.4847E+01	0.7878E+00	0.5915E+03	17.729
9003	4380	-0.3542E+02	0.1974E+02	0.3691E+00	0.5901E+03	17.685
9003	4810	0.1048E+03	-0.2987E+01	0.1156E+01	0.6155E+03	18.449

APPENDIX C - TABLE 3
LARGE SINGLE ROLL
DYNAMOMETER FORCE COEFFICIENTS

ID	WT (LB)	F0 (NT)	F1 (KG/SEC)	F2 (KG/M)	F50 (NT)	HP50 (HP)
5902	4330	0.4583E+02	0.2630E+02	0.3496E+00	0.8083E+03	24.226
5902	5020	0.2355E+03	0.1504E+01	0.1036E+01	0.7867E+03	23.577
5902	5500	0.1174E+03	0.1681E+02	0.5878E+00	0.7867E+03	23.579
7001	4620	0.2947E+03	0.1483E+02	0.4459E+00	0.8489E+03	25.444
7001	4930	0.3006E+03	0.6386E+01	0.8560E+00	0.8709E+03	26.102
7001	5530	0.3725E+03	-0.2756E+01	0.1101E+01	0.8609E+03	25.801
7001	6050	0.1959E+03	0.2000E+02	0.3673E+00	0.8265E+03	24.771
7001	6490	0.2066E+03	0.1570E+02	0.4333E+00	0.7740E+03	23.197
7001	8190	0.2029E+03	0.1667E+02	0.5747E+00	0.8626E+03	25.853
7101	4110	0.1379E+03	0.5763E+01	0.7936E+00	0.6631E+03	19.875
7101	4500	0.1456E+02	0.1872E+02	0.4260E+00	0.6458E+03	19.356
7101	5020	0.6346E+02	0.1555E+02	0.4793E+00	0.6504E+03	19.492
7101	5330	0.9162E+02	0.8049E+01	0.8141E+00	0.6782E+03	20.326
7201	4540	0.4720E+02	0.1502E+02	0.5397E+00	0.6526E+03	19.558
7201	5110	0.4910E+02	0.1555E+02	0.5514E+00	0.6721E+03	20.143
7201	5480	0.1347E+03	0.2133E+01	0.9955E+00	0.6796E+03	20.369
7201	6060	0.2517E+02	0.1591E+02	0.5473E+00	0.6542E+03	19.608
7301	4060	0.8357E+02	0.1151E+02	0.7422E+00	0.7117E+03	21.330
7301	4560	0.1684E+03	0.4764E+01	0.9434E+00	0.7461E+03	22.362
7301	5050	0.9657E+02	0.1522E+02	0.6076E+00	0.7404E+03	22.190
7301	6420	0.1871E+03	0.3722E+01	0.9219E+00	0.7307E+03	21.902
7402	5110	0.2058E+03	0.5184E+01	0.8931E+00	0.7678E+03	23.012
7402	5540	0.3032E+03	0.6090E+00	0.9776E+00	0.8052E+03	24.133
7402	5760	0.2107E+03	0.1421E+02	0.6139E+00	0.8349E+03	25.023
7502	4540	0.1841E+03	0.3401E+01	0.9580E+00	0.7386E+03	22.137
7502	5020	0.1163E+03	0.1694E+02	0.5567E+00	0.7730E+03	23.167
7502	5500	0.2385E+03	0.6915E+01	0.7128E+00	0.7491E+03	22.453
7502	6240	0.1134E+03	0.2187E+02	0.4298E+00	0.8168E+03	24.482
7602	4430	-0.2007E+02	0.2592E+02	0.2393E+00	0.6788E+03	20.344
7602	5030	0.3186E+02	0.1988E+02	0.3649E+00	0.6585E+03	19.738
7602	5540	0.5238E+02	0.2026E+02	0.3916E+00	0.7008E+03	21.004
7701	4500	0.1271E+03	-0.1358E+02	0.1932E+01	0.7887E+03	23.640
7701	4990	0.1304E+03	0.8088E+01	0.8276E+00	0.7246E+03	21.717
7701	5620	0.1113E+03	0.9080E+01	0.8228E+00	0.7253E+03	21.738
7701	5980	0.1233E+03	0.8122E+01	0.7382E+00	0.6735E+03	20.187
7901	5690	0.1755E+03	0.2400E-01	0.1084E+01	0.7175E+03	21.505
7901	6060	0.4281E+02	0.1562E+02	0.5439E+00	0.6635E+03	19.886
7901	7040	0.1964E+03	0.5454E+01	0.7315E+00	0.6837E+03	20.491
7901	7900	0.1054E+03	0.1503E+02	0.6061E+00	0.7441E+03	22.303
8002	4370	0.5093E+02	0.1428E+02	0.6179E+00	0.6787E+03	20.343
8002	5010	0.1830E+03	-0.3855E+01	0.1157E+01	0.6747E+03	20.223
8002	5530	0.1997E+03	0.2419E+01	0.9441E+00	0.7254E+03	21.741
8002	6320	0.1357E+03	0.1109E+02	0.7361E+00	0.7512E+03	22.515
8206	2750	0.7614E+02	0.5965E+01	0.4578E+00	0.4381E+03	13.132
8206	4050	0.1184E+03	0.3399E+01	0.5326E+00	0.4604E+03	13.799
8306	2700	0.3149E+03	0.1287E+02	0.2390E+00	0.4384E+03	13.141
8306	3490	0.1008E+03	0.6956E+01	0.4346E+00	0.4733E+03	14.187
8306	4140	0.8849E+02	0.6589E+01	0.4419E+00	0.4565E+03	13.682
8507	2560	0.9967E+02	0.8207E+01	0.3209E+00	0.4434E+03	13.289
8507	3370	0.1364E+03	0.8265E+01	0.3556E+00	0.4988E+03	14.950
8507	3900	0.1381E+03	0.8166E+01	0.3606E+00	0.5007E+03	15.008
9003	3970	0.1823E+03	0.4275E+01	0.7878E+00	0.6714E+03	20.122
9003	4380	0.6411E+02	0.1936E+02	0.3691E+00	0.6811E+03	20.413
9003	4810	0.2107E+03	-0.3418E+01	0.1156E+01	0.7117E+03	21.331

Appendix D - Table 1
EPA and Submitted Data

Vehicle		Type	Frontal Area (Ft**2)	Small Twin Roll Dynamometer Adjustment at 50 mph (horsepower)	Data Source
Ford	F100	PU	31.6	20.1	EPA Report
Ford	F100	PU (4X4)	33.8	21.8	↓
Ford	F100	PU	31.6	17.7	
Ford	F250	PU	34.4	21.8	
Ford	E150	Van	38.6	18.5	
GM	C20	PU (Step)	33.1	22.8	
GM	C10	PU	31.9	17.8	
GM	C10	PU	31.9	18.5	
GM	G10	Van	38.1	19.5	
GM	G20	Van	38.1	19.6	
GM	G30	Van	38.1	18.4	
Dodge	B100	Van	36.2	18.0	
Toyota	Hilux	PU	21.4	12.5	
Toyota	Hilux	PU	21.4	12.4	↓
Datsun		PU	21.2	13.8	
Ford	F150	PU	31.7	18.1	Ford letter, 9-17-76
Ford	F250	PU	33.3	17.4	Ford letter, 9-17-76
Chry	B200	Van	36.6	19.7	Chrysler letter, 9-1-76
Chry	AW100	PU(Util)	32.5	18.9	↓
Chry	D100	PU	31.6	18.0	
Chry	D200	PU	32.1	20.4	
GM	LUV	PU	22.2	12.5	GM letter, 9-22-76
GM	LUV	PU	22.2	10.8	
GM	C10	PU	32.7	19.1	
GM	C10	PU	32.7	16.7	
GM	C10	PU	32.7	18.6	
GM	C10	PU (SUB)	32.9	18.3	
GM	C20	PU	33.1	19.2	
GM	G10	Van	38.1	21.8	
GM	G20	Van	38.1	15.3	
GM	G20	Van	38.1	18.1	
GM	G30	Van	38.1	20.9	
GM	K5	PU(Blaz)	34.3	18.3	
GM	K5	PU(Blaz)	34.3	19.6	↓
GM	K10	PU (4X4)	34.3	19.9	

Vehicle Usage Factors Affecting Fuel Economy

Soak Time Effects on Car Emissions and Fuel Economy

Robert L. Srubar and Karl J. Springer
Southwest Research Institute
Martin E. Reineman
U.S. Environmental Protection Agency

IN PART 86 of the Code of Federal Regulations, test procedures used to determine the conformity of motor vehicles to statutory levels of gaseous emissions and fuel economy are outlined. The 1975 Federal Test Procedure(1)* specified that a minimum 12-hour soak period precede emission tests for light-duty vehicles. No maximum soak length was specified. The test procedure for 1978 and later model year vehicles(2) requires a soak period between 12 and 36 hours in length.

This raises the question of what effect soak period length has on emission rates and fuel consumption; and just as important, is there some soak period length that results in abnormal emission behavior? In an attempt to obtain answers and insight to these and related questions, a group of light-duty passenger cars were studied by Southwest Research Institute

for the Environmental Protection Agency (EPA). This experimental study resulted in a comprehensive final report(3) from which this paper was derived.

OBJECTIVE

The objective was to quantify the effects of soak period length on emission rates and fuel consumption for a group of light-duty vehicles. Of secondary interest was whether cold start emission results could be predicted from tests following abbreviated soak periods.

TEST PROCEDURES, EQUIPMENT, AND TEST PLAN

All emission testing was done following the 1975 Federal Test Procedure (FTP) for light-duty vehicles. Evaporative emission testing, which is part of the 1975 FTP, was of no interest in this project and was therefore deleted. The hot (start) transient phase of the FTP would have been of little use in

---

* Numbers in parentheses designate References at end of paper.

*Paper 780083 presented at the Congress and Exposition, Detroit, February 1978.

--- ABSTRACT ---

Five light-duty vehicles were used to investigate HC, CO, and NO_x emissions and fuel economy sensitivity to changes in the length of soak period preceding the EPA Urban Dynamometer Driving Schedule (UDDS). Emission tests were conducted following soak periods 10 minutes to 36 hours in length. Each of the first 8 minutes of the driving cycle was studied separately to observe vehicle warm-up. Several engine and fuel system temperatures were monitored during soak and run periods and example trends are illustrated. The extent to which emission rates and fuel consumption are affected by soak period length is discussed.

Fig. 1 - 1976 Honda Civic CVCC under dynamometer test

Fig. 2 - Under hood thermocouple installation for soak-run temperature measurement

this study and was also deleted. The test procedure was begun by driving the test vehicle over the 23-minute, 12 km (7.5 mi) long EPA Urban Dynamometer Driving Schedule (UDDS) one time, then allowing it to soak (be parked) for a period of time at a nominal ambient temperature of 24 ± 2°C (75 ± 3°F). The Federal Register permits the soak to be between 20° and 30°C (68° and 86°F).

The vehicle was then driven over the UDDS cycle from a cold engine start. The vehicle exhaust was diluted, mixed, and sampled during the test with a positive displacement pump type constant volume sampler (CVS). The first 505 seconds of the UDDS is referred to as the "cold transient" phase, and the remainder the "stabilized" phase.

Emission tests were conducted following soak periods of 36, 16, 8, 4, 2, 1, 1/2, 1/3, and 1/6 hours. Emissions of hydrocarbons (HC), carbon monoxide (CO), carbon dioxide (CO_2), and oxides of nitrogen (NO_x) were measured using the hydrogen flame ionization method for HC, non-dispersive infrared analyzers for CO and CO_2, and chemiluminescence analyzer for NO_x. Fuel consumption was computed by the carbon balance method (i.e., carbon bearing compounds in the exhaust account for carbon in the fuel burned).

The front wheel drive 1976 Honda Civic CVCC mounted on the chassis dynamometer is shown in Figure 1. The chassis dynamometer used was a Clayton Model ECE-50 with belt drive inertia, which has a 37 kw (50 hp) water-brake power absorber. This model has 0.219 m (8.625 in.) diameter rolls spaced 0.438 m (17.25 in.) apart. Vehicle engine cooling during emission tests was accomplished with a single 150 m³/min (5300 cfm) fan positioned approximately 0.305 m (12 in.) from the vehicle front bumper. This is also shown in Figure 1. The constant volume sampler (CVS) used was a positive displacement pump unit with a nominal capacity of 8.5 m³/min (300 cfm).

Each vehicle tested was instrumented to measure carburetor inlet air, engine water out, oil sump, fuel tank vapor, and fuel tank liquid temperatures. Carburetor inlet air temperature was measured with a thermocouple positioned inside the air cleaner above the carburetor air horn. Engine water out temperature was measured in the upper radiator hose, and oil sump temperature was measured using a dipstick thermocouple. Fuel tank temperatures were measured with two thermocouples inserted through the tank filler neck, with one thermocouple submerged in liquid fuel and one at the tank end of the filler neck to measure fuel vapor temperature. Figure 2 shows the under hood thermocouples for the 1976 Plymouth.

These temperatures, measured with iron constantan thermocouples, were continuously recorded on a 6-channel strip chart recorder during all soak periods and emission tests. The sixth channel on the recorder was used to record vehicle miles per hour. In addition to this temperature record, the output of each thermocouple, corrected for room temperature, was electronically integrated during all tests and soak periods so that average soak or run temperatures could be obtained.

The temperature measuring apparatus can be seen in Figure 3 adjacent to a 1976 Ford LTD. During all soak periods and emission tests, am-

Fig. 3 - 1976 Ford LTD during soak with temperature measurement system

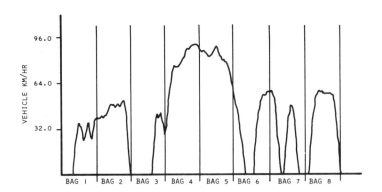

Fig. 4 - Transient portion of UDDS with sequential sampling periods indicated

bient temperature was continuously recorded with the multipoint recorder shown with the temperature measurement apparatus. This was done to insure that spurious data was not generated due to abnormal fluctuations in test area ambient temperature.

As an added investigative tool, a sequence of eight CVS sample bags was collected in the first 505 seconds of the UDDS. These samples, each representing approximately 63 seconds of the driving cycle, were taken in addition to the 505 second sample bag. Figure 4 shows the division of the first 505 seconds of the UDDS for purposes of the 8-bag sequential sampling. Each division of the UDDS speed-time trace in Figure 4 is 63 seconds and indicated by the vertical lines.

The CVS used was modified to allow a continuous sequence of dilute exhaust sample bags to be taken in addition to the two sample bags normally obtained. To simplify analysis, the values from the 505 second background sample were used in the calculations of mass emissions from the sequential samples. CVS flow calculations were corrected to compensate for the additional sample extracted. A schematic of the modified CVS is shown in Figure 5, and Figure 6 shows the sequential sampling system in operation.

For each automobile tested, three replicate emission tests were run following 16-hour soak periods; and two replicates were made following all other length soak periods. Minimal test-to-test variability was considered very important in the conduct of the laboratory tests. To this end, the same laboratory equipment, driver, analytical cart, and constant volume sampler were used throughout. Also, to avoid differences in the testing of each car, tests were performed in the sequence outlined in Table 1, grouping tests each day in the same manner for each car except the Ford LTD. In the event a test was repeated for any reason, the vehicle was preconditioned to a state comparable to that in the original test sequence.

In testing the Ford LTD, tests following the 36-hour and 8-hour and shorter soak periods were performed in a sequence opposite that in Table 1. Tests following shorter soak periods were run first, followed by tests using longer and longer soak periods. This was done to investigate the effects of different evaporative emission control system loading and to determine if there might be an unintentional bias to the results due to the order or history of tests. No obviously different trends in emissions or fuel consumption were linked to this change.

TEST VEHICLES

Five vehicles representing present vehicle populations and emission control systems were tested. In general terms, the vehicles were:

1. One 1976 GM automobile, large displacement engine, high inertia weight, pelleted oxidation catalyst.
2. One 1977 Ford automobile, large displacement engine, high inertia weight, monolith oxidation catalyst.
3. One 1976 Chrysler automobile, large displacement engine, high inertia weight, monolith oxidation catalyst.
4. One 1976 GM automobile, small displacement engine, low inertia weight, pelleted oxidation catalyst.
5. One 1976 Honda automobile, small displacement CVCC engine, low inertia weight.

The vehicles tested are described in Table 2; and several are shown in Figures 7, 8, and 9 under various stages of preparation, soak, or dynamometer test.

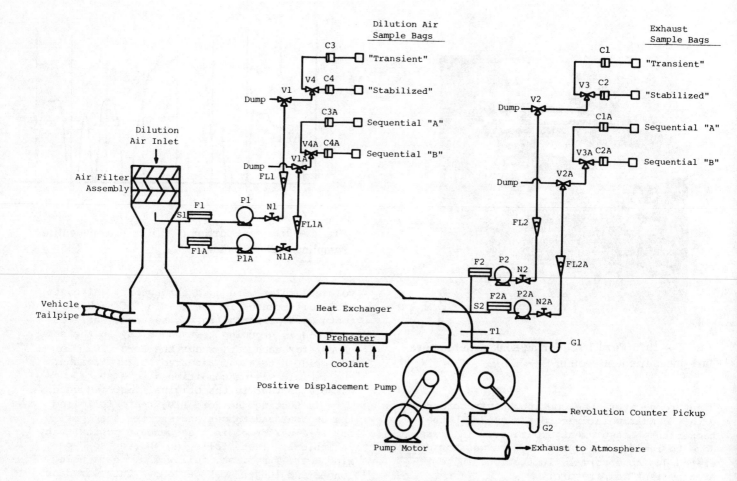

Fig. 5 - Flow schematic of modified CVS for 8-bag sequential sampling

Fig. 6 - 8-bag sequential sampling system in use during first 505 second portion of the UDDS

TABLE 1 - SEQUENCE OF LABORATORY TESTS FOLLOWING SPECIFIC LENGTH SOAK PERIODS

1. FTP following 16 hour soak period - 3 replicates
2. FTP following 36 hour soak period - 2 replicates
3. FTP following 8 hour soak period - 2 replicates
4. FTP following 4 hour soak period - 2 replicates
5. FTP following 2 hour soak period - 2 replicates
6. FTP following 1 hour soak period - 2 replicates
7. FTP following 1/2 hour soak period - 2 replicates
8. FTP following 1/3 hour soak period - 2 replicates
9. FTP following 1/6 hour soak period - 2 replicates

Table 2 - Description of Test Vehicles

Vehicle No.	1	2	3	4	5
Model Year	1976	1977	1976	1976	1976
Manufacturer	Chevrolet	Ford	Plymouth	Chevrolet	Honda
Model	Impala	LTD	Fury	Vega	Civic
Type of Vehicle	Sedan	Sedan	Sedan	Sedan	Sedan
Number of Doors	4	2	4	2	2
Odometer, km	20,515	7,787	20,590	14,320	17,168
Number of Cylinders	8	8	8	4	4
Displacement, litre (CID)	5.73 (350)	5.75 (351)	5.90 (360)	2.29 (140)	1.49 (91)
Carburetor Venturis	2	2	2	2	3
Transmission Type	Automatic	Automatic	Automatic	Automatic	Manual
Speeds	3	3	3	3	4
Tire Size	HR78-15	HR78-15	GR78-15	A78-13	600S12
Tire Type	Radial	Radial	Radial	Fiberglass Belted	Radial
Emission Controls	CAT EGR EFE CAN	CAT EGR AIR CAN	CAT EGR CAN	CAT EGR CAN	TR CAN
Test Weight (Inertia), kg (lbs)	2268 (5000)	2268 (5000)	2041 (4500)	1361 (3000)	907 (2000)

Fig. 7 - 1976 Chevrolet Vega under test

Fig. 8 - 1976 Chevrolet Impala under test

Fig. 9 - 1976 Plymouth Fury prepared for cold start 1975 FTP

EMISSION TEST RESULTS

Five vehicles were tested in this project, each with different emission levels. To be able to discuss trends only and not absolute emission rates, it is desirable to use a common baseline test and discuss deviations from that baseline. The 16-hour soak period is fairly representative of an overnight soak. Therefore, it will be used as the baseline. Values used in the following discussion are percent of values obtained following a 16-hour soak period.

$$\frac{\text{\% of}}{\text{baseline}} = \frac{\text{emission rate after given soak}}{\text{emission rate after 16 hr soak}} \times 100\%$$

Therefore, these values are normalized to the 16-hour values, which represent an overnight soak. For quantitative comparison of the emission levels of each vehicle, average emission rates and fuel consumption for each vehicle following a 16-hour soak period are listed in Appendix A.

COMPOSITE RESULTS - Emission rates and fuel consumption for each vehicle following specific length soak periods are shown in Figure 10. Normalized results for the 23-minute UDDS are plotted versus $\log_{10}$ of soak period length. The log-normal presentation of data was found to adequately define the changes early in the soak period as well as those after very long soak periods.

HC emission rates followed essentially the same trend for all vehicles, increasing with soak period length over the range of soak periods tested. The Honda Civic, the only non-catalyst car tested, had fairly stable HC rates following 2-hour and shorter soak periods but did follow the prevalent trend following longer soak periods.

CO emission rates were stable for four of the five cars following 2-hour and shorter soak periods and increased with increases in soak period length thereafter. The type of exhaust aftertreatment had a discernable effect on the level of CO following short soak periods. The Honda Civic, with no catalyst, had a minimum CO level of approximately 70 percent of the baseline soak length value. The cars equipped with pelleted catalysts (Impala, Vega) reached a minimum value of about 20 percent, and the monolith catalyst equipped cars (LTD, Fury) attained a minimum CO level of approximately 10 percent on a hot start. The CO emission rates following 4-hour and longer soak periods increased with increases in soak period length for all cars.

NO_x emission rates for the 23-minute cycle increased with increases in soak period length over the entire range of soak periods tested. Data for the Ford LTD is particularly scattered, but the scatter is accentuated by the comparatively low emission rate following a 16-hour soak period. The Plymouth Fury showed little variation in NO_x levels due to soak period length. For the other vehicles, NO_x following a 10-minute soak period was 80 to 90 percent of the baseline soak period value.

Composite fuel consumption for the UDDS is the most predictable of the items measured. For each of the five vehicles tested, the trend is a remarkably linear function of the $\log_{10}$ soak length, but the slopes of the functions for each vehicle are not similar enough to allow a good overall linear correlation.

A characteristic common to all emission rates and fuel consumption for the majority of the vehicles tested is higher levels following a 36-hour soak period than following the 16-hour soak. Except for NO_x, levels following an 8-hour soak period were lower than the 16-hour soak levels. This indicates that the shortest soak period allowable by the FTP would yield the lowest emission rates in most cases.

TRANSIENT AND STABILIZED PORTION RESULTS - For the transient phase of the UDDS, normalized emission rates and fuel consumption are plotted in Figure 11. For each emission rate and for fuel consumption, the behavior in the transient portion of the cycle is qualitatively identical to composite cycle results. The trends, however, are slightly exaggerated as compared to trends for the composite results.

Data from the stabilized portion of the UDDS for each car is plotted versus $\log_{10}$ of soak period length in Figure 12. CO, NO_x, and fuel consumption show weak trends with increased soak period length. HC emission rates are essentially constant in this portion of the cycle regardless of soak period length. It is impor-

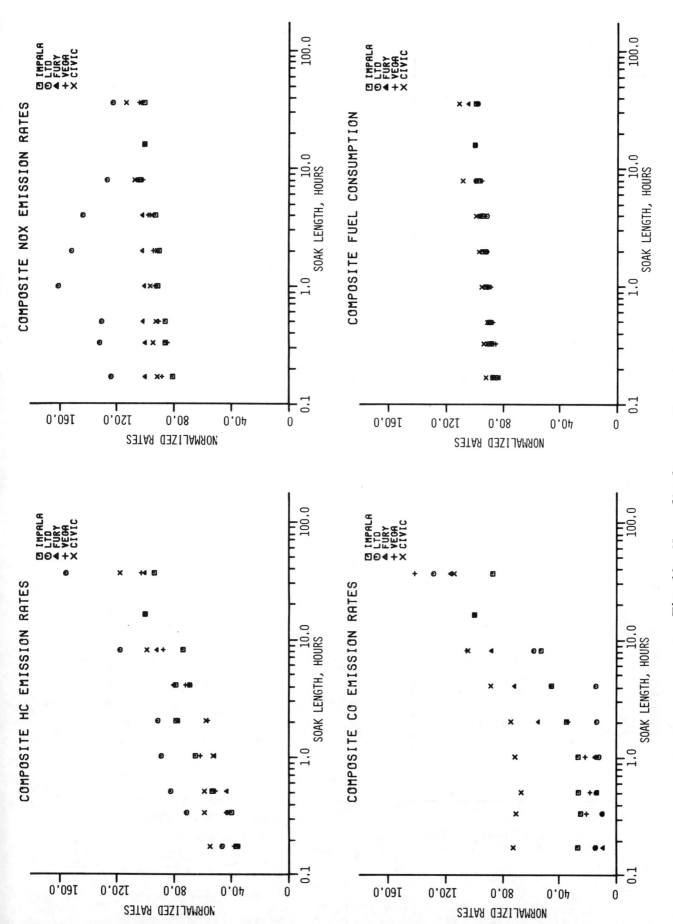

Fig. 10 - Normalized composite emission rates and fuel consumption

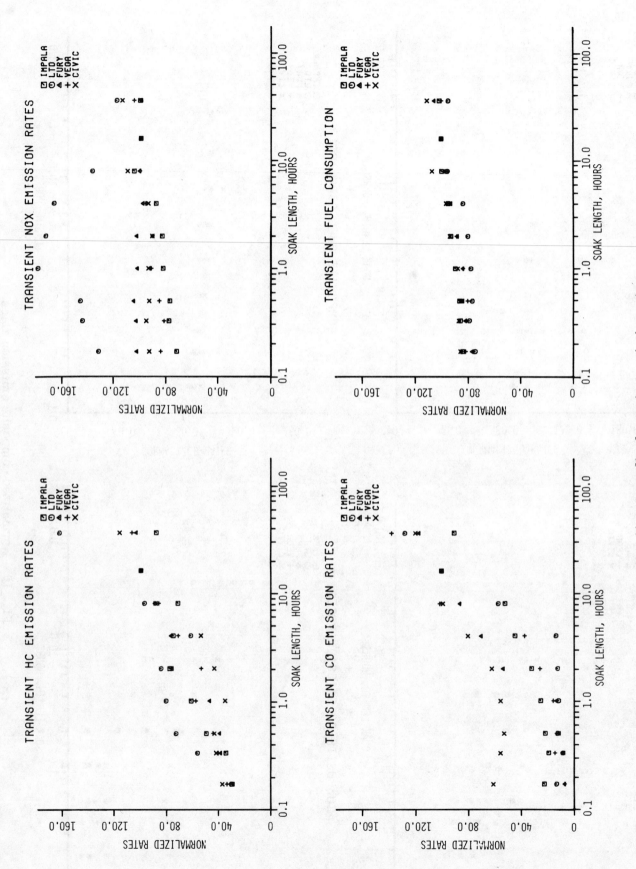

Fig. 11 - Normalized transient emission rates and fuel consumption

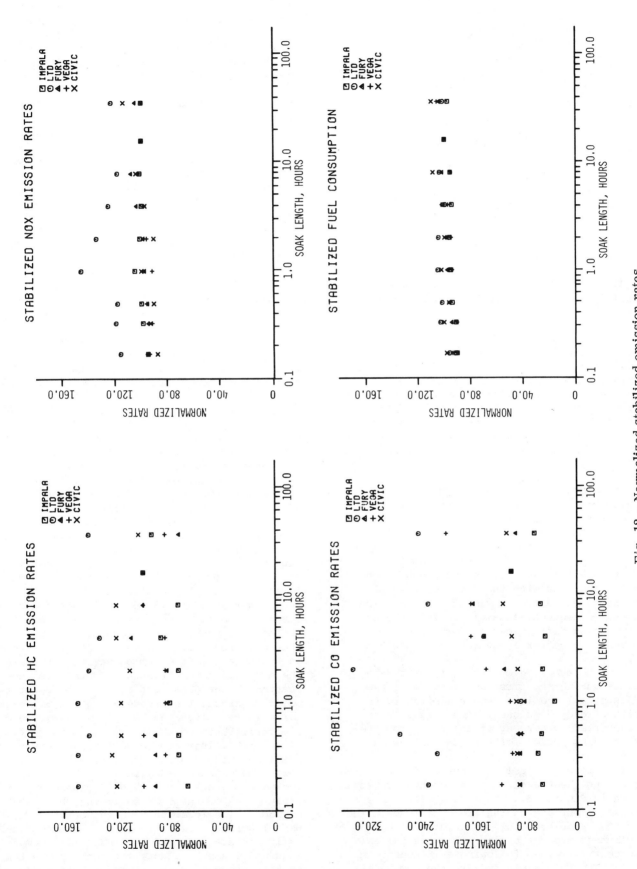

Fig. 12 - Normalized stabilized emission rates and fuel consumption

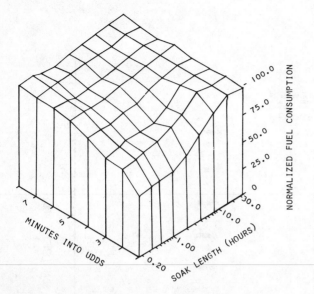

Fig. 13 - Fuel consumption during transient phase of the UDDS following specific length soak periods

tant to note the expanded scale of the plot of CO emission rates in Figure 12. Some of the catalyst equipped cars show considerable scatter in this portion of the cycle, but the amount of CO emitted in the stabilized portion is normally negligible compared to the magnitude of CO emissions in the transient portion. This accounts for the scatter and trends in the stabilized portion having little effect on the composite results.

SEQUENTIAL SAMPLING RESULTS - Mass emission rates and fuel consumption were calculated for each of the sequential sampling periods in the same way as for the transient and stabilized portions of the cycle. As a check on the validity of results from the eight sequential samples, the masses of each emission measured with the 505 second sample bag were compared to the sum of that emission measured in sequential samples. This check seldom showed differences of more than 5 percent and even better agreement was usually the case.

Fuel consumption measured with the 8-bag sequence during the first 505 seconds of the UDDS is normalized in a manner similar to the results already discussed. Absolute values are converted to percent of the value for the corresponding period for that car following a 16-hour soak period.

The trends in fuel consumption for the five cars during the transient phase of the UDDS can be seen in Figure 13. The figure is a three dimensional surface of the five car means. The horizontal axes are $\log_{10}$ soak period length and minutes into the cycle. The vertical axis is percent of values following a 16-hour soak period. The values plotted are the numerical averages of data for all five cars. The surface indicates that fuel consumption following a soak period as short as eight hours in length is fairly representative of the baseline values. Results of tests following shorter soak periods are not entirely representative of values following an overnight soak, even in the portions 8 minutes into the test.

Appendix B contains the tabulated data used to generate Figure 13. Also in Appendix B is the coefficient of variation for each mean used in Figure 13. These coefficients indicate good agreement between the five cars tested and justify the preparation of a single surface when presented in terms of percent of the baseline or 16-hour soak test results. The average of the coefficients of variation for all data points is less than 5 percent.

Generalizations similar to that for fuel consumption cannot be accurately made for HC, CO, or NO_x emission rates for the five cars. An analysis, similar to that made for fuel consumption, was made for HC, CO, and NO_x. As one would expect, the behavior of the emission rates with varying soak period lengths was unique for each car, even though it is qualitatively similar for all vehicles tested. As can be seen in Figure 11, the data for the transient portion of the UDDS indicates varied behavior for each automobile.

For the sake of illustration of some of the differences, Figures 14 and 15 are bar charts of emission rates following 10 minutes, 8 hours, and 36 hours of soak for the Plymouth Fury and the Chevrolet Vega. These three soak periods represent both extremes tested as well as a moderate soak period length. These two vehicles typify both full size and subcompact size cars with large V-8 and small I-4 engines, respectively. The Fury has a monolith oxidation catalyst while the Vega uses a pelleted catalyst. The road load and inertia settings are quite different because of the large difference in vehicle curb weight.

Figures 14 and 15 show fair agreement in trends in HC emission rates between the two vehicles, but the magnitude of changes is somewhat different for each car. The differences in the behavior of CO emission rates between the two cars is interesting. Both cars show scatter in the data; but even if the scatter is disregarded, the trends are quite different for each vehicle. The NO_x plots show some dissimilarity between the behavior of these two cars which illustrates a difference between the small and large cars tested. The high test weight cars tested (Impala, LTD, Fury) all showed comparatively high NO_x emissions in the early parts of the UDDS following short soak periods. The lower test weight cars (Vega, Honda) agreed qualitatively with the continually increasing trend in Figure 15. Complete sequential sampling data for each car tested in this program can be found in Reference 3.

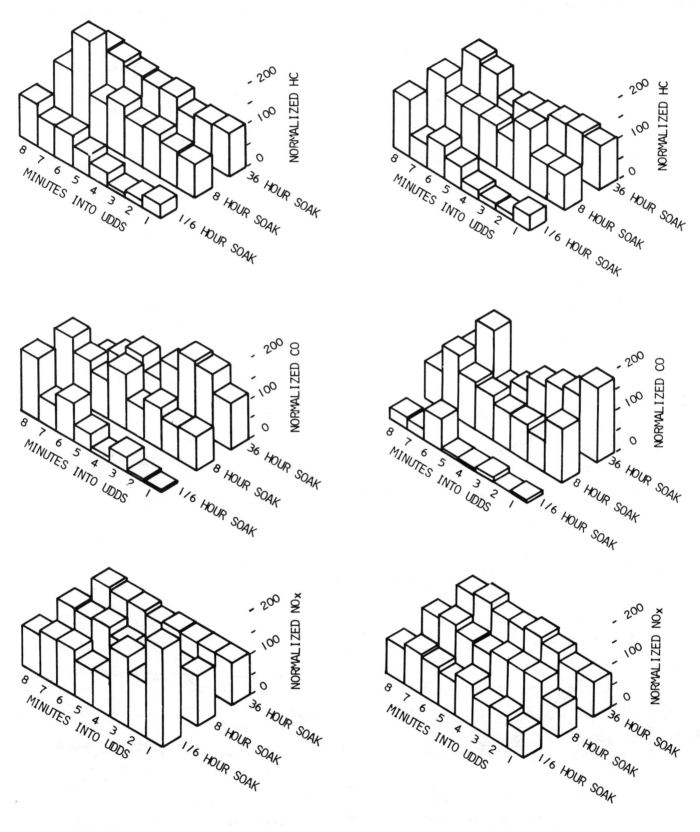

Fig. 14 – Sequential emission rates, 1976 Plymouth Fury

Fig. 15 – Sequential emission rates, 1976 Chevrolet Vega

TEMPERATURE MEASUREMENT RESULTS

As was the case of the sequential bag sample data, the temperature measurements made during each soak period and emission test indicate individual behavior for each vehicle. Overall, the three full-size sedans may be grouped together. For discussion purposes, the Ford LTD will be described as an example of what was observed. The two subcompact cars were found to have soak temperature profiles different from the full-size cars. Overall, the Vega had the lowest temperatures and is used as an example for illustration. Recall that the test sequence used with the Ford LTD was the reverse of that used on the other cars tested. The impact of this will be discussed later in this section.

Typical traces of vehicle temperatures monitored during a 16-hour soak period and during a subsequent 23-minute long emission test are shown in Figures 16 and 17 for the LTD and Vega, respectively. The 16-hour soak period was selected because it is most representative of an overnight soak normally used for emission testing. From the soak portions of these traces, the amount of soak time required for temperature stabilization can be seen. At the 12-hour point, complete stabilization of the various temperatures monitored had not occurred. Approximately 90 to 95 percent of the temperature transition occurred in the first 12 hours, but essentially complete stabilization appears to occur after 14 to 15 hours. After approximately eight hours of soak, the five temperatures monitored began to vary as a group; and no single component can be singled out as necessarily stabilizing first.

Figures 18 and 19 are bar charts of the average temperatures of each component during various soak period lengths. The maximum, minimum, and average temperatures measured are listed in Appendix C for each soak period. The minimum value is the temperature at the end of the soak period for all components measured except carburetor inlet air. This temperature

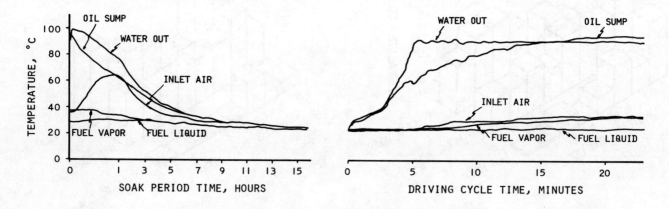

Fig. 16 - Vehicle temperature behavior during 16-h soak period and subsequent emission test, 1977 Ford LTD

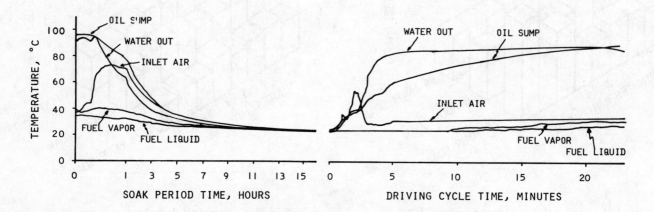

Fig. 17 - Vehicle temperature behavior during 16-h soak period and subsequent emission test, 1976 Chevrolet Vega

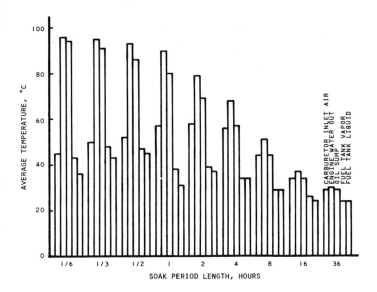

Fig. 18 - Average temperatures during specific length soak periods, 1977 Ford LTD

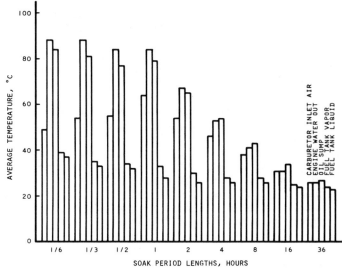

Fig. 19 - Average temperatures during specific length soak periods, 1976 Chevrolet Vega

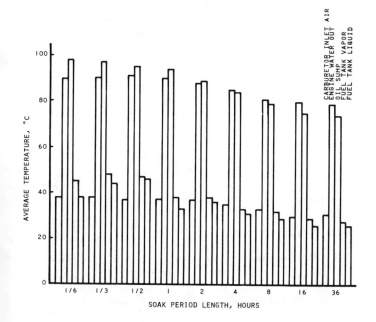

Fig. 20 - Average vehicle temperatures during emission tests following specific length soak periods, 1977 Ford LTD

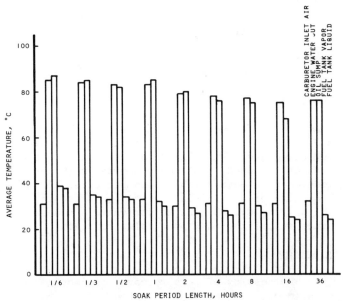

Fig. 21 - Average vehicle temperatures during emission tests following specific length soak periods, 1976 Chevrolet Vega

was found to be lower at the end of the preconditioning cycle for short soak periods but behaved like the other temperatures during long soak periods. The closer the minimum, or final, temperature was to 24° ± 2°C room temperature, the more complete was the soak stabilization.

Recall that the Ford LTD was tested using a sequence opposite that given in Table 1. The trends in fuel tank temperatures shown in Figures 18 and 19 reflect this change. The trends in average fuel tank temperatures seen in Figure 19 (increasing with decreases in soak period length up to the 10 minute points) are representative of the behavior of all cars tested in this sequence. The scatter in the Ford LTD temperature data during 1 hour and shorter soak periods suggests that one 23-minute cycle as preconditioning does not result in the same fuel tank temperatures as several cycles. This behavior is also apparent in the emission test temperature data discussed next.

The average temperature during each emission test is also a measure of the degree of stabilization which occurred during a soak. Figures 20 and 21 are bar charts of average temperatures during emission tests following each length soak period. For tests following soak periods of 1 hour and less, average temperatures during the 23-minute cycle are approximately constant for each car. Average temperatures during tests following 16-hour and 36-hour soak periods are also approximately equal but substantially lower, as expected. These observations agree closely with observations of emission rate and fuel consumption behavior following these length soak periods.

The temperature behavior shown in Figures 16 and 17 during emission tests suggests that the stabilized portion of the UDDS is affected by soak period length. Following a soak period long enough for vehicle temperature stabilization, only carburetor inlet air and engine water out temperature stabilized in the first 505 seconds of the UDDS. Following shorter soak periods, with elevated vehicle temperatures at the beginning of the driving cycles, the time into the driving cycle required for vehicle temperature stabilization is naturally reduced.

SUMMARY

The results of this research project pointed out some major differences in the behavior of emission rates, fuel consumption, and vehicular temperatures with soak period length of each vehicle; but many trends were common to all vehicles tested. The major findings in this study are as follows:

1. HC and NO_x emission rates and fuel consumption during the UDDS increased almost linearly with $\log_{10}$ of soak period length. HC emission rates following a 10-minute soak period are typically 40 percent of values following a 16-hour soak period but are not normally higher following a 36-hour soak than a 16-hour soak period. NO_x and fuel consumption following a 10-minute soak are approximately 90 and 85 percent of 16-hour soak values but are slightly higher following a 36-hour soak than a 16-hour soak period.

2. CO emission rates during the UDDS are stable following soak periods of 10 minutes to 1 hour in length, but the level ranges from 10 to 70 percent of the 16-hour soak value depending on the type of exhaust after treatment. Following 2-hour and longer soak periods, CO increased approximately linearly with $\log_{10}$ soak length.

3. Emission rates and fuel consumption during the transient portion of the UDDS show the same trends as composite cycle results with changes in soak period length, mentioned in 1. and 2. above.

4. Soak period length <u>does</u> affect emission rates and fuel consumption during the stabilized portion of the UDDS, but the magnitude of the effect is small compared to that observed during the transient portion.

5. The trends that HC, CO, and NO_x emission rates followed during each minute of the transient phase of the UDDS with varied soak period length were similar for each car tested. The magnitudes of changes for the five vehicles were so dissimilar, however, that a universal characterization was not possible.

6. Fuel consumption behavior during each 1-minute portion of the transient phase of the UDDS following different length soak periods is similar for each of the vehicles tested.

7. The soak time required for the vehicle temperatures measured to stabilize was from 14 to 15 hours for the five cars evaluated. Approximately 90 to 95 percent of the temperature transition occurred during the first 12 hours of soak, however.

8. Following a soak period long enough to allow vehicle temperatures to stabilize, only carburetor inlet air and engine water out temperature appear to reach equilibrium in the first 505 seconds of the UDDS.

ACKNOWLEDGMENT

The research upon which this publication is based was performed under Contract No. 68-03-2196, Task 10, for the Emission Control Technology Division of the U.S. Environmental Protection Agency.

REFERENCES

1. Federal Register, Vol. 40, No. 126, June 30, 1975, Section 86.177-11.
2. Federal Register, Vol. 42, No. 124, June 28, 1977, Section 86.135-78.
3. R. L. Srubar, "Emission and Fuel Economy Sensitivity to Changes in Light-Duty Federal Test Procedures." Final Report to the Environmental Protection Agency under Contract No. 68-03-2196, Task 10, May 1977.

Appendix A - Average Emission Rates and Fuel Consumption During UDDS Following 16-Hour Soak Period

		Chevrolet Impala	Ford LTD	Plymouth Fury	Chevrolet Vega	Honda Civic
UDDS Composite,	HC, g/km	0.60	0.45	0.60	0.39	0.83
	CO, g/km	10.92	8.78	7.36	4.62	4.10
	NO_x, g/km	1.16	0.76	2.01	0.80	1.00
	F.C., ℓ/100 km	21.02	18.39	18.02	11.94	8.48
Transient Phase,	HC, g/km	1.10	0.79	1.14	0.74	1.42
	CO, g/km	19.80	18.27	15.06	8.97	4.91
	NO_x, g/km	1.54	0.65	2.11	1.14	1.27
	F.C., ℓ/100 km	20.95	19.92	18.32	12.75	8.78
Stabilized Phase,	HC, g/km	0.15	0.12	0.11	0.06	0.29
	CO, g/km	2.78	0.07	0.29	0.61	3.35
	NO_x, g/km	0.80	0.87	1.93	0.49	0.74
	F.C., ℓ/100 km	21.17	17.07	17.82	11.24	8.33
Sequential Sampling, Minute 1	HC, g/km	6.04	11.88	10.23	6.88	12.65
	CO, g/km	123.85	304.06	190.15	61.56	15.62
	NO_x, g/km	3.40	0.46	1.49	1.44	1.94
	F.C., ℓ/100 km	43.65	59.22	43.06	26.00	21.10
Minute 2	HC, g/km	1.85	0.21	1.79	1.45	1.28
	CO, g/km	30.98	0.64	14.83	24.22	3.72
	NO_x, g/km	1.32	0.28	1.08	0.40	1.20
	F.C., ℓ/100 km	18.15	16.37	15.69	11.96	1.20
Minute 3	HC, g/km	5.14	0.50	5.00	1.96	5.23
	CO, g/km	67.42	2.41	19.82	20.82	30.07
	NO_x, g/km	0.68	0.25	1.99	1.85	1.11
	F.C., ℓ/100 km	50.14	37.63	41.15	25.03	15.64
Minute 4	HC, g/km	1.04	0.16	0.30	0.35	1.08
	CO, g/km	20.54	7.66	4.88	5.57	4.04
	NO_x, g/km	2.05	0.70	3.11	1.84	2.04
	F.C., ℓ/100 km	21.18	19.22	18.27	12.58	8.88
Minute 5	HC, g/km	0.15	0.07	0.14	0.09	0.05
	CO, g/km	2.19	0.10	0.25	1.00	1.15
	NO_x, g/km	0.75	0.33	1.60	0.67	1.01
	F.C., ℓ/100 km	12.04	11.14	10.37	7.65	5.82
Minute 6	HC, g/km	0.21	0.13	0.14	0.07	0.35
	CO, g/km	4.54	0.07	0.26	0.65	4.47
	NO_x, g/km	1.68	0.48	2.44	1.03	1.05
	F.C., ℓ/100 km	23.86	19.87	19.28	12.77	8.51
Minute 7	HC, g/km	0.23	0.15	0.21	0.09	0.48
	CO, g/km	5.72	0.09	1.52	1.17	5.24
	NO_x, g/km	1.45	0.97	2.12	1.03	1.17
	F.C., ℓ/100 km	23.23	20.59	20.82	12.56	9.22
Minute 8	HC, g/km	0.17	0.11	0.09	0.04	0.33
	CO, g/km	4.34	0.18	0.26	0.49	3.27
	NO_x, g/km	1.24	0.61	2.05	1.00	1.16
	F.C., ℓ/100 km	19.59	16.18	15.87	11.33	7.64

Appendix B - Means and Coefficients of Variation of Average Fuel Consumption from Sequential Data for All Five Cars

Means of Five Car Data

Soak Length	Min. 1	Min. 2	Min. 3	Min. 4	Min. 5	Min. 6	Min. 7	Min. 8
.17 hrs.	56.4	74.0	74.8	83.6	87.6	91.7	90.8	91.8
.33 hrs.	57.6	74.1	77.2	82.1	89.5	91.1	93.3	95.4
.50 hrs.	57.9	75.6	76.4	83.2	89.3	88.3	93.2	92.4
1.00 hrs.	59.3	77.6	77.2	85.6	90.3	89.9	94.0	91.0
2.00 hrs.	66.5	83.3	78.2	87.2	93.1	91.2	95.6	95.4
4.00 hrs.	78.9	88.2	85.2	92.0	95.0	93.8	97.4	97.1
8.00 hrs.	92.2	102.4	98.6	95.5	97.6	96.0	97.8	99.3
16.00 hrs.	100.0	100.0	100.0	100.0	100.0	100.0	100.0	100.0
36.00 hrs.	98.4	100.5	97.3	99.0	100.6	97.8	99.5	101.7

Coefficients of Variation of Five Car Means

Length	Min. 1	Min. 2	Min. 3	Min. 4	Min. 5	Min. 6	Min. 7	Min. 8
.17 hrs.	15.1	6.7	10.3	6.7	6.3	7.8	3.2	8.6
.33 hrs.	14.4	9.3	11.3	10.8	6.3	6.7	5.6	8.9
.50 hrs	14.9	10.4	10.4	6.6	5.5	9.3	2.8	4.6
1.00 hrs.	18.4	11.8	9.3	9.3	8.0	7.3	4.3	5.0
2.00 hrs.	19.5	9.9	8.3	6.8	6.3	4.7	5.6	7.0
4.00 hrs.	16.6	9.4	11.7	8.0	6.3	5.7	1.8	5.1
8.00 hrs.	9.0	13.7	15.8	8.7	6.7	6.2	5.3	3.6
16.00 hrs.	0.0	0.0	0.0	0.0	0.0	0.0	0.0	0.0
36.00 hrs.	9.0	5.5	6.9	7.3	5.8	5.2	3.4	8.4

Appendix C - Soak Temperature Data, °C
1977 Ford LTD and 1976 Vega

Soak Length		Carburetor Inlet Air		Engine Water Out		Oil Sump		Fuel Tank Vapor		Fuel Tank Liquid	
		LTD	Vega	LTD	Vega	LTD	Vega	LTD	Vega	LTD	Vega
10 min	max	53	59	101	92	99	88	44	39	37	37
	min	38	33	91	86	92	81	42	39	35	37
	avg	45	49	96	88	94	84	43	39	36	37
20 min	max	59	63	101	92	98	87	48	35	43	33
	min	37	32	91	82	56	77	48	35	42	32
	avg	50	54	95	88	91	81	48	35	43	33
30 min	max	60	61	101	91	98	83	48	35	46	32
	min	37	34	85	77	75	72	47	34	45	31
	avg	52	55	93	84	86	77	47	34	45	32
1 hr	max	63	67	103	97	99	93	39	35	31	29
	min	37	34	76	70	65	66	37	32	28	28
	avg	57	64	90	84	80	79	38	33	31	28
2 hrs	max	63	63	102	93	99	90	43	34	37	29
	min	39	31	63	45	51	48	37	28	36	26
	avg	58	54	79	67	69	65	39	30	37	26
4 hrs	max	69	62	103	95	104	89	45	34	37	29
	min	40	34	46	35	39	37	30	26	29	25
	avg	56	46	68	53	57	54	34	28	34	26
8 hrs	max	61	65	100	96	96	94	40	35	34	27
	min	32	28	33	28	29	29	26	27	25	25
	avg	44	38	51	41	44	43	29	28	29	26
16 hrs	max	61	67	99	71	97	93	38	34	32	32
	min	26	24	25	23	25	23	24	22	23	22
	avg	34	31	37	31	34	34	26	25	24	24
36 hrs	max	62	66	99	93	96	93	38	34	31	27
	min	24	23	24	24	24	23	24	22	23	21
	avg	29	26	30	26	29	27	24	24	24	23

Ambient Temperature and Trip Length—Influence on Automotive Fuel Economy and Emissions*

B. H. Eccleston and R. W. Hurn
U.S. Dept. of Energy,
Bartlesville Energy Research Center,
Bartlesville, OK

INTRODUCTION

It has been known for many years that ambient temperature affects both the rate of fuel consumption and the levels of exhaust emissions (1-6)*. However, recent major changes in engines and their fuel systems have resulted in significant decreases in fuel consumption and may have altered the manner or degree in which ambient temperature affects fuel consumption. These recent changes in automotive systems have resulted in the need for currently-applicable engineering information on the influence of temperature and trip length.

Short trips made from a cold start have a major influence on fuel economy because the rate of fuel consumption is the greatest during the first few miles that are driven. In fact, trips of 5 miles or less (estimated to account for about 62% of automobile trips) account for only 16% of vehicle miles traveled, but require an estimated 31% of the fuel consumed (7-8). This perspective emphasizes the importance of sufficient information for a reliable estimate of the short trip fuel economy and of the effect of temperature. It is also apparent that improvements in short trip fuel economy could have a significant impact on energy conservation.

Current federal fuel economy and emission measurement procedures use dynamometer tests made within a range of 68° to 86°F. This is a much

*Numbers in parentheses designate References at the end of the paper.

*Paper 780613 presented at the Passenger Car Meeting, Troy, June 1978.

higher temperature than the 30° to 40°F average daily minimum temperature in the most populous U.S. cities for six months of the year. As a need for more rigorous assessment of fuel consumption and pollutant effects develops, the fuel economy and emissions data from the federal test procedure (FTP) may require adjustment to reflect effects attributable to seasonal and geographic ambient temperatures. The present FTP fuel economy and emissions measurement partially accounts for the cold start effect by weighting the cold and hot start values according to an assumed hot/cold start ratio. The weighting that is used amounts to assuming that 3 of every 7 trips are made from a cold start within the 68° to 86°F temperature, but such an assumption cannot adequately factor the start and warmup effects on economy or emissions when the cold start is outside this temperature range. Nor does the use of the FTP weighted value permit accounting for regional differences where either the major portion of the trip is made at a cold ambient or conversely at very warm temperatures.

The present study used a limited sample of vehicles and the results are not presumed to provide adjustment factors for use in weighting FTP data for temperature and short trip effects. The study does, however, provide information on the magnitude of these and also information for a limited comparison of 1977 with earlier vehicles with regard to temperature sensitivity.

The work reported was sponsored by the Department of Transportation, Transportation Systems

ABSTRACT

Experimental work was done to examine the interrelationships among automotive fuel economy, ambient temperature, cold-start trip length, and drive-train component temperatures of four 1977 vehicles. Fuel economy, exhaust emissions, and drive-train temperatures were measured at temperatures of 20, 45, 70, and 100°F using the 1975 Federal test procedure and the Environmental Protection Agency's highway fuel economy test. Results showed that vehicles used for short cold-start trips consume fuel at a much greater average rate than during long trips, and the effect is magnified with decreasing ambient temperature.

Center, Kendall Square, Cambridge, MA 02142.

EXPERIMENTAL

Four 1977 vehicles were used in a series of experiments for which the vehicles were soaked for approximately 12 hours at the designated test temperature then operated through the federal test cycle for measurement of fuel economy and emissions. Data were obtained at test temperatures of 20, 45, 70, and 100°F using a chassis dynamometer in a fully-controlled environment. Emissions and fuel economy were determined on both a continuous basis and by the conventional "bag analysis" method. The temperature profile for each of several key engine and drive-train components was determined from thermocouple output during the test procedure. Additional data were gathered for steady-state modes of vehicle operation.

VEHICLES - Four 1977 vehicles were chosen to represent the classifications, sub-compact, compact, intermediate, and full-size. They were: Ford 140-CID, four-cylinder Pinto; Chevrolet 250-CID, six-cylinder Nova Concours; Oldsmobile 350-CID, eight-cylinder Cutlass; and Plymouth 400-CID Gran Fury. Descriptive information for the vehicles is provided in Table 1. All were low mileage leased vehicles. They were tuned to the manufacturer's specifications and 4,000 miles of city/highway road mileage accumulated to further stabilize the engines. Emissions of carbon monoxide (CO), hydrocarbon (HC), and oxides of nitrogen (NOx) were determined prior to assignment in the testing phase; this qualification testing showed that idle mixture adjustments were required on two of the vehicles to bring CO within specification.

No further adjustments were made during the test period.

FUELS - Fuels were selected to minimize the risk of an interaction of volatility and temperature with fuel consumption. A fuel with 8.6 psi Reid Vapor Pressure (RVP) and a 10% distillation temperature of 134°F was used at 45, 70, and 100°F ambients. A fuel with 12.9 psi RVP and a 10% distillation temperature of 108°F was used at 20°F ambient.

DRIVING CYCLES - A combination of the Environmental Protection Agency's (EPA) 1975 FTP and the highway fuel economy test (HWFET) driving cycles was used. This selection was made so that standard FTP and HWFET fuel economy data could be one part of the temperature, emissions, and fuel economy results. Five phases of the 1975 procedure constituted the test cycle: cold transient (CT), stabilized (ST), hot transient (HT), highway fuel economy warm-up (HWFE No. 1), and the highway fuel economy test (HWFE No. 2). A 10-minute soak interval was included between the stabilized and hot transients, 7 minutes of engine idle between HT and HWFE No. 1, and 3 minutes of engine idle between HWFE No. 1 and No. 2 during which time no emissions or temperature measurements were made. These periods of engine idle were to provide time for analytical measurements that required the free time. Data from the continuous analysis were used for real-time judgment of test validity and enabled the calculation of fuel economy at selected time intervals during the driving cycle. A description of the test cycle with times and distances is listed in Table 2.

INSTRUMENTS AND APPARATUS - The chassis dynamometer was a Clayton ECE-50-120 with direct-

Table 1 - Vehicle Description

DOE/FESR No.*	156	157	158	159
Manufacturer	Ford	Plymouth	Oldsmobile	Chevrolet
Model	Pinto	Gran Fury	Cutlass	Nova Concours
Engine	4-cylinder	V8	V8	6-cylinder
Displacement, cubic inch	140 (2.3 L)	400	350	250
Carburetor venturi	2V	4V	4V	1V
Compression ratio	9.0	8.2	8.0	8.3
Rated horsepower	89	190	170	110
Transmission	Auto	Auto	Auto	Auto
Axle ratio	3.18	2.71	2.41	2.73
Air conditioning	Yes	Yes	Yes	Yes
Curb weight	2428	4250	3913	3287
Actual weight	2640	4610	4000	3560
Emission controls:				
Air	No	No	No	No
Catalyst	Yes	Yes	Yes	Yes
EGR	Yes	Yes	Yes	Yes
Power:				
Power steering	Yes	Yes	Yes	Yes
Power brakes	No	Yes	Yes	Yes
Cruise control	No	Yes	Yes	No
Inertia weight used, lbs	3000	5000	4500	4000
Odometer first test	4046	4546	3980	3825
Odometer last test	5800	5486	6000	4350

Table 2 - Modifed 1975 FTP and HWFET Driving Schedule

Designation	Time Interval, Minutes	Cumulative Distance, Miles
Cold transient	0 - 8.4	3.6
Stabilized	8.4 - 22.9	7.5
Soak	22.9 - 32.9	7.5
Hot transient	32.9 - 41.3	11.0
Idle	41.3 - 48.3	11.0
Highway warmup (Highway No. 1)	48.3 - 61.0	21.3
Idle	61.0 - 64.0	21.3
Highway fuel economy (Highway No. 2)	64.0 - 76.8	31.6

NOTE.-No measurements were made during soak or idle periods.

Table 4 - Fuel Consumption for Discrete Test Segments -- Four Car Averages --

Trip, Miles*	Fuel Economy, mpg			
	20°F	45°F	70°F	100°F**
0 - 3.6	9.6	11.3	12.9	12.0
0 - 7.5	11.1	12.4	13.5	12.2
0 - 11.1	11.9	13.0	14.0	12.5
0 - 21.3	14.6	15.5	16.3	14.5
0 - 31.6	16.0	16.7	17.4	15.5
21.3 - 31.6	20.6	20.8	21.2	18.6

	Fuel Consumed, Gallons/100 Miles			
	20°F	45°F	70°F	100°F**
0 - 3.6	10.46	8.82	7.78	8.35
0 - 7.5	9.01	8.05	7.39	8.20
0 - 11.1	8.38	7.69	7.15	8.00
0 - 21.3	6.85	6.46	6.15	6.90
0 - 31.6	6.25	5.99	5.75	6.47

	Highway Fuel Consumed, Gallons/100 Miles			
	20°F	45°F	70°F	100°F**
21.3 - 31.6	4.85	4.80	4.72	5.38

*Trip miles represent the cumulative mileage in sequential completion of test segments as described in Table 2.
**100° F tests with air conditioner in operation.

drive inertia system and road load power control. Carbon monoxide, carbon dioxide (CO_2), HC, and NO_x analytical instruments were conventional with computer data acquisition and software used for continuous measurements of emissions and fuel consumption. The test cell temperature was controlled at 20, 45, 70, and 100°F. However, the 20°F could not be maintained throughout the 77-minute test so the starting temperature was reduced to approximately 10°F, resulting in approximately 20°F average for the test. The average conditions of temperature, humidity, and barometric pressure are listed in Table 3.

RESULTS AND DISCUSSION

Examination of the individual vehicle data established that the trends in fuel consumptions and emissions with trip length were basically unchanged among the four vehicles. Having noted this commonality, the results from the four vehicles are averaged for purposes of discussions to follow.

TRIP LENGTH AND FUEL CONSUMPTION - Results (Table 4) show that the shorter the trip the lower the fuel economy in mpg. This must result mainly from fuel-air mixture enrichment and lower drive-train efficiency during engine and drive-train warm-up period. An additional factor affecting fuel economy and emissions is the vehicle duty cycle or driving pattern. With respect to its effect on fuel economy, a "mild" cold start dynamometer duty cycle would be start cold with acceleration after start to a defined speed followed by cruise at constant load. The federal test procedure for emissions tests represents a more severe but more realistically "average" short trip duty cycle. This driving segment followed by the less severe highway fuel economy schedule may represent an appreciable portion of vehicle miles traveled--i.e., very short trips being almost all city with frequent stops represented by the emission test cycle and trips of 11 miles or more involving urban highway driving represented by the HWFET. While our rationale for choice of the combination of test modes is believed to be logical, it should be clearly understood that the choice was not derived from critical analysis of driving patterns. All discussion of results should therefore be taken as an analysis of elemental effect and not necessarily deductive of what happens in practice.

Fig. 1 shows the complete time/speed profile for the cycle used. To enable some visualization of the severity of the cycle, fuel economy values for 30 and

Table 3 - Test Conditions

Nominal Test Temp., °F	Average Temperatures					Full Test		Barometric Pres., mmHg		Moisture, Grains/lb	
	Test Phase										
	Cold Trans.	Stabilized	Hot Trans.	Hwy. #1	Hwy. #2	Avg.	Std. Dev.	Avg.	Std. Dev.	Avg.	Std. Dev.
20	14	18	18	25	29	20.8	6.1	745	4	12	2
45	46	47	46	48	48	47.0	1.0	741	2	23	6
70	71	70	71	73	73	71.6	1.3	743	2	48	5
100	100	101	101	102	103	101.4	1.1	742	2	73	9

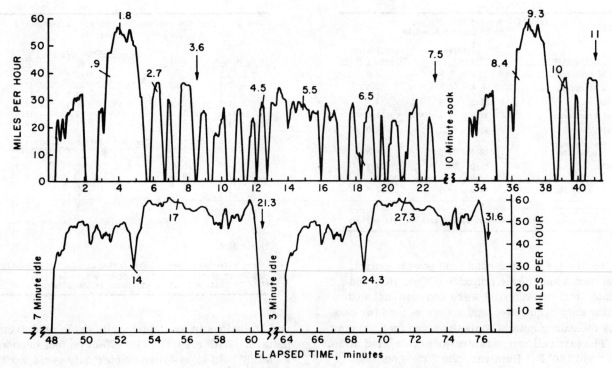

Fig. 1 - Driving schedule used - Numerical values with arrows toward x axis indicate miles at end of a FTP driving phase - Other numerical values on curves indicate trip distances in miles used in the report

60 mph cruise are shown, Fig. 2. From these data it is apparent that the cruise cycle at 60 mph is less severe than the test cycle. Temperature data from drive-train measurements also showed a slower rate of warmup for the constant speed tests.

Analyses of the results in the study show a solid relationship of fuel economy to trip length. Data for the four ambient temperatures (Fig. 3 and Table 4) show that at 70°F over a distance of 32 miles, average fuel economy at 70°F is 35% greater than the

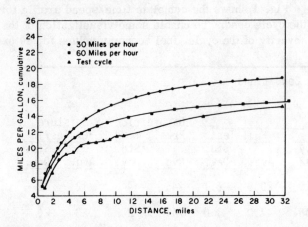

Fig. 2 - Fuel economy for the test cycle and 2 constant speed tests, both from cold start - vehicle No. 158 at 20°F ambient

value attained over a 3.6-mile trip. At 20°F ambient, the mpg value is 67% greater for the longer trip. Another useful comparison is provided by the data of Table 4 where values are shown for fuel consumption at varied trip lengths compared to the fuel consumption for the EPA highway test cycle. At the 20°F test temperature, 100 miles of driving composed of cold-start short trips of 3.6 miles would require over twice the fuel required for 100 miles of warmed-up highway driving. A similar comparison at 100°F shows that the fuel requirement for short trips is still over 1-1/2 times the requirement for high-way driving. (Averaged values of fuel economy measurements from which the above were calculated are summarized in Table 7 which appears later.)

The data of Table 4 for 100°F ambient were taken with the vehicles' air conditioners in operation and the fuel consumption therefore represents both the air conditioner energy requirement and temperature effect. The temperature effect between 70° and 100°F cannot be inferred directly; however, comparison of the 70° and 100°F data show a 7 to 12% increased fuel consumption at the 100°F ambient with the difference unchanged between city and highway driving. Below 70°F the temperature effect was much accentuated in city driving and it might be expected that if there were a 70 to 100°F ambient effect, the same test mode sensitivity would be seen. None was observed and it was therefore concluded

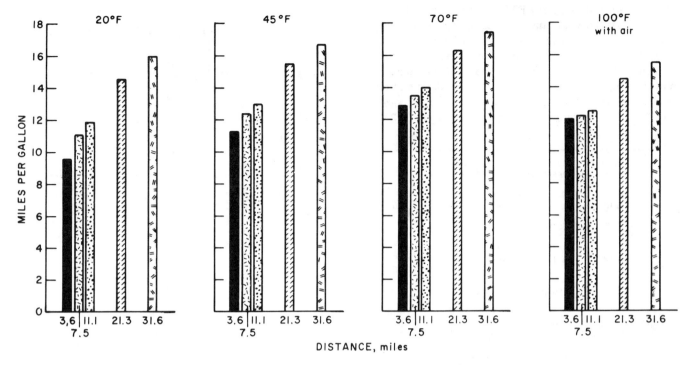

Fig. 3 - Fuel consumption rate as influenced by trip length, 4-vehicle average (100°F ambient with air conditioner in operation)

that fuel penalty in the 70 to 100°F change probably was attributable in large part to use of the air conditioner at the high temperatures. However, the fact is not established in the experimental data and more definitive information will be obtained for inclusion in a later report of the work.

RATE OF WARMUP FOR FTP CYCLES - Cumulative fuel economy values at one-mile intervals of the FTP were obtained from the continuous measurements; data are shown in Table 5 and Fig. 4. As shown in Fig. 4, fuel consumption is influenced by distance traveled and is strongly influenced by the driving modes as indicated by the slope changes of the curve. The data for the 100°F ambient (with the air conditioner operating) are believed to show an overriding effect of the air conditioner load as discussed earlier.

AMBIENT TEMPERATURE AND FUEL CONSUMPTION - The effect of temperature including integrated trip length effects can be shown in terms of the familiar FTP city and highway fuel measurement; data for the 70°F test provide a near-equivalent

Table 5 - Cumulative Fuel Economy at 1-Mile Intervals of the Federal Test Procedure--4-Vehicle Average

Distance, Miles	Fuel Economy, mpg			
	20°	45°	70°	100°*
0.9	6.0	8.1	10.5	11.1
1.79	7.8	9.6	12.2	12.6
2.69	9.4	11.3	13.4	12.6
3.59	10.0	11.9	13.5	12.4
4.53	10.3	11.9	13.4	12.4
5.54	11.2	12.9	14.2	12.9
6.53	11.6	13.2	14.4	12.8
7.50	11.6	13.1	14.2	12.5
8.44	11.7	13.1	14.2	12.5
9.24	12.0	13.3	14.4	12.7
10.14	12.4	13.7	14.7	12.9
11.04	12.5	13.7	14.7	12.9

*With air conditioner in operation.

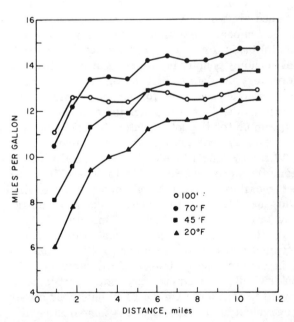

Fig. 4 - Cumulative fuel economy at 1-mile intervals of the FTP (See Fig. 1 for driving cycle)

of the specified standard FTP and is used as the base for comparison with other test temperatures (Table 6). The greatest temperature effect found in these tests for the FTP fuel economy measurements was a 15% increase in city driving at the 20°F ambient with a comparable 7% increase for the 45°F ambient. In the highway cycle the increase in fuel use from 70°F to 20°F ambient was found to be only 2 to 3%.

Table 6 - Ambient Temperature and Fuel Consumption--4-Vehicle Average

Test Cycle	Test Temperature, °F			
	20	45	70	100*
FTP Fuel Economy, Miles Per Gallon				
City	12.9	13.9	14.8	13.1
City/Highway	15.5	16.3	17.1	15.1
Highway	20.6	20.9	21.2	18.6
FTP Fuel Consumption, Gallons/100 Miles				
City	7.79	7.20	6.76	7.65
City/Highway	6.47	6.12	5.84	6.62
Highway	4.85	4.80	4.72	5.38
FTP Fuel Consumption, % Increase Over 70° Test				
City	15	7	Base	13
City/Highway	11	5	Base	13
Highway	3	2	Base	14

*With air conditioner in operation.

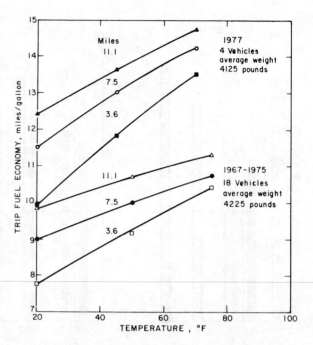

Fig. 5 - Sensitivity to ambient temperature

Findings in the current study were compared with those of previous years (Fig. 5) and the comparison shows fuel economy of the 1977 vehicles to have increased sensitivity to ambient. Whereas the short-trip fuel economy of the 1977 vehicles at standard ambient is shown to be increased by roughly 1/3 over the earlier models, the comparable increase at 20°F is no more than 15 to 20%. The data that were obtained are wholly inadequate to establish vehicle population characteristics but they do provide substantial basis for concern about a possibly increased fuel economy penalty in operating at reduced ambients. Moreover, the data and trends that appear to be observable suggest need for experimental work to determine the degree in which the several factors are responsible for the deterioration in fuel economy and the corollary need to find corrective measures.

VEHICLE TEMPERATURES - The following temperatures were recorded: Air-to-the-vehicle, air-to-the-carburetor, engine oil, coolant, transmission fluid, and differential lubricant. Thirty readings were taken during the elapsed 77 minutes of a test; no readings were taken for the soak and idle periods. A typical temperature warm-up profile is illustrated in the data for one vehicle (Fig. 6). Influence of the soak and idle periods on temperature profile development is apparent with the 10-minute soak causing the greatest disturbance. Of the temperatures monitored, the coolant and oil most rapidly approach equilibrium values; the transmission and differential

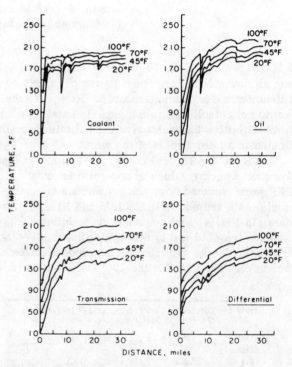

Fig. 6 - Vehicle temperatures and distance traveled during cycle operation from a cold start - Vehicle No. 156

approach most slowly. Of the temperatures monitored, the transmission and differential were affected to the greatest degree by a change in test temperature. Referring to maximum temperatures reached during the test cycles, the transmission showed highest sensitivity to ambient with its maximum increased 0.7°F per degree change in ambient; corresponding values were 0.5°F for the differential and 0.3°F and 0.2°F for oil and coolant. With a 70°F ambient, maximum temperatures in the test averaged for the transmission, 188°F--differential, 159°F--oil, 214°F--and coolant, 203°F. For these tests air flow was provided into the frontal area of the vehicles at velocity keyed to and following roll (i.e., vehicle) speed; radiant heat loads and road vehicle surface effects were not simulated. Therefore, it is not known how well the temperature profiles established in the tests would correlate with values to be found in actual road use. However, the general shape and character of the curves are expected to bear a reasonable correlation with road temperatures.

STEADY-STATE OPERATION - Measurements from a cold start with rapid acceleration to a constant speed were made to provide information on warm-up rate unaffected by load variation within the FTP cycle. Two vehicles were operated on the dynamometer at speeds of 30, 40, and 60 mph with inertial loading as specified for the EPA-FTP. Test ambients were 20, 45, and 70°F with air conditioner off; and 100°F with air conditioner on. Temperatures, fuel economy, and drive shaft torque were measured. Data were obtained at one-minute intervals to 9 minutes, then at 3-minute intervals to 60 minutes. Results are illustrated in Fig. 7; data

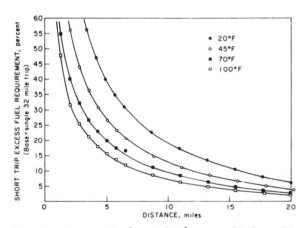

Fig. 7 - Excess fuel required for multiple cold start trips equivalent to a single 32-mile trip, steady-state cruise of 45 mph - Vehicle No. 156

are for the Ford Pinto operated at 100°F ambient and 45 mph. The unit on the ordinate, "Short Trip Excess Fuel Requirement, percent",* was chosen to emphasize the marked effect of cold start short trips on fuel economy. To obtain "percent excess fuel",* the cumulative fuel required for a trip of a given length was multiplied by the ratio of 32 miles to the trip length. This was subtracted from the

*Excess fuel = [(Fuel for trip $\frac{32}{\text{trip miles}}$ - fuel for 32 miles) ÷ fuel for 32 miles] x 100.

fuel requirement for the single 32-mile trip and the difference converted to a percentage. For the illustration of Fig. 7 the total trip length was taken at 32 miles. Similar data for comparing fuel economy at constant 30 mph with fuel economy for the test cycle used in this work are shown in the curves of Fig. 8. These data (100°F ambient) illustrate rather dramatically the very heavy penalty of the

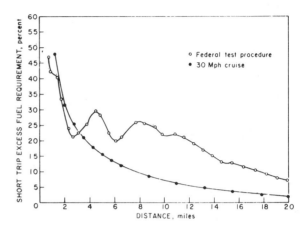

Fig. 8 - Excess fuel required for multiple cold start short trips totaling 32-miles over that required for a single 32-mile trip - vehicle No. 156 at 100°F ambient (See Fig. 1 for FTP driving cycle used)

combination of short trip and non-steady state operation. Patently, the curve for the FTP test data reflects the increased fuel demand of the acceleration modes of the city driving cycle; other factors in non-steady state operation also contribute to the fuel penalty and a more-definitive analysis of these appears to be needed.

Insight into the time (distance) required to stabilize fuel consumption rate was gained in calculating fuel economy values for successive trip segments. Typical trends are illustrated in Fig. 9; the data are for trips made from cold start and continued for thirty or more miles at steady cruise. These data for 70°F ambient show fuel economy to have reached around 95 to 98% of stabilized value within about 10 to 15 miles. Lower ambients result in somewhat extended trip lengths to stabilization, but the ambient effect was found to be less than expected.

EMISSIONS

The influences of trip length and ambient temperature on emissions were found to be quite strong in some cases. These emissions data together with fuel economy values--all representing 4-car averages --are summarized in Table 7. Findings are illustrated in the curves of Figs. 10 thru 12 (values given in Table 8) in which emission rates are expressed as a ratio of the rate over a given trip length to the rate measured for the 1975 FTP. The curves in Figs. 10

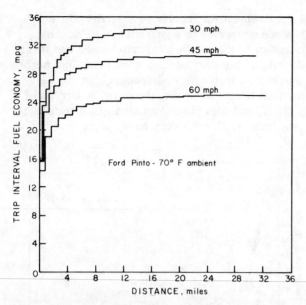

Fig. 9 - Fuel economy for steady-state operation from a cold start - Vehicle No. 156 at 70°F

Table 7 - Summary Fuel Economy and Emission Values
-- Four Car Averages --
(Non-Steady State Tests)

Time Interval, Minutes	Test Cycle	Fuel Economy, mpg		
		20°F	45°F	70°F	100°F*
0 - 3.6	Cold transient	9.9	11.8	13.5	12.5
3.6 - 7.5	Stabilized	13.5	14.2	14.9	12.9
7.5 - 11.1	Hot transient	14.8	15.2	15.8	13.8
11.1 - 21.3	Highway No. 1	19.8	20.2	20.6	18.1
21.3 - 31.6	Highway No. 2**	20.6	20.8	21.2	18.6
	Weighted***	12.8	13.9	14.8	13.1

		Carbon Monoxide, Grams/Mile			
		20°F	45°F	70°F	100°F*
0 - 3.6	Cold transient	141.1	77.1	25.9	18.6
3.6 - 7.5	Stabilized	12.1	7.7	5.0	12.3
7.5 - 11.1	Hot transient	7.8	4.8	4.4	14.7
11.1 - 21.3	Highway No. 1	1.1	.7	.6	2.8
21.3 - 31.6	Highway No. 2**	.8	.6	.5	3.1
	Weighted***	37.5	21.2	9.1	14.3

		Hydrocarbons, Grams/Mile			
		20°F	45°F	70°F	100°F*
0 - 3.6	Cold transient	7.45	3.82	1.89	1.60
3.6 - 7.5	Stabilized	.86	.71	.53	.85
7.5 - 11.1	Hot transient	.52	.44	.42	1.04
11.1 - 21.3	Highway No. 1	.11	.11	.11	.17
21.3 - 31.6	Highway No. 2**	.08	.09	.08	.16
	Weighted***	2.12	1.28	.78	1.06

		Oxides of Nitrogen, Grams/Mile			
		20°F	45°F	70°F	100°F*
0 - 3.6	Cold transient	3.1	2.7	2.0	2.6
3.6 - 7.5	Stabilized	1.8	1.4	1.1	2.0
7.5 - 11.1	Hot transient	2.2	2.1	1.9	2.6
11.1 - 21.3	Highway No. 1	2.1	1.8	1.8	2.3
21.3 - 31.6	Highway No. 2**	2.1	1.7	1.9	2.4
	Weighted***	2.2	1.9	1.5	2.3

*With air conditioner in operation.
**Highway No. 2 at 70°F is equivalent to EPA-HWFE test.
***Calculated from first three phases and at 70°F is equivalent to EPA FTP city fuel economy.

Table 8 - Relative Fuel Economy and Emissions

	Relative Fuel Economy				Relative Emission Rate**		
Miles	Cycle*	30 mph**	45 mph**	60 mph**	CO	HC	NOx
		20°F					
3.6	0.469	0.477	0.540	0.540	15.4	9.5	2.1
7.5	.544	.602	.659	.674	8.1	5.1	1.7
11.1	.585	.664	.712	.704	5.7	3.7	1.6
21.3	.716	.751	.795	.788	3.0	2.0	1.5
31.6	.784	.792	.841	.832	2.1	1.4	1.5
		45°F					
3.6	0.599	0.614	0.672	0.712	8.5	4.8	1.9
7.5	.610	.734	.772	.800	4.5	2.8	1.4
11.1	.641	.784	.821	.844	3.2	2.1	1.4
21.3	.761	.851	.881	.904	1.7	1.2	1.3
31.6	.823	.883	.914	.928	1.2	.8	1.3
		70°F					
3.6	0.634	0.728	0.755	0.716	2.8	2.4	1.3
7.5	.666	.822	.841	.812	1.6	1.5	1.0
11.1	.689	.865	.881	.852	1.3	1.2	1.1
21.3	.802	.924	.934	.912	.7	.7	1.2
31.6	.855	.950	.957	.940	.5	.5	1.2
		100°F, w/a***					
3.6	0.590	0.617	0.732	0.668	2.1	2.0	1.7
7.5	.600	.722	.801	.732	1.6	1.5	1.5
11.1	.614	.766	.831	.764	1.6	1.5	1.6
21.3	.713	.822	.868	.800	1.0	.9	1.5
31.6	.759	.851	.887	.820	.8	.7	1.6

*Relative fuel economy = fuel economy measured from start to the end of the segment divided by the average fuel economy for the 21.3 to 31.6 mile increment. The latter (21.3 to 31.6) segment economy is equivalent to the EPA highway fuel economy and is assumed to be a stabilized and maximum fuel economy for the given speed.
**Relative emission rate = average emission rate from start of test to end of a phase (miles) divided by the weighted emissions calculated for 1975 FTP using the 70°F data.
*** With air conditioning.

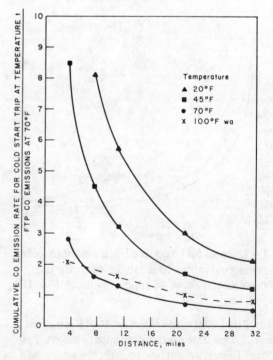

Fig. 10 - Trip length and temperature effect on average carbon monoxide emissions rate relative to FTP emissions rate - 100°F test conducted with air conditioner in operation

thru 12 are drawn through the 5 points representing the cumulative mileage at the end of each of the five phases of the test cycle. The rates taken over smaller increments of distance deviate from the curves, depending upon the distance increment and load and speed changes of the cycle (see Fig. 4).

The relative rate of CO emissions is shown to be highly sensitive to trip length and temperature. For example, the 4-car average 3.6 mile CO relative emission rate was over 15 times the FTP value for 20°F ambient but only about 3 times the FTP value for the 70°F ambient. To illustrate the trip length effect, it may be noted that the relative CO emission rate increased by a factor of about 7 as the trip distance was shortened from the 31.6 miles to 3.6 miles, at 20°F. For the 70°F ambient the comparable value was found to be about 5.

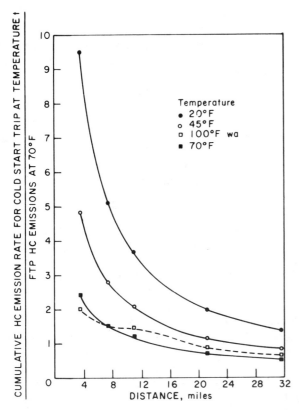

Fig. 11 - Trip length and temperature effect on average hydrocarbon emission rate relative to the FTP emission rate - 100°F test conducted with air conditioner in operation

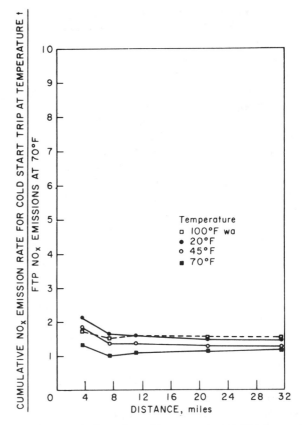

Fig. 12 - Trip length and temperature effect on the average oxides of nitrogen emission rate relative to the FTP emission rate - 100°F test conducted with air conditioner in operation

The effect of trip length on HC emissions is similar to that for CO, but the sensitivity is reduced. Details are available from the data given in Table 8.

Levels of NO_x emissions were found to be only slightly influenced by the cold start trip length or ambient temperature. Emissions for the short trip appeared somewhat increased over tests of longer trips but absolute differences were small.

SUMMARY

Four vehicles ranging in size from a compact to standard were used in experimental work to measure fuel economy, CO, HC, and NO_x emissions for operating temperatures between 20°F and 100°F. A temperature-controlled chassis dynamometer was used for the work. In each test the vehicles were driven 31.6 miles using a combination of driving duty cycles of the FTP and federal highway fuel economy test (HWFET) with all tests replicated three times. In addition, data were obtained using two of the vehicles at the four temperatures and in each test operated for a distance of 32 miles at constant speeds of 30, 45, and 60 mph. During these steady-state tests measurements were made for comparing the steady-state with cycle data.

Findings in the study are summarized following;

the generalizations that are drawn are believed to be reliable indications of directional effect, but absolute values should be used with proper appreciation for the very limited sample reported by the vehicles that were used.

- As ambient temperature decreases from about 70°F, fuel economy decreases. Using the familiar FTP city fuel economy measured at 70°F as a reference, the fuel economy measured at 20°F is 86% of the 70°F value, i.e., 14% "penalty" for cold operation. This temperature effect (20°F vs 70°F) varies from a 27% penalty for the 3.6 mile initial cold transient phase to only 3% penalty for the highway fuel economy interval from 21 to 32 miles.

- The four car average fuel economy for a 3.6-mile trip from cold start at 20°F was found to be 47% of the standard HWFET value; 56% of HWFE for a 3.6-mile trip at 45°F, and 64% of HWFE for the 3.6-mile trip at 70°F.

- CO and HC pollutant emissions also are significantly influenced by ambient temperature and by trip length. Using the 70°F FTP weighted emissions rate as a comparison reference, the CO emissions rate during a 3.6-mile trip from cold start at 20°F was more than 15 times the reference rate. However, at 100°F the CO emissions rate for the short trip was only twice the reference rate. At 70°F ambient CO emissions rate for a 3.6-mile trip was found to be about three times the reference value—with trip length increased to between 20 and 30 miles, the CO rate dropped to about 1/2 the FTP reference value. Similar trends were noted for HC emissions, but with reduced sensitivity to both ambient temperature and trip length. Only slight sensitivity to temperature or trip length was noted in measurements of NO_x emissions.

- Tests that minimize duty cycle effects provide a demonstration of the very substantial loss of fuel economy during start and warmup. Starting cold at 20°F and accelerating rapidly to and maintaining 60 mph, the average fuel economy in a 3.6-mile trip is only 54% of the fuel economy realized when fully warm. Fuel economy averaged over the 31.6-mile cold start trip was found to be 83% of the value reached by the fully warmed-up vehicle.

- Weighted FTP fuel economy was reduced roughly 10% between operating at 70°F with the vehicle air conditioning off and operating at 100°F with the air conditioner on. Via deductive analysis of the data the loss was attributed primarily to air conditioner operation but more definitive data are required to establish purely ambient effects in the 70° to 100°F range.

- Engine oil, transmission fluid, and differential lubricant temperatures recorded during the test cycle showed an effect of ambient temperature in both rate of warm-up and equilibrated temperatures. Engine oil reached 95% of equilibrated temperature (for a given ambient) in approximately 15 miles, the transmission in 16 miles, and the differential in 22 miles of cycle driving.

- In general, the data indicate that after great improvements in carburetion over the past few years, current-production autos suffer as much or more cold-operation penalty in fuel economy and emissions (relative to warmed-up operation) as was found with models dating back several years. It therefore appears that reduction of cold weather sensitivity still offers a good field for overall fuel economy improvement.

REFERENCES

1. C. E. Scheffler and G. W. Niepoth, "Customer Fuel Economy Estimated from Engineering Tests." Paper 650861 presented at the SAE National Fuels and Lubricants Meeting, Tulsa, OK, November 1965.

2. H. A. Ashby, R. C. Stahman, B. H. Eccleston, and R. W. Hurn, "Vehicle Emissions Summer to Winter." Paper 741053 presented at the SAE Automobile Engineering Meeting, Toronto, Canada, October 1974.

3. Society of Automotive Engineers, Inc., "Automotive Fuel Economy Selected SAE Papers, 1965-1975."

4. R. R. Cirillo, T. D. Wolsko, and J. E. Norco, "The Effects of Cold Start on Motor Vehicle Emissions and Resultant Air Quality." Presented at the 67th Annual Meeting of the Air Pollution Control Association, Denver, Colorado, June 1974.

5. L. Grinberg and L. Morgan, "Effect of Temperature on Exhaust Emissions." Paper 740527 presented at the SAE Combined Commercial Vehicle and Lubricants Meetings, Chicago, Ill., June 1974.

6. J. C. Polak, "Cold Ambient Temperature Effects on Emissions from Light-Duty Motor Vehicles." Paper 741051 presented at the SAE Automobile Engineering Meeting, Toronto, Canada, October 1974.

7. R. H. Asin, "Nationwide Personal Transportation Study--Purposes of Automobile Trips and Travel." U.S. Department of Transportation Report No. 10, May 1974, pp. 16.

D. B. Shonka, et al, "Transportation Energy Conservation Data Book: Edition I.5." U.S. Energy Research and Development Administration, Washington, DC.

8. T. C. Austin and K. H. Hellman, "Passenger Car Fuel Economy as Influenced by Trip Length." Paper 750004 presented at the SAE Automotive Engineering Congress and Exposition, Detroit, Michigan, February 1975.

Urban Traffic, Fuel Economy and Emissions — Consistency of Various Measurements*

Leonard Evans
Traffic Science Dept.,
General Motors Research Labs.
Warren, MI

THE FUEL CONSUMPTION AND EXHAUST EMISSIONS of a given vehicle in urban and suburban traffic are determined by the complex interaction of many factors. These include the detailed control inputs the individual driver makes to his vehicle, and how his vehicle interacts with neighboring vehicles and with a complex traffic control system. As a consequence of these factors, vehicles imbedded in urban traffic undergo frequent changes in speed. Some examples of how the speed of one vehicle varied with time for samples of driving in actual traffic are shown in Figure 1 (driving in the Detroit CBD stop-go traffic) and Figure 2 (suburban driving). Fuel consumption and exhaust emissions of a vehicle in urban traffic are determined not only by the physical characteristics of the automobile, but also by the manner in which individual and collective human behavior interacts with a large complex system.

A casual observation of congested urban traffic might suggest that it has so many seemingly-haphazard features that it would not be amenable to a reliable quantitative description. It is the purpose of the present paper to review a number of investigations carried out by the Traffic Science Department of the General Motors Research Laboratories which, in fact, obtained rather consistent descriptions of how fuel consumption and hydrocarbon emission depend on traffic conditions. A general finding from these studies is that urban traffic conditions are well quantified by a single characteristic, namely, the average speed of the traffic. A vehicle which is driven "with the traffic" has an average trip speed, defined as distance of

*Paper 780934 presented at the International Fuels & Lubricants Meeting, Toronto, November 1978.

— ABSTRACT —

A number of studies of the effect of traffic conditions on fuel consumption and emissions are reviewed. A model based on driving vehicles in traffic is described in which the fuel consumption of a vehicle in urban traffic is expressed as a simple function of trip speed. Data from a variety of sources, including additional field data, detailed computer simulation, the same vehicle tested on different fixed urban driving schedules, and small segments of the Federal Test Procedure (FTP) have been all shown to fit the model. A similar model of HC emissions as a simple function of trip speed is derived from analyzing small segments of FTP data. Data from a variety of sources, including published EPA relations, detailed computer simulation, and dynamometer replication of street data have been all shown to fit this model. No simple models were found for CO and NO_x. In general, it is found that urban traffic, despite its seemingly disorganized appearance, yields rather consistent relations between fuel consumption and average speed, and between HC and average speed.

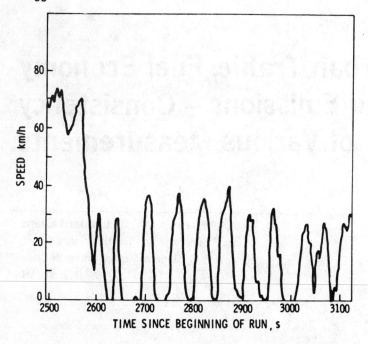

Fig. 1 - Sample of speed-time history for Detroit central business district stop-go traffic. From Evans, Herman and Lam (2)

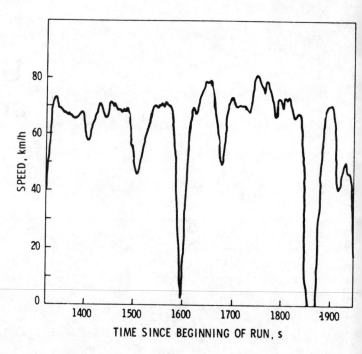

Fig. 2 - Sample of speed-time history for driving in free flowing suburban roads. From Evans, Herman and Lam (2)

travel divided by total time including stopped time, that is the same as the average speed of the traffic.

The variable used in the studies to be described was generally the average trip time per unit distance, $\bar{t}$, rather than the average trip speed, $\bar{v}$, where $\bar{t} = 1/\bar{v}$. This variable was used because relations involving it are both simpler and more readily interpreted in physical terms than relations in $\bar{v}$.

We first summarize a simple model in which urban fuel consumption is expressed as a linear function of $\bar{t}$ and then discuss other studies which have shown that different types of data from a variety of sources can be explained in terms of this model. A similar approach is successfully applied to hydrocarbon emissions. This approach, however, is not successful for carbon monoxide and oxides of nitrogen. Only the general findings of the various studies reviewed will be mentioned -- details on experimental methods, statistics, etc. may be obtained in the reference cited.

FUEL CONSUMPTION

THE MODEL - In late 1973 and early 1974 a study (1,2)* was carried out in which an instrumented 1973 subcompact car was driven "normally, with the traffic" for 383 km in the Detroit metropolitan area. The goal of the study was to identify what quantitative aspects of urban traffic are most useful in determining urban fuel consumption. The approach was to divide long speed-time histories, (examples of portions are shown in Figures 1 and 2) into smaller segments which were still of sufficient length for each to contain some of the essential characteristics of urban driving. For each segment, various traffic measures, such as the average speed, the maximum acceleration, the maximum braking, the number of stops per unit distance and the amount of stopped time per unit distance were computed from the recorded speed-time and acceleration-time histories. In all, 16 different traffic measures were examined in detail (1). For each of four different methods of dividing the data into segments, it was found that the same single variable, the average trip time per unit distance, $\bar{t}$, best explained the average fuel consumption per unit distance, $\bar{\phi}$, leading to the equation

$$\bar{\phi} = k_1 + k_2 \bar{t}, \quad (\bar{v} < \sim 60 \text{ km/h}) \qquad (1)$$

or $1/\bar{E} = k_1 + k_2/\bar{v}$,

where $\bar{E} = 1/\bar{\phi}$ is the average fuel economy. As Equation 1 was found (2) to apply with essentially the same values of the parameters k_1 and k_2 for all the methods of sampling the data, the simplest sampling method was used in succeeding studies. This method was to divide the data into segments of travel between consecutive

---

*Numbers in parentheses designate references at end of paper.

stops of the vehicle, so that the only measurements required to give $\bar{\phi}$ and $\bar{t}$ were the total fuel consumed, distance traveled and elapsed time each time the vehicle came to a stop (1,2).

Note that the variables $\bar{t}$, $\bar{v}$, $\bar{\phi}$ and $\bar{E}$ in Equation 1 are averages over segments of travel (for example, between consecutive stops of the vehicle) that include a variety of speeds. They must not be interpreted as instantaneous values.

By comparing Equation 1 with an analytical model of the engine vehicle system developed by Amann, Haverdink and Young (3) of the Engine Research Department of General Motors Research Laboratories, the following physical interpretations of k_1 and k_2 were given (1,2).

The parameter k_1 corresponds to the fuel consumed per unit distance to overcome rolling resistance, and is hence approximately proportional to the vehicle mass, M. That is,

$$k_1 = c_M M, \quad (2)$$

where c_M is a constant of proportionality.

The parameter k_2 is associated with time-dependent losses, and may therefore be approximately associated with the idle fuel flow rate, I, giving

$$k_2 = c_I I, \quad (3)$$

where c_I is a constant of proportionality.

These interpretations of the parameters are supported by observed relations between k_1 and M, and between k_2 and I (4,5).

Equations 1 - 3 constitute our model of fuel consumption in urban traffic. Equation 1 applies only to low speed traffic ($\bar{v} < \sim 60$ km/h). At higher speeds aerodynamic effects became dominant and $\bar{\phi}$ increases as $\bar{t}$ decreases. If $\bar{\phi}$ and $\bar{t}$ are related according to Equation 1 for each segment of a long trip, then the values of $\bar{\phi}$ and $\bar{t}$ for the entire trip are also related according to Equation 1 (1,2).

Note that the model applies only to driving fully-warmed vehicles "normally, with the traffic." Modifications to the model when these restrictions do not apply have been reported (4,6) but will not be discussed here.

We now discuss published studies in which the model has been successfully applied to data of the following types:

Additional Field Data

Detailed Computer Simulation

Same Vehicle Tested on Different Test-Track and Dynamometer Driving Schedules

Small Segments of FTP Dynamometer Data.

ADDITIONAL FIELD DATA - A number of studies in addition to those (1,2) which led to the model have been conducted in which fuel consumption and average speed have been measured in actual traffic.

Chang, Evans, Herman and Wasielewski (4) obtained data for six different vehicles driven normally in Detroit metropolitan area traffic. The data for these and other vehicles fitted Equation 1 with parameters k_1 and k_2 in accord with the interpretation represented in Equations 2 and 3. Data for one vehicle from this study (4) is plotted in Figure 3. Data from two additional vehicles (one also plotted in Figure 3) are reported by Evans and Herman (5) to also fit the model.

In another field study (7), it was found that some of the fuel observed to be saved by the introduction of a permissive right turn on red in Michigan could be explained in terms of reduced trip times, in accord with Equation 1.

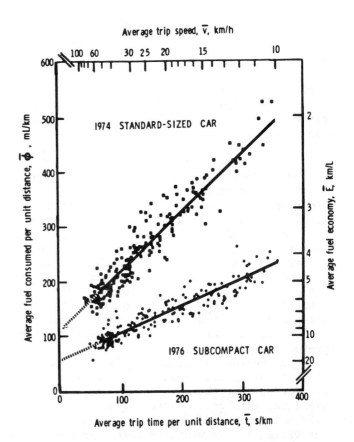

Fig. 3 - Examples of the fit to Equation 1 of fuel consumption data obtained by driving instrumented vehicles in traffic. Each point is for a segment of travel from the beginning of one stopped period to the beginning of the next stopped period during the course of a long trip which encompassed a variety of traffic conditions. From Evans and Herman (5). Data for the standard-sized car from Chang, Evans, Herman and Wasielewski (4)

In addition to the above GM studies, linear relations between urban fuel consumption and trip time have also been reported in the literature for urban traffic in Britain (8,9), Australia (10) and India (11). For the cases for which vehicle test mass (8-10) and idle fuel flow rates (8) are reported, the parameters of the linear relations have been successfully interpreted (4) in accord with Equations 2 and 3.

DETAILED COMPUTER SIMULATION - One approach to estimating fuel consumption in urban traffic is to use computer simulation to calculate speed-time behavior of a vehicle in a simulated traffic situation. The fuel consumption is then calculated by adding the fuel required to execute each small individual maneuver, as inferred from the dependence of instantaneous fuel consumption on speed and acceleration.

Lieberman and Cohen (12) calculated the effect on fuel economy of permitting a right turn on red at signalized intersections in a simulation of a network of streets in Washington, D.C. Three levels of flow were simulated, each with and without a right-turn-on-red policy, and the average speeds for these six cases were calculated. These data were converted (13) to the six $(\bar{t}, \bar{\phi})$ pairs shown in Figure 4. In this representation, the data lie close to the least squares linear fit shown in Figure 4, showing that Equation 1 interprets these data well.

In another computer simulation study, Honeywell Traffic Management Center (14) estimated fuel economy and other traffic quantifiers for a variety of traffic control scenarios (e.g. pre-timed versus fully-actuated signal systems; one way versus two way streets). Data for 40 different cases were reported, which were converted to $(\bar{t}, \bar{\phi})$ pairs (15), as shown in Figure 5. Six of the data, represented by open circles in Figure 5, refer to a simulation of the effect of different pedestrian strategies at a single isolated intersection rather than to a network. The fit of these 40 data to the line shows that the data may be successfully interpreted in terms of Equation 1.

The regression lines in Figures 4 and 5 have similar values of the parameters k_1 and k_2, reflecting that the same "composite vehicle" was simulated in both the original computer studies (12,14). These values are similar to the average values of the vehicles tested in traffic (4).

SAME VEHICLE TESTED ON DIFFERENT TEST-TRACK AND DYNAMOMETER DRIVING SCHEDULES - Urban fuel economies of different vehicles are often measured and compared using test procedures based on fixed urban driving schedules. The EPA's "city" fuel economy and emissions are determined from the 1975 Federal Test Procedure (FTP), which will be discussed in the next section. In this procedure the vehicle executes the LA-4 driving schedule on a chasis dynamometer. Other urban schedules include the SAE urban and the GM city-suburban, which are used

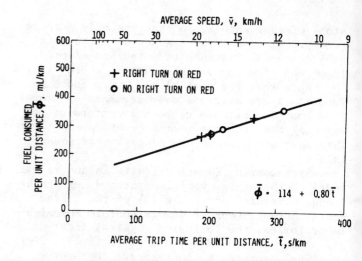

Fig. 4 - The six data points derived by computer simulation in Reference 12 transformed to fuel consumed per unit distance versus trip time per unit distance. From Chang, Evans, Herman and Wasielewski (13)

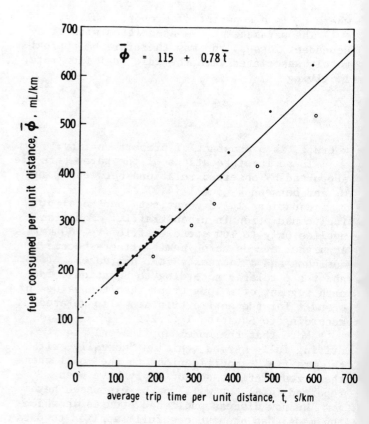

Fig. 5 - The 40 data points derived by computer simulation in Reference 14 transformed to fuel consumed per unit distance versus trip time per unit distance. The line is a least squares fit to all 40 data points shown. The six open circle points refer to a simulation of the effect of different pedestrian strategies at a single isolated intersection rather than to a network. From Evans and Herman (15)

for test-track measurements. The advantage of the fixed urban driving schedule approach is that each vehicle executes the same speed-time history. Although the different urban schedules rank the fuel economies of different vehicles in a consistent way, the actual measured urban fuel economy of a vehicle depends systematically on which particular schedule is chosen. However, it has been shown (5) that much of the difference between the fuel economies measured on the different schedules may be explained in terms of their different average speeds, in accordance with Equation 1.

In Figure 6 the fuel consumption for four vehicles measured on three different schedules is shown plotted versus the average trip time per unit distance inferred from the definitions of the schedules. The fit of these data to linear relations (Equation 1) shown in Figure 6 is typical of that observed for 111 vehicles studied (5). Because each regression (Equation 1) was based on only three data points, the parameters k_1 and k_2 for individual cars were obtained with less reliability than for field data (see Figure 3). However, the relations Equation 2 and Equation 3 yielded average values of c_M and c_I similar to those derived from the field tests (4,5).

SMALL SEGMENTS OF FTP DYNAMOMETER DATA - The LA-4 fixed urban driving schedule is defined by a table of 1373 values of speed and time, at one second intervals (16). The schedule contains 18 stop- (in one case a near stop) to-stop cycles (see Figure 7).

The 1975 FTP requires a 12 hour soak at normal room temperature before starting the vehicle. The first five cycles, from time zero to 505 s, are referred to as the cold-start phase. Cycles 6 through 18 (505 to 1372 s) are called the stabilized phase. At the end of cycle 18, the engine is switched off for 10 minutes. The first five cycles are then repeated (i.e., cycles 19 through 23); this third phase is called the hot-start. To obtain the official FTP emissions and fuel economy values, the data from the cold-start and hot-start phases are weighted by factors of 0.43 and 0.57, respectively, to reflect estimated proportions of in-use cold and hot starts.

Evans (17) showed that data from one FTP test could be interpreted in terms of the model (Equations 1 - 3) by treating each of the 18 cycles as portions of travel between consecutive stops in the same manner as was done for field data (1,2,4,5,7).

Fuel consumption (and exhaust emissions) data for 12 arbitrarily-chosen 1975 and 1976 model-year cars run on the FTP by General Motors were analyzed (17).

In an attempt to obtain data for vehicles after they had reached the stable operating conditions for which Equation 1 applies, the cold-start phase (i.e., cycles 1 through 5) was excluded from the analysis. However, because

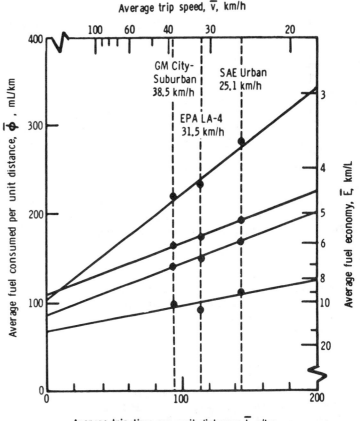

Fig. 6 - Data for four automatic transmission vehicles run on three fixed urban driving schedules. The scatter of the three points from the least squares linear fits shown is typical of that for 111 vehicles studied. From Evans and Herman (5)

the test includes a hot-start repeat of cycles 1 through 5, (cycles 19 through 23 in Figure 7) "non-cold" data were available for each of the 18 cycles.

The fuel consumed in each of the 18 "non-cold" cycles was determined from the exhaust emissions using the carbon balance technique. The average trip time per unit distance and other variables for each of the 18 cycles were calculated from the LA-4 speed-time history (16). The stopped portion was included at the beginning of the segment.

The resulting data were subjected to a multivariate analysis similar to that applied to field data (1,2). It was found that $\bar{t}$ was the best explanatory variable for $\bar{\phi}$ for the data from 10 of the 12 cars, and the second best variable for the other two.

Values of $\bar{\phi}$ are shown in Figure 8 plotted versus $\bar{t}$ for one of the 12 cars. The line is a least squares fit to the 17 data points plotted with a solid symbol. Cycle 15 (open symbol), which has a distance of travel of 0.11 km has been excluded in keeping with the practice of excluding data representing less than 0.2 km

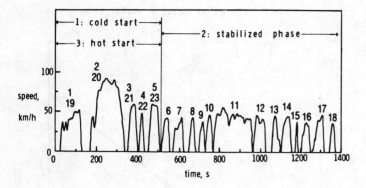

Fig. 7 - Speed-time history of the LA-4 driving schedule, with the three phases of the 1975 Federal Test Procedure also indicated.

travel which was used in analyzing the field data (4). Of the 12 cars studied, 6 fitted Equation 1 better than the one shown in Figure 8 and 5 fitted Equation 1 less well than the one shown in Figure 8.

HYDROCARBON EMISSIONS

THE MODEL - Exhaust emission data (i.e., g/km hydrocarbons, carbon monoxide and oxides of nitrogen) for each of the 18 non-cold cycles for the 12 cars run on the FTP were analyzed (17) in the same way as described in the previous section for fuel. All 12 vehicles were equipped with catalytic converters, and emissions both before (engine-out) and after (tailpipe) were studied. The goal was to obtain relations between exhaust emissions and simple measures of traffic conditions, as was done for fuel. Because absolute exhaust emissions do not depend simply on, e.g., vehicle mass, as is approximately so for fuel consumption, relations between traffic characteristics and relative, rather than absolute, exhaust emissions were sought. A simple relationship was found only for the case of hydrocarbons (HC).

The engine-out hydrocarbons, being larger and not subject to additional statistical variability associated with passing through the catalyst, yielded more consistent car-to-car relations than the tailpipe emissions. The variable $\bar{t}$ explained more of the variance in HC than any other variable studied for 9 of the 12 cars, and in fact explained more than half the variance in HC for 11 of the cars. No traffic variable explained a significant fraction of the variance in HC emissions for one of the cars. The average variance explained by $\bar{t}$ was more than 70% for these 11 cars. Because of their higher correlations with traffic variables, engine-out hydrocarbons for individual cars will be discussed first (17). Later, it will be shown that similar results apply to tailpipe values averaged over all the cars.

For low speed urban traffic a relation between engine-out HC and average trip time per

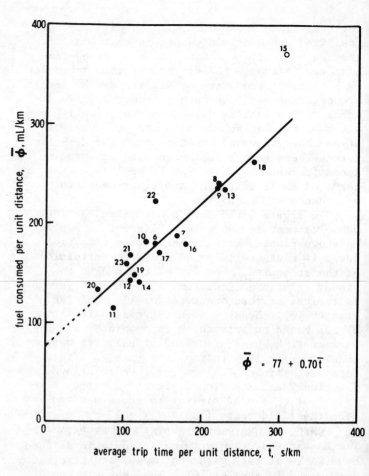

Fig. 8 - Fuel consumption per unit distance in mL/km versus $\bar{t}$ for the car with correlation coefficient rank 7 out of 12. The number of the cycle (see Figure 7) which provided each datum is indicated beside it. From Evans (17)

unit distance similar to that found for fuel therefore applies, namely,

$$y(\bar{t}) = a + b\bar{t}, \qquad (4)$$

where $y(\bar{t})$ is hydrocarbon emissions in g/km at average trip time per unit distance $\bar{t}$.

A plot of $y(\bar{t})$ versus $\bar{t}$ is shown in Figure 9 for one of the cars. Six of the cars fitted Equation 4 better than the case illustrated in Figure 9, and five fitted less well than the case in Figure 9. The broken line extending beyond the data is to indicate the intercept, a, in Equation 4. This relation is not valid for high speeds ($\bar{v} > \sim 60$ km/h) for which hydrocarbon emissions per unit distance, in fact, increase with speed.

The time rates of HC emissions at idle were available for three cars, and were 6.2 mg/s, 8.4 mg/s and 22.5 mg/s. These may be compared to the corresponding values of the parameter b derived by fitting Equation 4 to the data for these three cars, namely 6.4 mg/s, 8/4 mg/s and

16.8 mg/s respectively. We see that, as in the case of fuel consumption, the coefficient of $\bar{t}$ in the linear relation can be associated with the time rate at idle.

A relation of the form of Equation 4 enables us to express the fractional change, f, in HC emissions corresponding to a change in trip time from $\bar{t}_o$ to $\bar{t}$ as a simple multiple of the change in trip time. The dependence of f on $\bar{t}$ can be expressed in terms of a single quantity, α, as follows:

$$f = \frac{y(\bar{t}) - y(\bar{t}_o)}{y(\bar{t}_o)} = \alpha(\bar{t} - \bar{t}_o), \quad (5)$$

where α is given explicitly by

$$\alpha = \frac{b}{a + b\bar{t}_o}. \quad (6)$$

If R is the ratio of emissions at the two trip times $\bar{t}$ and $\bar{t}_o$, i.e., $y(\bar{t})/y(\bar{t}_o)$, then

$$R = 1 + f = 1 + \alpha(\bar{t} - \bar{t}_o). \quad (7)$$

We take as a reference value $\bar{t}_o = 114.4$ s/km, the average value for the LA-4 driving schedule. Note that this value is identical to the average value for the FTP test because the total weighting factor for phases one and three is unity. It is inappropriate to use the unweighted average speed actually experienced by the driver of the test as the average speed for the FTP, as has been suggested in the literature (18,19). To do so gives double weighing to the first five cycles.

The expression for R (Equation 7) for the 11 cars with signficant correlations with $\bar{t}$ are contained in the shaded region indicated in Figure 10. The average of the 11 values of α is

$$\bar{\alpha} = 0.0059 \text{ km/s}. \quad (8)$$

Note that if $y(\bar{t})$ were directly proportional to $\bar{t}$ (i.e., a = 0 in Equation 6), then α would have the value 1/(114.4 s/km) = 0.0087 km/s.

In the above analysis engine-out HC emissions were used because the larger variability of the tailpipe emissions reduced correlations with traffic variables for individual cars. However, by averaging over all the cars the effect of this variability can be reduced. For each car the ratio, p, of g/km HC for each cycle to the average value for that car for all cycles except #19 and #20 was determined. These two cycles, which occur just after the 10-minute soak, sometimes had much higher HC

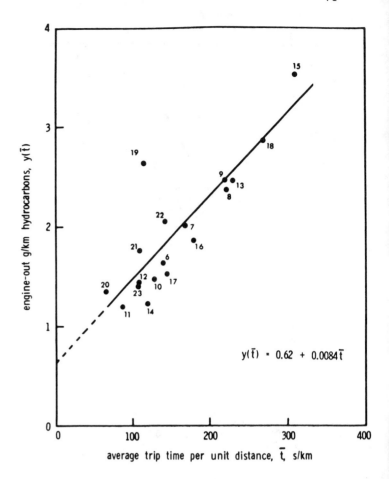

Fig. 9 - Engine-out hydrocarbon emissions in g/km (= $y(\bar{t})$) versus $\bar{t}$ for the car with correlation coefficient rank 7 out of 12. The number of the cycle which provided each datum is indicated beside it (see Figure 7). From Evans (17)

emissions than the other cycles. The average value of the ratio p over the 12 cars is shown plotted versus $\bar{t}$ for all 18 cycles in Figure 11. The least squares fit to the data for 16 cycles (i.e., excluding #19 and #20) shown in Figure 11 leads, through Equation 6, to α = 0.0067 km/s (c.f. Equation 8). Thus, the overall tailpipe HC emissions exhibit a relative dependence on trip speed essentially similar to that found in the engine-out case.

The following simple statement approximately summarizes the above findings on hydrocarbon emissions:

> In low speed urban driving, for each one second increase in trip time per kilometer, the hydrocarbon emissions for a given vehicle increase by about 0.6% of their value at the average trip speed of the LA-4 driving schedule.

The above statement summarizes the model represented in Equations 5 through 8. We discuss below other HC data which have also been shown (17) to be linear functions of $\bar{t}$. The

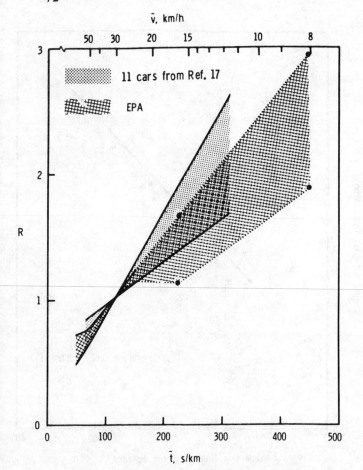

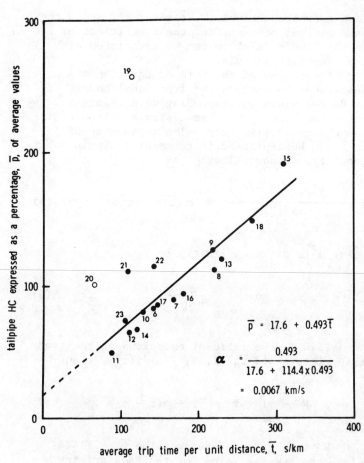

Fig. 10 - Hydrocarbon emissions (mass per unit distance) versus trip time per unit distance, $\bar{t}$, expressed as a multiple, R, of their value at the average trip time per unit distance of the LA-4 fixed driving schedule. One of the shaded areas contains the 11 linear relations for the cars studied in Reference 17. The other shaded area contains all 11 curves given by the EPA, as well as their estimates at the average speeds of 8 km/h and 16 km/h. From Evans (17)

Fig. 11 - Dependence of tailpipe hydrocarbons on average trip speed. The plotted data were obtained by first expressing the g/km HC for each car as a percentage, p, of the average of the values for 16 cycles (i.e., all except #19 and #20), and then computing the average p for 12 cars. The quantity α is defined by Equation 6. From Evans (17)

values of α computed by Equation 6 from these derived linear relations are compared to the value of 0.0059 km/s given by Equation 8. The extent to which the values agree indicates the consistency of the relative way in which hydrocarbon emissions per unit distance increase with decreasing traffic speed in the different studies.

EPA RELATIONS - The ratio of emissions at different speeds, R, (see Equation 7) are given in the form

$$R = \exp(A + B\bar{v} + C\bar{v}^2) \qquad (9)$$

in EPA document AP-42 for the speed range 24 km/h to 72 km/h (20). Values of the parameters in Equation 9 for 11 different cases, (e.g.

different locations and altitudes) together with separately determined values at the particular average speeds of 16 km/h and 8 km/h are also given (20).

When each of the 11 curves given by Equation 9 for HC is plotted over its stated range of validity as a function of $\bar{t}$, none show other than small departures from linearity. All 11 curves are contained within the indicated area in Figure 10, which also shows the maximum and minimum of the specific values given for 16 km/h and 8 km/h.

Figure 12 shows Equation 7 plotted with $\alpha = 0.0059$ km/s (Equation 8). A curve representing the average values for the EPA data is also shown. The reasonable agreement of these two curves supports the simple model. The degree of complexity of the EPA's Equation 9 would appear to be difficult to support with presently available data.

DETAILED COMPUTER SIMULATION - In addition to the previously-mentioned fuel consumption results, Lieberman and Cohen (12) also calculated emissions at each of the six conditions simulated in their right-turn-on-red study. Their results for HC for a "composite" vehicle are shown plotted versus $\bar{t}$ in Figure 13, which may be compared to Figure 4.

The near perfect fit of this data to Equation 4 appears to result from the smooth analytic dependence of emissions on acceleration and speed which is build into the simulation model. The value $\alpha = 0.0050$ km/s may be compared to the value 0.0059 km/s in Equation 8.

DYNAMOMETER REPLICATIONS OF STREET DRIVING - Data given for HC emissions for trips at the four urban speeds of 12, 18, 19 and 31 miles per hour in Figure 3 of Reference 21 are plotted versus $\bar{t}$ in Figure 14. These data yield $\alpha = 0.0065$ km/s, which may be compared to the value $\alpha = 0.0059$ km/s given by Equation 8.

CARBON MONOXIDE AND OXIDES OF NITROGEN

No consistent relations between CO and NO_x and average trip speed emerged from the analysis of the FTP data (17). Another variable, the work used to accelerate the vehicle in each of the 18 segments of travel, was a better predictor of these emissions than $\bar{t}$. However, this finding does not preclude the possibility that speed characteristics alone, through their correlations with acceleration characteristics, might provide an adequate description of CO and NO_x emissions for a more extensive data set.

For the case of CO, some other studies do suggest relations with average speed. The right-turn-on-red simulation data (12) yields values of CO that fit a linear relation in $\bar{t}$ with as little scatter as that shown in Figure 13 for HC. Using computer simulation, Watson and Milkins (22) and Watson, Milkins and Bulach (23) produced data which lay close to hand drawn curves relating CO and average speed and HC and average speed. They concluded (23) that, for these emissions, models that accounted for variations in traffic speed alone were as reliable as more complex models. The analysis of actual FTP emissions data yielded this same conclusion for HC, but a less definitive result for CO (17).

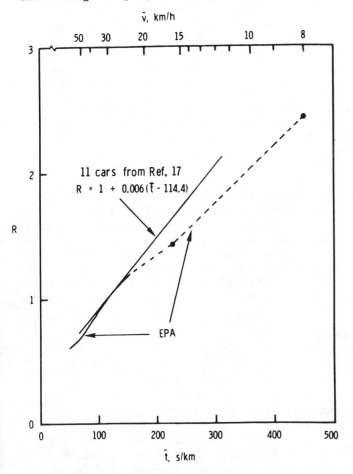

Fig. 12 - The average values of the Figure 10 representations. From Evans (17)

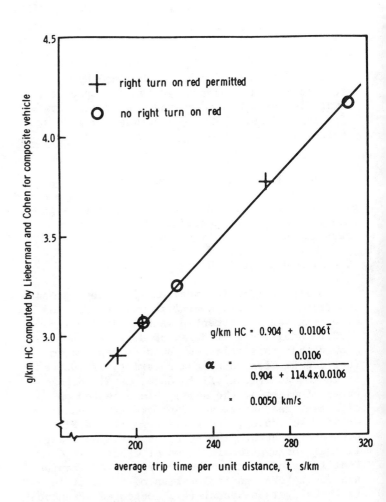

Fig. 13 - The six data points derived by computer simulation in Reference 12 plotted versus trip time per unit distance. The quantity α is defined by Equation 6. From Evans (17)

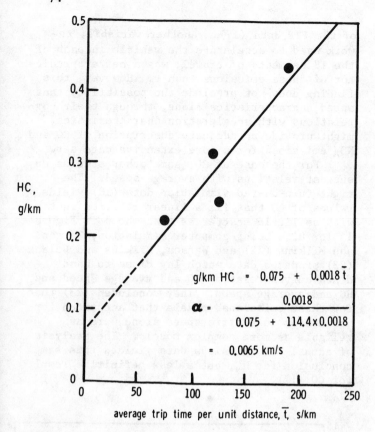

Fig. 14 - Values of g/km HC obtained by replicating on a chassis dynamometer speed-time histories of urban trips under different traffic conditions. Data from Herman, Jackson and Rule (21)

An analysis of the measured CO emissions corresponding to trips at the four urban speeds given in Figure 5 of Reference 21 gave no significant relation between g/km CO and $\bar{t}$, in contrast to the result (Figure 14) for HC.

SUMMARY AND DISCUSSION

A linear relation between fuel consumption and trip time in urban traffic was established by driving an instrumented vehicle in traffic (1,2). Additional studies, as described above, showed that this relation could be applied to fuel consumption data from a variety of sources, including additional field data, detailed computer simulation, the same vehicle tested on different test-track and dynamometer fixed urban driving schedules, and small segments of FTP dynamometer data. The relation explains changes in fuel consumption in terms of changes in traffic speed whether the traffic speed changes are due to changes in location of the traffic (suburban versus urban) (4), congested or uncongested (2,4), changes in traffic ordinances (right-turn-on-red) (7,13) or various alternate traffic engineering strategies (one-way versus two-way streets) (15).

As has been pointed out (1) part of the reason that so simple a relation applies is that other quantities important in determining fuel consumption (e.g. acceleration characteristics, stop time) are themselves largely determined by average trip speed. When urban speeds are relatively high, the traffic is "smooth" (21), whereas when speeds become low due to congestion, quantities such as acceleration, stop time, stops/km, etc. all increase.

A relation similar to that found for fuel was found to apply between relative hydrocarbon emissions and trip time per unit distance for small segments of FTP data. Because the absolute emissions do not depend simply on e.g., vehicle mass, as is approximately so for fuel consumption, the relative rather than absolute values were studied. It was found that for each one second increase in trip time per kilometer (s/km) the g/km HC emissions increase by about 0.6% of their value at the average trip speed of the LA-4 schedule. Hydrocarbon data from the following other sources were found to fit this relation: published EPA relations; detailed computer simulation; dynamometer replication of street data.

This simple relation enables one to compute how changes in overall traffic systems affect hydrocarbon emissions, as has previously been done for the case of fuel consumption (24). For example, it has been estimated (25) that the average circulation speed of traffic in the New York City CBD is 16 km/h. In terms of the reported fuel consumption model (24) and above relation for HC, it is estimated that if the circulation speed could be increased by, e.g. traffic engineering improvements, to, say 20 km/h, then system fuel economy would increase by 14% and system HC emissions would decrease by 16%.

The analysis (17) of the segments of FTP data did not yield simple relations in $\bar{t}$ for carbon monoxide and oxides of nitrogen.

In general, it is found that urban traffic, despite its seemingly haphazard appearance, yields rather consistent relations between fuel consumption, hydrocarbon emissions and average traffic speed, and that these same relations apply to field, chasis dynamometer and computer simulation data.

REFERENCES

1. L. Evans, R. Herman and T. N. Lam, "Multivariate Analysis of Traffic Factors Related to Fuel Consumption in Urban Driving," Transportation Science, Vol. 10, No. 2, 205-215, May 1976.

2. L. Evans, R. Herman and T. N. Lam, "Gasoline Consumption in Urban Traffic," Society of Automotive Engineers, SAE Paper No. 760048, February 23, 1976.

3. C. A. Amann, W. H. Haverdink and M. B. Young, "Fuel Consumption in the Passenger Car

System," General Motors Corporation, Research Publication GMR-1632, 1975.

4. M.-F. Chang, L. Evans, R. Herman and P. Wasielewski, "Gasoline Consumption in Urban Traffic," Transportation Research Board 599, 23-30, 1976.

5. L. Evans and R. Herman, "Automobile Fuel Economy on Fixed Urban Driving Schedules," Transportation Science, Vol. 12, No. 2, 1978.

6. L. Evans, "Driver Behavior Effects on Fuel Consumption in Urban Driving," General Motors Corporation, Research Publication GMR-2769, June 1978.

7. M.-F. Chang, L. Evans, R. Herman and P. Wasielewski, "Observations of Fuel Savings Due to the Introduction of Right Turn on Red," Traffic Engineering and Control, Vol. 18, No. 2, 475-477, October 1977.

8. G. J. Roth, "The Economic Benefits to be Obtained by Road Improvements, with Special Reference to Vehicle Operating Costs, Department of Scientific and Industrial Research," Road Research Laboratory, Research Note No. RN/3426/-GJR, Harmondsworth, 1959 (unpublished).

9. P. F. Everall, "The Effect of Road and Traffic Conditions on Fuel Consumption," Road Research Laboratory, Report LR 226 Crowthorne, England, 1968.

10. E. Pelensky, W. R. Blunden and R. D. Munro, "Operating Costs of Cars in Urban Areas," Proceedings of the Fourth Conference of the Australian Road Research Board, Vol. 4, Part 1, 475-504, 1968.

11. N. S. Srinivasan, P. S. Shetty and S. B. Sripathi Rao, "Fuel Consumption on Roads with Interrupted Traffic Flow," Highway Research Bulletin No. 1, New Dehli, India, 1975.

12. E. B. Leiberman and S. Cohen, "New Technique for Evaluating Urban Traffic Energy Consumption and Emissions," Transportation Research Board 599, 41-45, 1976.

13. M.-F. Chang, L. Evans, R. Herman and P. Wasielewski, "Fuel Consumption and Right Turn on Red: Comparison Between Simple Model Results and Computer Simulation," Transportation Science, Vol. 11, No. 1, 1977.

14. Honeywell Traffic Management Center, Fuel Consumption Study - Urban Traffic Control System (UTCS) Software Support Project, Report No. FHWA-RD-76-81 prepared for Federal Highway Administration, February 1976.

15. L. Evans and R. Herman, "Urban Fuel Economy -- Computer Simulation Calculations Interpreted in Terms of Simple Model," Transportation Research, Vol. 12, No. 3, 163-165, 1978.

16. EPA Urban Dynamometer Driving Schedule (Speed versus Time sequence), Appendix 1, Federal Register, Vol. 37, No. 221, Wednesday, November 15, 1972.

17. L. Evans, "Exhaust Emissions, Fuel Consumption and Traffic: Relations Derived from Urban Driving Schedule Data," General Motors Corporation, Research Publication GMR-2599, December 1977.

18. D. J. Simanaitis, "Emission Test Cycles Around the World," Automotive Engineering, 36-43, August 1977.

19. M. Kuhler and D. Karstens, "Improved Driving Cycle for Testing Automotive Exhaust Emissions," Society of Automotive Engineers, SAE Paper No. 780650, June 6, 1978.

20. U.S. Environmental Protection Agency Publication AP-42, "Compilation of Air Pollutant Emission Factors," December 1975.

21. R. Herman, M. W. Jackson and R. G. Rule, "Fuel Economy and Exhaust Emissions Under Two Conditions of Traffic Smoothness," Society of Automotive Engineers, SAE Paper No. 780614, June 6, 1978.

22. H. C. Watson, E. E. Milkins and V. Bulach, "How Sophisticated Should a Vehicle Emissions Source Model Be?" Smog '76, Proceedings Supplement, Clear Air Society of Australia and New Zealand, 110-119, 1976.

23. H. C. Watson and E. E. Milkins, "Prediction of CO Concentrations in Street Canyons," Australian Environmental Council, Symposium on Air Pollution Diffusion Modelling, Canberra, Paper 15, 1976.

24. L. Evans and R. Herman, "A Simplified Approach to Calculations of Fuel Consumption in Urban Traffic Systems," Traffic Engineering and Control, Vol. 17, Nos. 8 and 9, August/September 1976.

25. M.-F. Chang and R. Herman, "An Attempt to Characterize Traffic in Metropolitan Areas," Transportation Science, Vol. 12, No. 1, 58-79, February 1978.

Vehicle Design Factors Affecting Fuel Economy

Effects of the Degree of Fuel Atomization on Single-Cylinder Engine Performance*

William R. Matthes and Ralph N. McGill
Research Labs., General Motors Corp.

IT IS GENERALLY ACCEPTED that the degree of fuel atomization and preparation can affect the performance of conventional engines. For example, we know that the degree of fuel atomization can affect cylinder-to-cylinder air-fuel ratio distribution and the degree of mixing between the fuel and air in multicylinder engines. Furthermore, it has been conjectured for some time that the degree of atomization can affect the formation of exhaust emissions, even in the absence of geometric maldistribution.

For example, the degree of homogeneity of the fuel-air mixture at the time of ignition can influence combustion and emissions. A homogeneous charge yields emission results that are quite different from a stratified charge. And, the degree of fuel atomization influences the degree of homogeneity of the charge through the mechanisms of vaporization and mixing. Small droplets can be vaporized and mixed more fully in the intake manifold and during intake and compression strokes than large droplets.

*Paper 760117 presented at the Automotive Engineering Congress and Exposition, Detroit, 1976.

ABSTRACT

An investigation has been made to determine the effects of the degree of fuel atomization on exhaust emissions, fuel consumption, lean limit, MBT spark timing, and cyclic variations in peak cylinder pressure. A single-cylinder engine was used to isolate the effects of atomization on combustion from the additional effects of maldistribution that would be present in a multicylinder engine.

Three degrees of gasoline atomization were investigated, along with the case of a well-mixed charge of gaseous propane. The degrees of atomization investigated varied from "Good" (10-20 μm droplets) to "Bad" (400-700 μm droplets) to "Wall-Wetted" (400-700 μm droplets deposited on the intake-port walls).

Results from this investigation show that the degree of atomization can have considerable effect on exhaust emissions, but little effect on fuel consumption. Generally, as atomization deteriorated, hydrocarbon emissions increased; nitric oxide and carbon monoxide emissions increased for certain air-fuel ratio ranges; MBT (Minimum for Best Torque) spark advance decreased; lean limit was extended; and cyclic variations in peak cylinder pressure decreased.

It was hypothesized that the case of Good atomization resulted in an essentially homogeneous charge of vaporized gasoline and air at the time of ignition, while the cases of Bad and Wall-Wetted atomization resulted in inhomogeneous charges; that is, some form of stratification of the fuel-air mixture existed. All of the results are discussed in light of this hypothesis and are shown to be consistent with it.

Thus, we might reasonably expect that very finely atomized fuel should lead to a more homogeneous charge at combustion and, consequently, different emissions and engine performance than poorly atomized fuel.

BACKGROUND

Even though we have suspected that the degree of fuel atomization must surely affect combustion and the formation of emissions, we have not known just how great the effects might be. The problem has been that if we are to study rigorously the effects of the degree of fuel atomization, we first need to know the details of the spray that an atomizer produces (droplet size distributions, velocities, geometric distribution, etc.). We then need to know what happens to that spray through the induction system, past the intake valve, and during the intake and compression strokes. Finally, we would like to know the entire description of the charge that exists immediately prior to ignition. Needless to say, this is difficult knowledge to obtain, and for this reason the question of the effects of the degree of fuel atomization has been an elusive one. Consequently, previous fuel-atomization investigations generally have examined comparative engine performance with vaporized gasoline or a gaseous fuel, and with gasoline supplied by a conventional carburetor.

Multicylinder engine tests have been made by Robison and Brehob (1)* and by Lindsay, et al. (2). A vaporized-gasoline system and conventional carburetion were compared on a six-cylinder engine by Robison and Brehob, while Lindsay, et al. compared performance with a mixture generator (vaporized gasoline allowed to recondense into a fog) to that of a conventionally carbureted mixture on a four-cylinder engine. The two investigations had similar results. Both found geometric air-fuel ratio (A/F) distribution improvements and lean limit extensions with vaporized gasoline. Additionally, Robison and Brehob found less cyclic variations in peak cylinder pressure but little change in fuel consumption with vaporized gasoline. In these multicylinder engine experiments, it is difficult to tell whether the lean limit extension was due primarily to the improvement in geometric distribution between cylinders or to changes in the combustion characteristics with the more homogeneous mixtures attained with vaporized gasoline.

Single-cylinder engine experiments are more appropriate for that determination; and there have been several such investigations in the past. Dodd and Wisdom (3) studied the following four means of introducing gasoline in two different single-cylinder engines: fully vaporized, port fuel injection, carburetor, or a very coarse type of drip feed. The major result of their experiments dealt with the lean limit. They found that the engine would operate at the leanest A/F with vaporized gasoline. Port fuel injection was next in lean limit, followed by the carburetor and drip feed which were nearly the same. Their emission results were not conclusive because of repeatability problems.

Quader (4) studied propane and iso-octane with and without special fuel-air mixing provisions on a single-cylinder engine. With the special mixing provisions he felt that he had a much more homogeneous mixture of iso-octane than without it. Again, a significant extension of the lean limit was found with the more homogeneous mixtures. But, he also found that nitric oxide (NO) and hydrocarbon (HC) emissions were higher at a given equivalence ratio for iso-octane with good mixing than without.

Mizutani and Matsushita (5) studied the effects of a small amount of light diesel oil injected into the combustion chamber of a single-cylinder engine running on a propane-air mixture. Because of the low volatility of the diesel oil and the very short time available for fuel-air mixing, they may very well have had heterogeneous combustion. They found a significant extension of the lean limit with the addition of the diesel oil. They also found emissions of NO and carbon monoxide (CO) for a given equivalence ratio to be reduced with light-oil injection, while HC emissions increased.

The brief review above serves to illustrate that there is some confusion about the effects of the quality of mixture preparation. On the one hand it appears that more homogeneous charges should lead to leaner limits and lower emissions. On the other hand, there is some evidence (Quader) that indicates that homogeneous charges lead to higher NO and HC emissions. And, there is some evidence (Mizutani and Matsushita) which indicates that heterogeneous charges can lead to leaner limits.

The present investigation was undertaken in an effort to add to this body of

---

*Numbers in parentheses designate References at end of paper.

knowledge. Our major concern was, just how great are the effects of the degree of fuel atomization. This entails creating a fuel spray, with some knowledge of the character of the spray, then trying to estimate the character of the charge that might exist immediately prior to combustion.

We were able to determine the range of drop sizes produced by a given atomizer. Then, with analytical estimates of vaporization, it was possible to make reasonable estimates of droplet sizes at the intake valve. But, the state of the charge after the intake valve remains unknown, and any description of the charge in the combustion chamber must rely on estimates and deduction.

The experiments were performed on a single-cylinder engine so as to isolate the effects of atomization on combustion from the additional complications of cylinder-to-cylinder maldistribution. Furthermore, the tests were limited to steady-state engine operation. Thus, the effects of atomization on transient engine performance are not revealed by these experiments. Certainly, in considering the total effects of atomization on engine performance, one would have to include the effects of atomization on both geometric distribution and transient engine operation. But, those effects are not included in this research.

Three degrees of atomization were studied, and they covered a very wide range. Thus, there is a gap between the best atomizer and the worst; and it would have been desirable to investigate some atomizers within that range. However, it was felt that, by comparing the effects of large differences in fuel atomization, this research would contribute to our understanding of this very important facet of conventional engine technology.

TEST FACILITIES AND PROCEDURES

The single-cylinder engine used in this study was one cylinder of an Oldsmobile 7.46 ℓ (455 CID) V-8 engine. The crankshaft of an Oldsmobile 5.73 ℓ (350 CID) V-8 engine was used, giving the single cylinder a swept volume of 744 cc (45.4 in^3). A schematic diagram of the overall experimental setup is shown in Fig. 1. Several unique and specialized subsystems were required for these experiments. The details of these subsystems along with other pertinent descriptions are given below.

FUEL PREPARATION SYSTEMS - In this

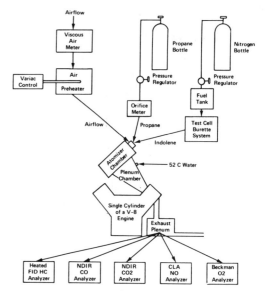

Fig. 1-Schematic of test facility

study three degrees of liquid fuel atomization, plus propane, were evaluated. These included Good atomization (10-20 μm droplets), Bad atomization (400-700 μm droplets), and Wall-Wetted atomization (400-700 μm droplets deposited on the cylinder-head port wall just above the intake valve). It was thought that this would blanket the range of possible charge conditions that could exist within an induction system and thus maximize differences in engine performance.

Good Atomization - Good atomization was produced by a converging-diverging nozzle (shown in Figs. 2 and 3) in which

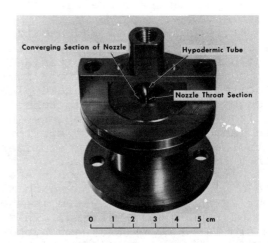

Fig. 2-Atomizer for Good atomization

the fuel was introduced at the high velocity throat section through a small hypodermic tube. This atomizer produced droplets in the size range from 10 to 20 μm and had a critical pressure ratio of 0.92. Droplet sizes for the atomizers were determined by photographing and

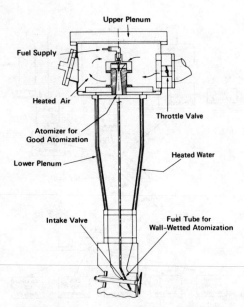

Fig. 3-Cross-sectional drawing of atomizers and induction system

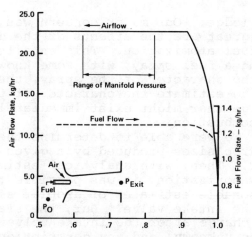

Fig. 4-Air flow and fuel flow characteristics of the atomizer for Good atomization

analyzing the sprays.

All engine airflow passed through the nozzle, thus promoting mixing of the liquid fuel and air and minimizing stratification. Airflow and fuel flow characteristics of the nozzle are shown in Fig. 4. Several sizes of nozzles were fabricated to cover the airflow range tested. This was necessary in order to maintain critical-flow nozzle operation and assure uniform atomization over the airflow range tested. Small changes (±15%) in airflow for a given nozzle were accomplished by varying the nozzle upstream pressure with a small throttle valve.

Bad Atomization - Bad atomization was produced by removing the nozzle section from the Good atomizer and injecting fuel into a relatively low-velocity section of the intake plenum through the same hypodermic tube. This resulted in very large droplets in the size range of 400-700 μm. Airflow was controlled upstream of the atomizer with a small throttle valve.

Wall-Wetted Atomization - Another type of charge preparation which may occur in an induction system is the case of severe wall-wetting. This case was simulated by extending a hypodermic tube down into the cylinder head intake port and spraying 400-700 μm fuel drops directly onto the port wall just above the intake valve. Therefore, this case is similar to the situation that may occur during cold engine operation or with some types of manifold fuel injection. The hardware to produce this type of preparation is illustrated in Fig. 3.

Propane - The apparatus used for propane metering was very similar to that used for Good gasoline atomization. For propane the hypodermic tube of the Good atomizer was replaced with a critical flow orifice. Mixing of fuel and air occurred within the converging-diverging nozzle, as with the case of Good atomization. This feature is desirable since the degrees of mixing are then similar, and comparisons between Good gasoline atomization and propane are more meaningful.

INDUCTION SYSTEM - The induction system consisted of an upper plenum and a lower plenum as shown in Figs. 3, 5, and 6. The upper plenum (Fig. 6) housed the atomizers and airflow control throttle valve. The clear plastic cover plate facilitated rapid changing of nozzles and provided for limited visual observations of the fuel spray within the lower plenum. The atomizer holder was spring-loaded and sealed with an "O" ring to the top of the lower plenum in order to relieve excessive pressures that might be caused by a backfire. The upper plenum also contained a pressure relief valve so that excessive pressures could be safely relieved. Upper plenum pressure and temperature were measured through a connection in one side of the plenum. Provisions were also made for the use of either liquid or gaseous fuel. The upper plenum volume was 3.7 ℓ (226 in.^{3}), and the lower plenum volume was 1.475 ℓ (90 in.^{3}). The interior surface of the lower plenum was heated to 52°C (125°F) by means of a water jacket surrounding the interior surface. This temperature prevented condensation of vaporized fuel and was helpful in estimating the character of the fuel-air charge just prior to entering the

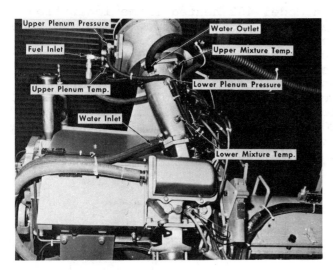

Fig. 5-Single-cylinder induction system

Fig. 6-Upper plenum

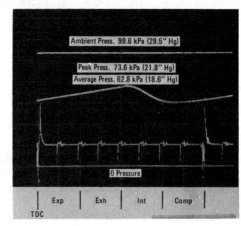

Fig. 7-Lower plenum pressure for motored engine

Fig. 8-Dish-shaped piston

cylinder head intake port. The lower plenum volume was carefully chosen such that pressure pulsations resulting from the intake stroke did not significantly affect metering and atomization. Pressure measurements within the plenum were recorded under motored engine conditions as shown in Fig. 7. For typical operating conditions, the average manifold pressure was 62.8 kPa (18.6 in. Hg), while the maximum pressure reached 73.6 kPa (21.8 in. Hg) just after intake valve opening. This results in a maximum pressure ratio of 0.74 across the nozzle, well below the critical pressure ratio of 0.92. Thus, the metering of fuel and air, as well as the quality of atomization, were unaffected by lower-plenum pressure pulsations.

COMBUSTION CHAMBER - The combustion chamber design chosen for these experiments was a relatively quiescent type. It was thought that the effects of atomization might be more readily revealed with a quiescent chamber as opposed to a more turbulent chamber, since turbulence may enhance mixing and vaporization. The piston, shown in Fig. 8, was of a rounded dish design. The cylinder head had a pancake-shaped combustion chamber and a second access hole to the combustion chamber which was used in this study for a cylinder pressure transducer. The combination of the dished piston and the pancake chamber cylinder head yielded a compression ratio of 7.84.

FUEL SYSTEMS - Specialized fuel metering systems were required for both propane and gasoline.

Propane - For the experiments involving propane, Pure Grade propane which is at least 99% pure on a molal basis was used. The propane flow was metered using a calibrated orifice meter and was controlled by modulating the pressure upstream of the orifice meter.

Gasoline - For the gasoline tests, Indolene clear was used. This is an unleaded gasoline having a hydrogen-to-carbon ratio of 1.81 and a specific gravity of 0.744 at 15°C (60°F). To meter and control the gasoline flow it was necessary to construct a fuel handling system that incorporated two important features. First, the gasoline flow had to be extremely steady. It was felt that such a requirement, from practical considerations, would rule out a system employing a fuel pump. Second, it was necessary to be able to measure accurately the low flow rates encountered. The basic components of the system are included in Fig. 1. A fuel tank was pressurized from a nitrogen bottle, thus delivering fuel at an extremely stable pressure. Fuel flow was varied by regulating this pressure. The gasoline flowed from the fuel tank through the standard test cell burette system and then to the engine.

AIRFLOW MEASUREMENT AND PREHEATING - Engine airflow was measured with a Meriam laminar airflow meter. In order to maintain a more controlled experiment, the intake air was heated, as required, to maintain a constant intake air temperature. The temperature chosen for the experiments will be discussed in the section on test procedure. The air heating was accomplished in a large tank with an electric heater controlled by a Variac rheostat. The air flowed first through the airflow meter, then into the preheating tank, and finally into the upper plenum of the induction system as shown in Fig. 1.

IGNITION SYSTEM - A Delco-Remy high-energy ignition system was used throughout the testing. This unit typically produces a spark duration of 1.8 msec at a voltage of 36 kV. Standard ignition systems have a spark duration of 1.0 msec at 24 kV. Standard spark plugs were used with a spark gap of 1.5 mm (0.060 in.).

CYLINDER PRESSURE MEASUREMENT - Cylinder pressure was monitored with a Kistler Model 649 water-cooled pressure transducer. This was done in order to compare the cyclic dispersion of the combustion events for the various degrees of atomization. The output of the transducer, along with the output of a crank angle indicator, were displayed on an oscilloscope.

EXHAUST EMISSION MEASUREMENT - Exhaust emission concentrations of CO and carbon dioxide (CO_2) were measured with Model 315A Beckman NDIR analyzers. HC concentration was measured with a Beckman Model 402 heated flame ionization detector, and concentrations of NO were measured with a Thermoelectron Chemiluminescent analyzer. Oxygen (O_2) concentration was measured with a Beckman Model 715 Process Oxygen Monitor. A separate exhaust sample line for the HC emissions was used, in addition to the CO, CO_2, O_2, and NO sample line. A diagram of the exhaust sampling system is shown in Fig. 9.

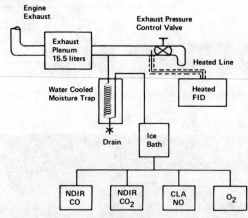

Fig. 9-Exhaust sampling system

TEST CONDITIONS - All tests were run at the same engine speed and indicated mean effective pressure. Fuel preparation and equivalence ratio (ϕ)* were the independent variables; and MBT (Minimum for Best Torque) spark advance, exhaust emissions, fuel consumption, and lean limit were the dependent variables. The operating conditions are listed in Table 1.

Table 1 - Operating Conditions

Speed	1600 rpm
Load	345 kPa (50 psi) IMEP*
Spark	MBT
Upper-Plenum Air Temperature	38°C (100°F)
Lower-Plenum Surface Temperature	52°C (125°F)
Exhaust Pressure	0.34 kPa (0.1 in Hg) gage
Coolant Temperature	85°C (185°F)
Oil Temperature	91°C (195°F)
Indolene Clear and Propane Fuels	

*IMEP was calculated from the sum of brake power and motoring power.

This speed and load approximates an 80 km/h (50 mph) road load condition. Upper-plenum inlet air temperature was held constant, although this resulted in different mixture temperatures depending on the type of fuel preparation. Originally, tests were run at constant mixture temperature, but it was con-

---

*Equivalence ratio (ϕ) is defined as the stoichiometric air-fuel ratio divided by the actual air-fuel ratio.

cluded that different degrees of fuel preparation should result in different mixture temperatures because of varying amounts of fuel vaporization. Therefore, a better control temperature for comparison of fuel preparation would be inlet air temperature. This resulted in mixture temperatures of 39.4°C (103°F) for Good atomization and 45.6°C (114°F) for Bad and Wall-Wetted atomizations. Mixture temperature for propane was controlled to 39.4°C (103°F).

TEST PROCEDURE - A period of about one hour was required to stabilize engine conditions after initial engine start-up. Upper-plenum air temperature was controlled to within ±0.5°C (1°F), and engine coolant temperature was controlled to within ±3.0°C (5°F). Lower-plenum skin temperature was controlled to within ±1.5°C (3°F) of the desired value. MBT spark advance was defined as the minimum spark advance necessary to provide the maximum brake output at a particular air-fuel ratio. A period of about 20 minutes was required to stabilize a test point prior to recording data. Fuel flow, airflow, and exhaust emission levels were monitored throughout this period in order to ensure stabilization.

LEAN MISFIRE LIMIT CRITERIA - In this paper the lean misfire limit is defined as that lean air-fuel ratio at which one misfire was detected in 100 to 200 engine cycles. This number of misfires is somewhat arbitrary. Nevertheless, it was chosen because, for greater misfire frequencies, it was difficult to control operating conditions and accurately measure exhaust emission levels. Also, misfire frequencies greater than one in 100 to 200 cycles for a vehicle would probably result in objectionable driveability. Misfires were detected by monitoring exhaust gas HC concentrations and cylinder pressure. A misfire usually appeared as a motored cylinder-pressure trace and an abrupt increase in HC concentration. Typical pressure traces at the lean limit are shown in Fig. 10. Because of large cyclic pressure variations under some conditions, it was difficult at times to detect a motored trace from a fired trace. Under these conditions, an abrupt change in exhaust gas HC concentration of about 200 ppm was used to detect a misfire.

RESULTS

In this section the results of the study will be presented without discussion. In the following section the sig-

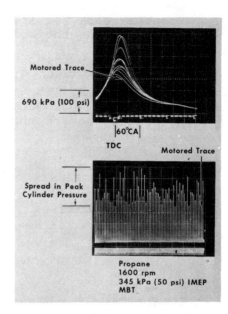

Fig. 10-Cylinder pressure traces at lean misfire limit

nificance of the results will be discussed.

MBT SPARK ADVANCE - MBT spark advance requirements for the three degrees of gasoline atomization and propane are shown in Fig. 11*. Large differences

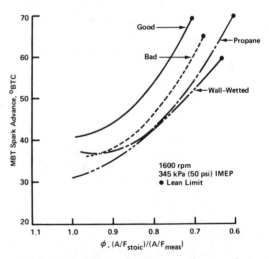

Fig. 11-Comparison of MBT spark requirements

in MBT spark advance were noted for the three cases of gasoline atomization. Good atomization required the largest spark advance while Wall-Wetted atomization required the least. Typically, this difference was 10 to 15 crank-angle

---

*The curves in Figs. 11 to 18 are least-squares curve-fits of the actual data points. Figures 11, 12, 13, 15, and 17 are shown again in the appendix with the actual data points included.

degrees. Propane spark requirements were nearly the same as the Wall-Wetted case of atomization.

LEAN MISFIRE LIMIT - The lean limit varied considerably for the three cases of gasoline atomization. The lean limit equivalence ratios for each degree of atomization and propane are shown in Table 2.

Table 2 - Lean Limit Equivalence Ratios (ϕ)

	ϕ
Good atomization	0.70
Bad atomization	0.68
Wall-Wetted	0.63
Propane	0.61

FUEL CONSUMPTION - Indicated specific fuel consumption (ISFC) was nearly identical for all three cases of gasoline atomization, as shown in Fig. 12. The equivalence ratios for minimum specific fuel consumption varied slightly, depending on the degree of atomization. The equivalence ratios at minimum ISFC for Good, Bad, and Wall-Wetted atomizations were 0.79, 0.77, and 0.75, respectively. The ISFC's for the three cases of atomization were nearly equal at the lean misfire limits. Fuel consumption for propane was about 8% lower than gasoline at all equivalence ratios tested, probably because of the greater heating value for propane (6). Minimum ISFC for propane occurred at ϕ = 0.73.

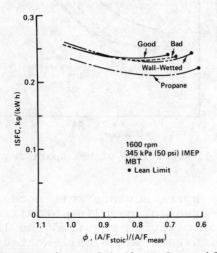

Fig. 12-Comparison of indicated specific fuel consumption

HC EMISSIONS - The exhaust HC emissions for each degree of gasoline atomization and propane are shown in Figs. 13 and 14 on a concentration and indicated

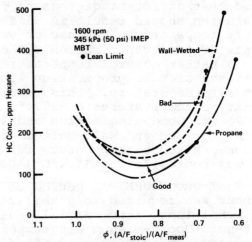

Fig. 13-Comparison of exhaust HC concentration

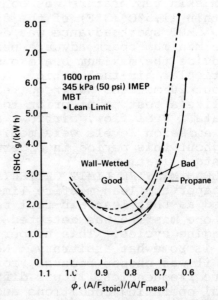

Fig. 14-Comparison of indicated specific HC emissions

specific basis, respectively. For equivalence ratios leaner than ϕ = 0.95, on either basis Good atomization resulted in less HC emissions than either Bad or Wall-Wetted atomization. The minimum HC point occurred at about ϕ = 0.85 for all cases of gasoline atomization. At the minimum HC point, Good atomization was about 20% lower than Wall-Wetted atomization. The difference in HC levels at the lean limit was large. HC concentration for Good atomization was 180 ppm at the lean limit while the HC concentration for the Wall-Wetted case was 490 ppm. Propane HC emission levels were lower than gasoline for all equivalence ratios tested. The minimum HC point for propane occurred at ϕ = 0.86 and the lean limit HC level was 380 ppm.

CO EMISSIONS - The CO emission results are shown in Figs. 15 and 16 on a concentration and indicated specific basis, respectively. Exhaust concentrations of CO for equivalence ratios richer than $\phi = 0.85$ were significantly lower for Good atomization than for the other two degrees of atomization. The equivalence ratio for minimum CO occurred at $\phi = 0.91$ for Good atomization and at $\phi = 0.85$ for Bad and Wall-Wetted atomizations. CO emissions for all cases of atomization were nearly the same for

Typically, propane CO emissions were 30% lower than gasoline for equivalence ratios leaner than $\phi = 0.90$.

NO EMISSIONS - NO emissions, shown in Figs. 17 and 18 (by convention, the molecular weight of NO_2 is used in calculating the mass emissions shown in Fig. 18), were significantly different between the three types of atomization investigated. Good atomization resulted in a peak NO value of 2125 ppm at $\phi = 0.92$, and Bad atomization resulted in a peak NO concentration of 2050 at $\phi = 0.87$. Wall-Wetted atomization resulted in a peak value of 1900 ppm at $\phi = 0.88$. For leaner than $\phi = 0.87$, Bad atomization resulted in NO emissions which were 10 to 70% higher than Good atomization, depending on the equivalence ratio. NO emissions with propane were generally lower than gasoline NO emissions. The peak NO value for propane was 1600 ppm at an equivalence ratio of 0.92.

CYCLE-TO-CYCLE VARIATIONS IN PEAK CYLINDER PRESSURE - Both peak cylinder pressure and mean peak pressure* varied considerably between the different degrees of atomization and propane as shown in Fig. 19 and Table 3 for $\phi =$

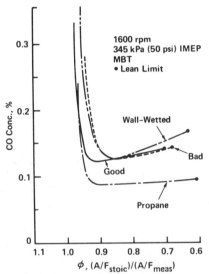

Fig. 15-Comparison of exhaust CO concentration

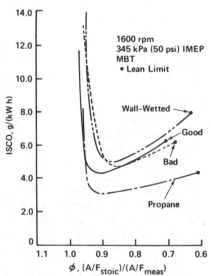

Fig. 16-Comparison of indicated specific CO emissions

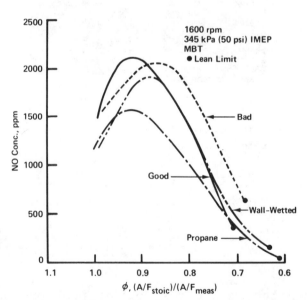

Fig. 17-Comparison of exhaust NO concentration

equivalence ratios leaner than $\phi = 0.85$. CO emissions with propane were lower than gasoline for all equivalence ratios leaner than $\phi = 0.95$. Minimum CO emissions for propane occurred at $\phi = 0.91$.

0.84. Generally, Good atomization resulted in the largest spread in peak cylinder pressure and the lowest value for mean peak pressure, while propane

---

*Cylinder pressure was measured with respect to the lowest recorded cylinder pressure. The lowest pressure observed for these conditions corresponds to an absolute lower-plenum pressure of 59.1 kPa (17.5 in. Hg).

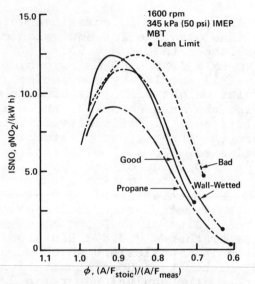

Fig. 18-Comparison of indicated specific NO emissions

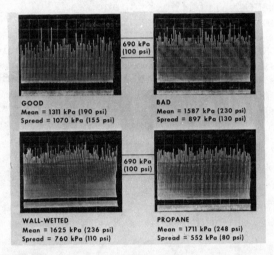

Fig. 19-Comparison of cycle-to-cycle peak cylinder pressures

Table 3 - Cylinder Pressure Data, $\phi = 0.84$

	Spread in Peak Cylinder Pressure	Mean Peak Pressure
Good atomization	1070 kPa (155 psi)	1311 kPa (190 psi)
Bad atomization	897 kPa (130 psi)	1587 kPa (230 psi)
Wall-Wetted	760 kPa (110 psi)	1625 kPa (236 psi)
Propane	552 kPa (80 psi)	1711 kPa (248 psi)

had the smallest spread in peak pressure and the highest mean peak pressure.

DISCUSSION

The results presented in the previous section generally show that emissions and MBT spark requirements are significantly affected by the degree of fuel atomization. The poorer degrees of atomization, i.e., Bad and Wall-Wetted, generally resulted in higher emissions and lower MBT values, i.e., shorter combustion durations*, for lean equivalence ratios than those resulting from Good atomization. In this section we will discuss the mechanisms and processes by which these results might have come about.

The manner in which the degree of fuel atomization influences emissions and combustion duration is not well understood because there are many complicated, interrelated processes that occur between the point of initial atomization and the time of ignition in the combustion chamber. Even though droplet sizes just after atomization can be determined accurately, albeit not trivially, and the amount of vaporizing that occurs prior to the intake valve can be estimated, the state of the charge (drop sizes, distribution, and degree of mixing) just prior to ignition is relatively unknown.

This is because the intake and compression processes are extremely complex, involving additional mixing and vaporization of liquid fuel. So, even if we know a lot about the charge at the intake valve, we know very little about it after the intake and compression events.

In spite of a lack of detailed understanding of the processes involved, it is reasonable to expect that the degree of atomization influences the amount of mixing between the fuel and air. We believe that differences in mixture uniformity resulting from differences in atomization are responsible for the results observed in these experiments. The discussion that follows is intended to justify this belief and explain the experimental results in light of the mixing phenomenon.

Consider the case of Good atomization where initial droplet sizes of 10-20 μm were observed. In this case all of the engine airflow passed through the throat of the atomizer where the fuel was introduced. This should lead to good initial mixing between the air and liquid fuel droplets because the two are brought into close contact in the atomizer throat. Furthermore, the normal-shock front developed in the diverging section of the nozzle aided in mixing the fuel droplets and air. The fuel and air then diffused together

---

*In this paper combustion duration is defined as the sum of the ignition delay period and the burn time. MBT spark timing is taken to be a qualitative, but not precise, indicator of combustion duration.

throughout the remainder of the nozzle and induction system.

Based on the work of Kent (7), and Law (8,9), we estimated that 10-20 μm gasoline droplets, as in the case of Good atomization, will be 70 to 80% (by mass) vaporized prior to the intake valve in our induction system, and that the droplets remaining will be 5-10 μm in diameter. Droplets of the 5-10 μm size generally will follow the airstream (10) and are less likely to impinge on the intake valve and combustion chamber walls than larger droplets. All of this favorable preparation of the charge prior to the intake and compression strokes leads us to believe that Good atomization probably resulted in a nearly homogeneous, vaporized charge in the combustion chamber immediately prior to ignition.

In contrast, the cases of Bad and Wall-Wetted atomization likely resulted in fuel-air stratification and inhomogeneity prior to ignition. There are several considerations involved in this conclusion in each case.

Consider the case of Bad atomization, where 400-700 μm droplets were introduced in a stream down the center of the induction system. Initial mixing of the fuel and air within the induction system was not good because the large droplets tended to remain in a stream for much of the induction system length. Furthermore, 400-700 μm droplets will not vaporize readily within the induction system, thus inhibiting good mixing between vaporized fuel and air; and the trajectories of the large droplets are not significantly affected by the air motions in the induction system and around the intake valve. Thus, it is likely that many droplets were impacted on the induction system walls and intake valve. When this deposited fuel vaporizes from these surfaces, stratification of the fuel and air can result. In the case of the large droplets that do happen to negotiate the intake passage without being deposited, they will most likely either be further atomized by passing through the intake valve or impinge on the hot combustion chamber walls. The impinged droplets will vaporize from the surfaces and contribute to additional stratification.

While it is possible that a small amount of liquid fuel existed at the time of ignition in the case of Bad atomization, we believe that most of the fuel was in the vapor phase. We have estimated that a 25 μm droplet of gasoline will just barely vaporize completely before ignition during the compression stroke at 1600 rpm engine speed for part load operating conditions. It is also estimated that nearly 100% droplet impaction will occur for 50 μm and larger droplets arriving at the intake valve (10). Thus, any liquid fuel droplets still existing at the time of ignition probably entered the combustion chamber in the 25-50 μm size range. When one considers that a large part of each droplet in this size range will vaporize during the intake and compression strokes, the conclusion is that Bad atomization probably had very little fuel in the liquid state just prior to ignition.

In the case of Wall-Wetted atomization, initial mixing of fuel and air again was poor because the fuel was squirted directly onto the intake-port walls. Also, it is possible that a larger amount of liquid fuel than in the case of Bad atomization passed through the intake valve, particularly during the period of maximum valve lift. This liquid fuel may not be well-atomized because the velocities through the valve are relatively low during this period; and the droplets could impact the combustion chamber walls and produce stratification. Also, it might be reasonable to expect that Wall-Wetted atomization produced a greater amount of fuel in droplet form at the time of ignition than did Bad atomization.

It should be emphasized that the amount of fuel impacted on the combustion chamber walls and the amount of fuel in droplet form prior to ignition in both the Bad and Wall-Wetted cases depend heavily on the flow through the intake valve and the fluid motions within the chamber, and these factors are not well characterized. Also, recall that the combustion chamber was of a relatively quiescent design. Thus, in-cylinder mixing may not have been as great as in a more turbulent combustion chamber.

In light of all of the above discussion, one might expect the results for Good atomization to exhibit characteristics of a homogeneous, vaporized charge, such as propane. Results for the Bad and Wall-Wetted cases might be expected to show some of the characteristics of a stratified charge engine.

POSSIBLE DESCRIPTIONS OF THE STRATIFICATION - The character or description of the stratification that may have occurred in the cases of Bad and Wall-Wetted atomization is, of course, unknown. But, we feel that one of the three descriptions below is probably realistic.

First, these two cases of atomization may have resulted in fuel-air charges composed of small pockets of rich mixture with lean mixture filling in the areas between the rich pockets. This description is illustrated in Fig. 20A. The rich pockets could result either from liquid droplets vaporizing but not mixing well before ignition, or from any other vaporized fuel which has not been mixed well with air. For example, any of the fuel which might have impacted the cylinder walls might vaporize but remain localized resulting in rich pockets. It is also possible in these two cases of atomization that there was some amount of liquid fuel existing at the time of ignition, although, as we have mentioned, that amount probably was small. This small-scale kind of stratification might be called "distributed" stratification.

The second possible description, illustrated in Fig. 20B, might be a larger-scale stratification with relatively large rich and lean pockets. This might result from the fuel taking a preferential direction through the intake valve and remaining concentrated in a particular area of the cylinder. This might be especially true in the case of Wall-Wetted since the fuel was introduced

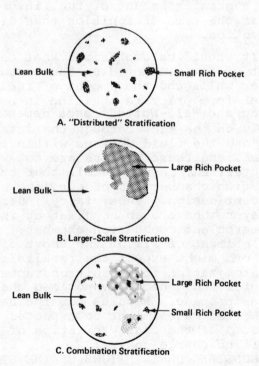

Figs. 20A,B&C-Illustrations of possible kinds of stratification

on the port wall in a very localized area. In the case of Bad atomization the fuel could have become quite localized by virtue of the fact that the large droplets generally could not negotiate the bend in the intake port.

The third possible description of the stratification might be visualized as a combination of the above two descriptions as shown in Fig. 20C. This case might consist of a fairly localized and larger rich pocket in part of the cylinder along with other small, distributed rich pockets in the remainder of the cylinder.

In the discussion that follows the results of this study will be shown to be consistent with the hypothesis that Good atomization resulted in an essentially homogeneous charge while Bad and Wall-Wetted atomizations resulted in some form of stratification and heterogeneous combustion.

HC EMISSIONS - Referring again to Figs. 13 and 14, Bad and Wall-Wetted atomizations resulted in higher HC emissions than Good atomization. Stratification or inhomogeneity of the charge in the Bad and Wall-Wetted cases could have caused this as a result of variable quench layer thicknesses due to pockets of rich and lean mixture. These rich and lean pockets could be the result of stratification within the bulk gas that might exist even next to the wall; or they could result from droplet impingement on the combustion chamber walls. Thus, for the same overall A/F, Good atomization, being a more homogeneous charge, would generally result in lower HC emissions.

Lower HC emissions could also have resulted with Good atomization from more reactions during expansion and exhaust due to longer combustion durations. Propane resulted in lower HC emissions than Good atomization primarily because of a thinner quench layer for propane fuel (11), not necessarily because of differences in fuel preparation.

CO EMISSIONS - The primary difference in CO emissions between Good, Bad, and Wall-Wetted atomizations is the shift in equivalence ratio at which minimum CO occurs, as shown in Figs. 15 and 16. Similar shifts in the CO curves were observed by Purins (12) with a prechamber stratified charge engine and by Dodd and Wisdom (3) with two single-cylinder engines. The shift in CO is consistent with the idea of A/F stratification within the cylinder, because CO emission levels are very nonlinear with A/F in the vicinity of the minimum and are, therefore, sensitive to small changes in A/F about an average. A/F stratification within the cylinder would cause a shift in the minimum CO emissions to leaner equivalence ratios. CO emission levels would be relatively insensitive to stratification for very rich and very lean mixtures, since the curve is nearly linear in these ranges.

Propane CO emissions are lower than Good atomization, probably as a result of less partial reaction of HC to CO in the exhaust system.

NO EMISSIONS - Refer again to Figs. 17 and 18 and consider, for the moment, only the differences in NO between Good and Bad atomizations. The major difference between these two appears to be a displacement or shifting of the NO curve to leaner mixtures for Bad atomization when compared to Good. The peak value of NO was about the same for both degrees of atomization. As a result of the shift, the NO values for Bad atomization at leaner than $\phi = 0.88$ were significantly greater than for Good atomization. The shift observed here can result from stratification and inhomogeneities within the charge and, in fact, has been seen to occur in stratified charge engines.

A similar shift in the NO curve was reported by Lavoie and Blumberg (13) in studies of an open-chamber stratified charge engine. They found both theoretically and experimentally that a linear rich-to-lean stratification (linear variation in equivalence ratio from rich to lean) resulted in a peak NO value at a significantly leaner overall equivalence ratio than a homogeneous, premixed case. Consequently, linear stratification resulted in lower NO concentrations for equivalence ratios near stoichiometric but higher NO emissions for leaner equivalence ratios -- similar to the results for our Bad versus Good atomization.

Lavoie and Blumberg also considered another form of stratification consisting of small-scale inhomogeneities superimposed on a uniform mixture. This is very pertinent because, perhaps, their model of stratification in this case begins to approximate the kind of stratification that we might have had with Bad atomization. Their model showed that NO emissions even for this small-scale stratification were very similar to those for large-scale stratification for overall lean mixtures.

Additionally, Blumberg (14) shows theoretically that, for an overall equivalence ratio of $\phi = 0.8$, an increase in NO emissions of 40% over the homogeneous case is possible with linear rich-to-lean stratification limits of $\phi = 0.85$ to $\phi = 0.75$. This shows that even for rather small differences in equivalence ratio, stratification can produce large effects on NO.

In light of these previous results, it is reasonable to expect that NO emissions for our Bad atomization case would be different from the Good atomization case, assuming that Bad atomization resulted in some kind of stratification -- either distributed, large-scale, or a combination of the two as described previously.

Now consider the results for Wall-Wetted atomization. Intuitively, one might expect the Wall-Wetted case to result in even more inhomogeneity than Bad atomization, and in line with the previous discussion, the NO for Wall-Wetted might be displaced to the lean side of Bad atomization. But that was not the case. Instead, the NO curve for Wall-Wetted lies to the rich side of Bad atomization and, in fact, almost lies entirely within the curve for Good atomization. However, the Wall-Wetted peak is shifted somewhat to the lean side of the peak for Good atomization, and its highest NO value is lower than either Good or Bad atomization. All of this does not necessarily mean that Wall-Wetted resulted in less inhomogeneities than Bad atomization.

Another part of Blumberg's paper (14) relates well to this point. He showed that for a given overall equivalence ratio, the amount of NO can vary significantly depending on the range or width of stratification about the average. In fact, he showed that for $\phi = 0.8$ NO emissions, when compared to the homogeneous case, can increase up to a peak value for small widths of stratification and then decrease as the width of stratification becomes even larger. Perhaps Bad atomization has a width of stratification such that NO is increased compared to Good atomization; while Wall-Wetted may have a larger width of stratification than Bad and is past the peak identified by Blumberg so that NO is less than for Bad atomization.

At this point we should re-emphasize that precise interpretation of our results in terms of stratification is difficult because of the many complex phenomena involved and the fact that there are still many unknowns about the fuel-air charges resulting from different degrees of atomization. However, we do feel that our results are sufficiently consistent and in agreement with the results of others, to justify the belief that different degrees of atomization caused different kinds of stratification and mixing of the fuel-air charge which, in turn, produced different emission results.

A final note for this section concerns the NO emissions of propane, which were significantly lower than Good atomized gasoline at nearly all equivalence ratios tested. This is thought to be due to lower combustion temperatures with propane (15), because propane has a larger hydrogen-to-carbon ratio and requires more air at a given equivalence ratio. Furthermore, Harrington and Shishu (16) showed that for fixed equivalence ratios, NO emissions decreased as hydrogen-to-carbon ratio was increased. Consequently, the differences in NO emissions between propane and Good atomization are interpreted to be the result of differences in fuel composition rather than differences in fuel preparation.

FUEL CONSUMPTION - The results for fuel consumption, Fig. 12, primarily show shifts in the equivalence ratios at which minimum fuel consumption occurs for the Bad and Wall-Wetted cases. Other than the shifts, the differences between the three degrees of atomization are very small. A sufficient explanation for these results has not been formulated as yet. However, it may be that the shorter combustion durations for Bad and Wall-Wetted, as indicated by the smaller MBT spark advances for the two, result in higher efficiencies at lean mixture ratios. A shorter combustion duration indicates that more of the charge burns close to top dead center and is, therefore, more efficient.

It is interesting to note that Dodd and Wisdom (3) have in their paper an amazingly similar set of curves. One of the parts of their investigation was the study of their "drip feed" atomization in three different sizes of mixing chambers. One can argue that the smallest mixing chamber would allow the least time for mixing of the rather poorly atomized fuel, while the largest mixing chamber might offer the most time. Consequently, the charge should be most homogeneous with the largest mixing chamber, with more inhomogeneities resulting from the two smaller chambers. Their fuel consumption results were very similar to ours, with the least homogeneous charge having the leanest minimum fuel consumption point. Furthermore, other than the shift of the minimum points, there were very few differences between their three curves.

MBT SPARK REQUIREMENTS AND CYCLIC VARIATIONS IN PEAK CYLINDER PRESSURE - MBT spark requirements and the cyclic variations in peak cylinder pressure will be discussed together because the two are intimately related. Figure 19 shows that as atomization of gasoline was improved the cyclic variations in peak cylinder pressure increased. At first this result seems startling, because one might think that more homogeneous charges should produce more consistent combustion. But, in order to understand this phenomenon we must consider the cyclic peak pressure results in conjunction with the results for MBT spark timing. Figure 11 shows that as atomization of gasoline improved, the combustion duration -- as indicated by MBT spark timing -- increased. The shorter combustion durations in the cases of Bad and Wall-Wetted can result from inhomogeneities in the charge, and here we have to draw a distinction between the possible descriptions of the stratification.

If the stratification were the small-scale, distributed kind, the inhomogeneities might have affected the overall combustion rate in two ways. First, the rich pockets would have a faster flame speed than lean pockets, with a resulting effect of increased wrinkling and distorting of the flame front. The more-wrinkled flame front will have a greater flame front area resulting in a faster combustion rate. Second, the rich pockets, if in the vicinity of the spark plug, will enhance the initial flame kernel development and result in decreased ignition delay time.

If the stratification in the cases of Bad and Wall-Wetted were more of the large-scale kind with a more concentrated and localized rich region in the vicinity of the spark plug, then the initial flame kernel development would again be enhanced with resulting decreases in the ignition delay period.

The above descriptions are merely an attempt to describe how the shorter combustion durations -- smaller MBT spark advances -- might have resulted from the poorer degrees of atomization. The question now is, what does that have to do with decreased cyclic variations which also resulted from the poorer atomizations? Patterson (17) states that "changes which increase the combustion rate will always reduce peak pressure and cyclic work differences." He was concerned primarily with the effects of <u>charge motions</u> on the combustion rate, and particularly with the effects during the initial flame development period. It was his conclusion that, when the flame front was wrinkled and distorted by charge motions just after ignition, the cyclic variations were reduced. It is our belief that with Bad and Wall-Wetted atomizations a similarly beneficial effect on cyclic variations occurred because of decreases in combustion durations, except that in our case this effect resulted from inhomogeneities, not charge motions.

A seemingly contradictory point is that propane, which certainly was more homogeneous than either Bad or Wall-Wetted atomization, had the least variation in peak cylinder pressure. This seems contradictory to the entire reasoning above. But, propane also had a combustion duration, as deduced from MBT spark requirements, that was about the same as the Wall-Wetted case. Thus, the premise tha shorter combustion durations lead to reduced cyclic variations is still supported with the results for propane. The question then becomes, why does propane have a shorter combustion duration than the Good atomization case? Although a complete answer to this question has not been formulated as yet, the explanation probably lies primarily in differences in fuel properties, not differences in fuel preparation. For example, part of the answer might be that propane has a slightly greater laminar flame speed than gasoline (11). (Propane flame speed must be compared to flame speeds of pure hydrocarbon components that make up gasoline such as hexane, heptane, iso-octane, etc. (7)).

LEAN MISFIRE LIMIT - The lean misfire limit was extended with Bad and Wall-Wetted atomizations when compared to Good atomization. This appears to be a consequence of the shorter combustion durations for Bad and Wall-Wetted. Quader (4) found that combustion duration reaches a limiting value at the lean misfire limit and that the lean limit correlates well with flame speed. It is possible, then, that mixtures with shorter combustion durations will not encounter this limiting combustion duration until leaner mixtures.

COMPARING EMISSION RESULTS AT FIXED SPARK ADVANCE - Because of the significantly different MBT spark timings, comparisons of exhaust emissions at a fixed spark timing for the three cases of fuel atomization would be considerably different than those reported at MBT spark advance. For example, if a fixed spark setting is chosen that is near MBT for Bad atomization, then the results for Bad atomization would be nearly the same as those for MBT. However, if Good atomization is tested at the same fixed spark setting, then emission results would exhibit characteristics of retarded spark for Good atomization; HC and NO emissions would be lower than at MBT. This would emphasize even more the differences between Good and Bad atomizations. Thus, when comparing different types of fuel atomization, one should recognize that there may be differences in MBT spark timing between the different types of atomization which could influence emission results.

SUMMARY AND CONCLUSIONS

In this single-cylinder engine study three degrees of gasoline atomization plus propane were investigated. Air-fuel ratios were varied from stoichiometric to the lean limit. The experiments were limited to a single engine operating condition -- 1600 rpm, 345 kPa (50 psi) IMEP, and MBT spark timing -- and were carried out using a relatively quiescent combustion chamber. The results and conclusions of the study are summarized below.

(1) The degree of fuel atomization significantly affected engine performance. It is suggested that the reason for the effects is that Good atomization probably resulted in an essentially homogeneous fuel-air mixture, while Bad and Wall-Wetted atomizations resulted in some form of stratification or inhomogeneity within the charge.

(2) HC emissions varied by about 20% between degrees of atomization with Good atomization being the lowest.

(3) CO emission levels for fuel-lean mixtures ($\phi < 0.85$) were not affected by the degree of atomization. For richer mixtures ($\phi > 0.85$) Good atomization resulted in the lowest CO emissions by as much as 40%.

(4) NO emission levels were significantly affected by the degree of atomization with up to 70% variation between the three degrees of atomization.

(5) Fuel consumption was not significantly affected by the degree of atomization.

(6) Both Bad and Wall-Wetted atomizations extended the lean misfire limit over that of Good atomization, with Wall-Wetted having the leanest misfire limit.

(7) Bad and Wall-Wetted atomizations decreased MBT spark advance requirements.

(8) Bad and Wall-Wetted atomizations resulted in less variation in cycle-to-cycle peak cylinder pressure than Good, probably the result of shorter combustion durations.

(9) The results for propane differed significantly from those of Good atomization of gasoline even though the two cases probably resulted in very similar fuel-air charges, i.e., vaporized and

homogeneous mixtures of fuel and air. Therefore, it was concluded that the differences in results between the two were, for the most part, the result of basic differences in fuels such as heating value, quench distance, and hydrogen-to-carbon ratio, and not the result of differences in fuel preparation.

We feel that at this stage of our understanding of the effects of atomization we still have a lot to learn. In all of the investigations of the phenomenon, including ours, the descriptions of the fuel-air mixtures that resulted from various charge preparation devices were more qualitative in nature than quantitative. As long as this situation exists, it might be reasonable to expect that different investigators would obtain different results depending upon the characteristics of the fuel-air mixtures resulting from their charge preparations. The problem is that we do not have a good basis for comparison. For example, the charge preparation that someone else calls "Bad" may bear little resemblance to what we may call "Bad." And, we have seen how engine performance can be quite different for what may actually be only moderate differences in fuel-air mixtures. Perhaps any apparent conflicts in results between different investigators are merely an indication of the complexity of the problem, and emphasize the need for further research.

Based on the results of the present investigation, we can suggest some areas for further research which may lead to a better understanding of the problem. For example, as we mentioned previously, the range of atomization between our Good and Bad cases should be investigated to fill in the gap between those two. Other engine operating conditions should also be investigated. And, an important area for further investigation involves the combustion chamber. We used a relatively quiescent combustion chamber. It is possible that a much more turbulent combustion chamber might contribute to significantly different results since, one would think, mixing of air and fuel would be enhanced. Thus, a more turbulent combustion chamber should be included in future work.

ACKNOWLEDGMENTS

The authors gratefully acknowledge all those at General Motors Research Laboratories who contributed to this work. Special thanks are due Dr. Julian M. Tishkoff and Dr. David L. Harrington for their contributions in determination of atomizer spray characteristics and in analysis of results, and to Harvey C. Mooer for test setup and data collection.

REFERENCES

1. J.A. Robison and W.M. Brehob, "The Influence of Improved Mixture Quality on Engine Exhaust Emissions and Performance," Journal of the Air Pollution Control Association, Vol. 17, No. 7, July 1967.

2. R. Lindsay, A. Thomas, J.A. Woodworth, and E.G. Zeschmann, "Influence of Homogeneous Charge on the Exhaust Emissions of Hydrocarbons, Carbon Monoxide and Nitric Oxide from a Multicylinder Engine," SAE Paper No. 710588 presented at the Mid-Year Meeting, Montreal, Canada, June 1971.

3. A.E. Dodd and J.W. Wisdom, "Effect of Mixture Quality on Exhaust Emissions from Single-Cylinder Engines," Proceedings of the Institution of Mechanical Engineers, Vol. 183, Part 3E, 1968.

4. A.A. Quader, "Lean Combustion and the Misfire Limit in Spark Ignition Engines," SAE Paper No. 741055 presented at the Automobile Engineering Meeting, Toronto, Canada, October 1974.

5. Y. Mizutani and S. Matsushita, "Fuel Vapor-Spray-Air Mixture Operation of a Spark-Ignition Engine," Combustion Science and Technology, Vol. 8, 1973.

6. C.F. Taylor and E.S. Taylor, "The Internal Combustion Engine," International Textbook Co., Scranton, Penn., 1961, p. 34.

7. J.C. Kent, "Quasi-Steady Diffusion Controlled Droplet Vaporization and Condensation," Applied Science Research, Vol. 28, 1973.

8. C.K. Law, "A Theory for Monodisperse Spray Vaporization in Adiabatic and Isothermal Systems," International Journal of Heat and Mass Transfer, Vol. 18, 1975.

9. C.K. Law, "Quasi-Steady Droplet Vaporization Theory with Property Variations," Physics of Fluids, Vol. 18 (No. 11), 1975.

10. D.A. Trayser, F.A. Creswick, J.A. Gieseke, H.R. Hazard, A.E. Weller, and D.W. Locklin, "A Study of the Influence of Fuel Atomization, Vaporization, and Mixing Processes on Pollutant Emissions from Motor-Vehicle Powerplants," Report No. PB185886, Battelle Memorial Institute, Columbus, Ohio, April 30, 1969.

11. "Basic Considerations in the Combustion of Hydrocarbon Fuels with Air," NACA Report 1300, 1959.

12. E.A. Purins, "Pre-Chamber Stratified Charge Engine Combustion Studies," SAE Paper No. 741159, presented at the International Stratified Charge Engine Conference, Troy, Michigan, 1974.

13. G.A. Lavoie and P.N. Blumberg, "Measurements of NO Emissions from a Stratified Charge Engine: Comparison of Theory and Experiment," Combustion Science and Technology, Vol. 8, 1973.

14. P.N. Blumberg, "Nitric Oxide Emissions from Stratified Charge Engines: Prediction and Control," Combustion Science and Technology, Vol. 8, 1973.

15. P.N. Blumberg and J.T. Kummer, "Predictions of NO Formation in Spark-Ignited Engines -- An Analysis of Methods of Control," Combustion Science and Technology, Vol. 4, 1971.

16. J.A. Harrington and R.C. Shishu, "A Single-Cylinder Engine Study of the Effects of Fuel Type, Fuel Stoichiometry, and Hydrogen-to-Carbon Ratio on CO, NO, and HC Exhaust Emissions," SAE Paper No. 730476, May 1973.

17. D.J. Patterson, "Cylinder Pressure Variations -- A Fundamental Combustion Problem," SAE Paper No. 660129, January 1966.

APPENDIX

Figures A-1 through A-5 show the actual data points along with the least-squares curve-fits that were presented in Figures 11, 12, 13, 15, and 17.

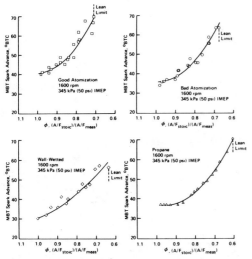

Fig. A-1-MBT spark requirements

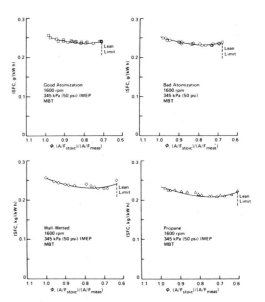

Fig. A-3-Exhaust HC concentration

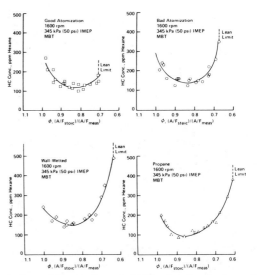

Fig. A-2-Indicated specific fuel consumption

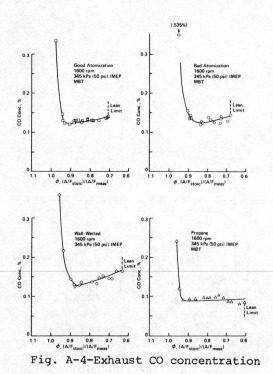

Fig. A-4-Exhaust CO concentration

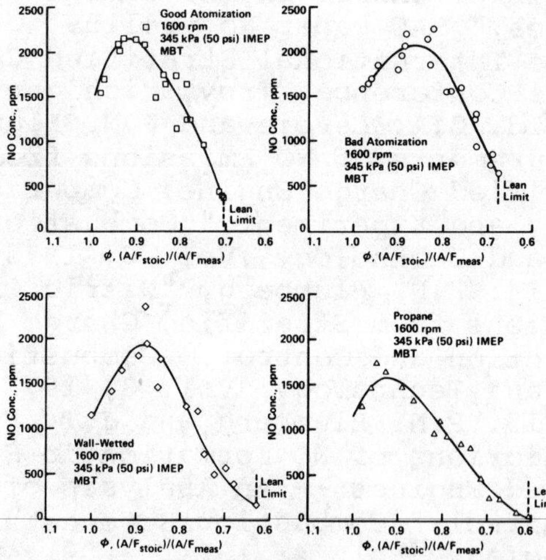

Fig. A-5-Exhaust NO concentration

DISCUSSION

I. N. BISHOP
FORD MOTOR CO.

THE AUTHORS IMAGINATIVELY have undertaken to study a most difficult subject. It has been my experience that each time a difficult experimental program is run, unforeseen information is uncovered which poses new questions and suggests that the experiment should be redesigned and rerun. The experiment described by the authors is consistent with this experience. Experienced and skilled hardware designers readily accept the fact that good design is not suddenly created but rather must be evolved by a tedious process of design and redesign. Thus, skilled and experienced researchers need not feel embarrassed when the analogous situation is encountered in a complex and difficult experimental study program.

In my opinion, there are several experimental details that should be altered or added in order to control engine performance and emission effects that are attributable to factors other than fuel vaporization. It is my further opinion that in the absence of these modifications, the significance of the data presented and the explanations postulated as the underlying causes of the results are questionable.

Inasmuch as spark timing has a very pronounced effect on NO_x level and a significant effect on HC emissions, the extreme difficulty of accurately defining MBT spark timing makes the conclusions drawn by the authors from their data on NO_x effects and, to some extent, on HC effects, highly suspect. It may be seen in the authors' Fig. A-1 that spark timing data varied from the trend line by as much as 5 crank deg. I have reproduced the authors' NO_x data in my Fig. 1 and have superimposed on this graph the variability in NO_x level that I find characteristic for the engines I have studied over a range of ± 2 crank deg near MBT. It can be seen from this Fig. that the error band shown easily covers the vast majority of the experimental data. I would propose, therefore, that if the program were to be rerun, the spark timing be set at some percentage torque loss from the best torque level, say 5%, which is a setting that can be reproduced more reliably than can MBT.

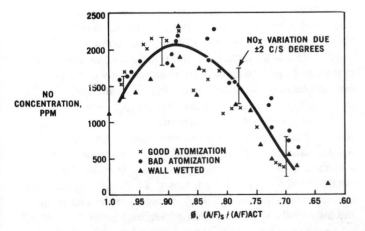

Fig. 1 - Comparison of atomization effect data and spark advance effect data on NO_x concentrations

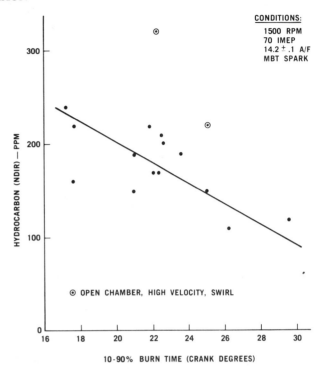

Fig. 2 - Effect of burn time on hydrocarbon emissions

The authors' Fig. 11 shows that the MBT spark timing for the cases of good, bad and wall-wetted progressively varied 5 deg for each level of atomization decrease which, as the authors state, shows slower combustion as the level of atomization improves. If this is indeed true, it is within the power of the engine designer to modify the engine design in such a way to provide a constant rate of combustion and, therefore, the effects of combustion rates can be isolated in an experimental program that is designed to identify the inherent advantages and disadvantages of atomization level. In support of this statement, I offer Fig. 2 which is taken directly from a paper by Mayo[1] which shows that combustion rate has a pronounced effect on HC level. Incidentally, the paper also shows that combustion rate has a significant effect on NO_x level. The HC data shown by Mayo is in a direction and of a magnitude that can more than explain the HC effect shown by the authors.

According to studies conducted at Ford to determine the factors that influence lean limit, gas temperature at the time of spark has a profound effect on the lean limit. Thus, the increased spark advance requirement observed by the authors for good atomization is significant since it results in a lower temperature at the time of spark due to a lower heat of compression. As a matter of fact, the changes in lean limit that were observed to occur as a function of vaporization correspond closely to the values we would predict on the basis of spark timing alone. (See Fig. 3.)

Recent single cylinder combustion/emission experiments at Ford using a fast response air/fuel ratio meter have shown a surprisingly large time resolved fluctuation in air/fuel ratio even though a vaporization tank was introduced between the fuel system, in our case a carburetor, and the engine. It is conceivable

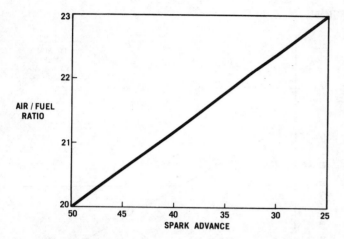

Fig. 3 - The effect of spark advance on lean limit

that cyclic or time dependent variations in air/fuel ratio could influence the experiments described by the authors. Thus, it might be appropriate to monitor air/fuel ratio variations in future experiments, even though what is believed to be a relatively stable fuel metering system is employed.

I wonder if the authors could comment on any discrepancy between air/fuel ratio values computed from flow measurements as opposed to estimates made from the analysis of exhaust gas chemistry. Since they did not publish CO_2 data, I was not able to determine this for myself. At Ford, we have adopted the redundant measurement of air/fuel ratio as an inviolate practice in order to help assure experimental accuracy.

The mixture temperature for bad and wall-wetted atomization was shown to be about 11° higher than for good atomization. According to my calculations, one could deduce, therefore, that percent vaporization was about 30% less than for the good atomization case. Thus, it could be estimated that, since good atomization is estimated to provide 70 - 80% vaporization, the other two cases provided 40 - 50% vaporization. Do the authors concur that this estimate is a valid one? Also, have the authors considered trying to monitor vaporization level experimentally?

After spending 6 - 8 years of my professional life trying to create charge stratification, I came to the conclusion that it is a goal that is nearly impossible to attain without drastic measures, namely, direct cylinder fuel injection. I, therefore, have a natural reluctance to accept the authors' contention that charge stratification has all the effects on their experimental results that they postulate, and I have discussed other possible explanations that can be related to experimental evidence. Thus, I would suggest that future studies along this line include the following:

1) Monitor 0 - 10 - 90% burn rates by analyzing pressure time records.

2) Provide some method to maintain 10 - 90% burn time constant, if required. Variable spark gap location would be an easy and acceptable experimental method to employ.

3) Ensure repeatable spark timing adjustment. Two methods appear valid.

[1] J. Mayo, "The Effect of Engine Design Parameters on Combustion Rate in Spark Ignited Engines." Paper 750355 presented at SAE Automotive Engineering Congress, Detroit, February 1975.

a) Set spark so that the average cycle peak pressure occurs at a fixed time in the cycle (15 deg is about normal for MBT).

b) Set spark timing for a fixed amount of torque loss from the best obtainable (2 - 5% of indicated torque should provide good control).

4) If spark timing requirements vary, run cyclic variation and lean limit tolerance tests at constant timing in order to isolate the effect of compression density and pressure.

Inasmuch as the authors observed pressure time traces in their study, it is unfortunate that they did not include an analysis of combustion rate in the paper. However, the data they have shown on peak cylinder pressure effects convincingly argue that the flame speed decreases with good atomization throughout the complete combustion period. Thus, the evidence that flame speed is affected significantly by fuel mixture preparation as presented in the paper appears to be irrefutable and their hypothesis that this effect results from stratification appears plausible. I sincerely hope that the authors will continue their very interesting experiments and that they again will share their findings with us in a future technical publication.

AUTHOR'S CLOSURE TO DISCUSSION

LET ME BEGIN by thanking Mr. Bishop for a very thorough and thought-provoking review of our paper. Mr. Bishop has raised a number of important points which apparently need further explanation and clarification. I think the main points that need comment are the following:

1) The variability in MBT data and its effect on our HC and NO_x emissions results.

2) The effects of combustion duration of NO_x and HC emissions.

3) The relation between spark advance and lean limit.

4) The influence of cyclic variations in A/F on our results.

5) And finally, a brief response to Mr. Bishop's direct questions about A/F measurement and percent fuel vaporization.

To start with point 1, we believe that the observed variability in MBT spark advance data from the nominal curve was a result of the very slow combustion rates for the bowl-shaped combustion chamber used in this study. This makes the job of determining MBT more difficult, and, it can be expected that larger variations in MBT data would be present in these tests.

Much of the variation in NO data from the nominal least-squares curve-fit can be attributed to the experimental variations in MBT spark timing, but this is not a random scattering as indicated by Mr. Bishop. Careful examination of the NO data and corresponding MBT data for good atomization, as shown in Fig. 1, convincingly shows that NO data points which lie above the curve-fit can be attributed to overly advanced spark timing. This is shown by the positive numbers next to the data points, indicating deviations in MBT from the nominal curve. In a similar manner, many of the points which lie below the curve can be linked to retarded spark timing. I have also included two dotted curves which form a band around all of the data points and indicate the maximum observed variation in the data. Identical correlations between NO and spark timing can be demonstrated for the case of bad atomization, as shown in Fig. 2.

I think these two Figs. demonstrate that the variation in NO

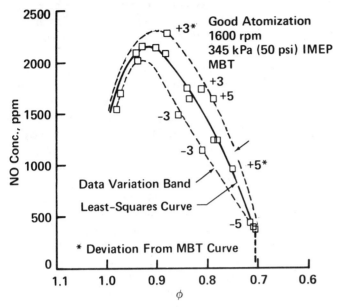

Fig. 1—Effects of deviations from MBT spark advance on NO emissions for good atomization

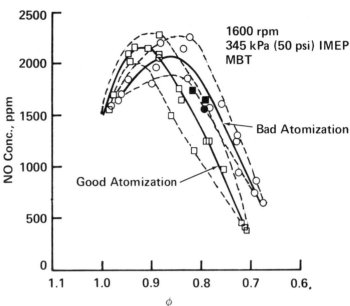

Fig. 3—Comparison of NO emissions for good and bad atomization including data variation bands

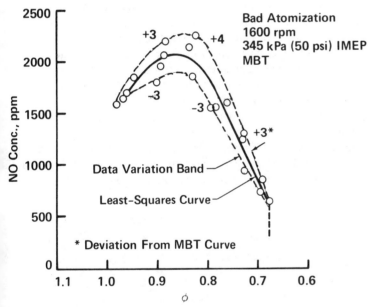

Fig. 2—Effects of deviations from MBT spark advance on NO emissions for bad atomization

data is consistent with spark timing differences and is not a random scattering as Mr. Bishop has implied. To represent all of the NO data points for good, bad and wall-wetted atomization with a single curve, as Mr. Bishop did in his first Fig., is incorrect.

To further demonstrate this, I have replotted the NO data for good and bad atomization including curve-fits and the corresponding data variation bands, as shown in Fig. 3. I think you will agree that, for at least lean air-fuel ratios, the NO data for good atomization is obviously lower than the case for bad atomization. Furthermore, when one considers that the few points that do overlap for good and bad atomization are accounted for by experimental variations from MBT, it is evident that the degree of fuel atomization has a pronounced effect on NO emissions which is not related to errors in MBT spark timing.

Another question has been raised concerning the change in combustion duration and whether this might account for the differences in NO and HC emissions. Of course, a change in combustion duration in itself is a direct result of the changes in atomization and any subsequent effects on emissions should be considered a direct result of fuel atomization. However, the question still remains whether the changes in emissions were a result of inhomogeneities in the change or a change in combustion duration.

We believe that the changes in HC emissions may well be the result of combustion duration, and we make this point in the paper; however, we do not think this is the case for NO emissions. In support of this, I would like to show some additional data, which is not in the paper, for the case of good atomization with and without a shrouded intake valve. This is shown in Fig. 4. The shrouded valve decreased MBT spark timing about 20 CA deg, thus indicating a significant decrease in the combustion duration. As you can see, NO emissions were not substantially affected by the change in combustion duration. Therefore, we have concluded that the differences in NO emissions were the result of some degree of stratification within the charge resulting from fuel atomization.

Mr. Bishop contends that the changes in lean limit that we observed can be predicted on the basis of spark timing alone, as he has shown in a previous Fig. According to his data, a change in lean limit of 3 A/F's would require a spark advance change of nearly 25 deg. Our data shows that a lean limit extension of 3 A/F's was observed between good and wall-wetted atomization with only a 10 deg change in spark timing. Thus, a majority of the change in lean limit was caused by other factors and we believe that some degree of charge stratification could easily account for the difference.

Mr. Bishop would like us to comment on the effects of cyclic A/F variations on our experimental results. One would expect

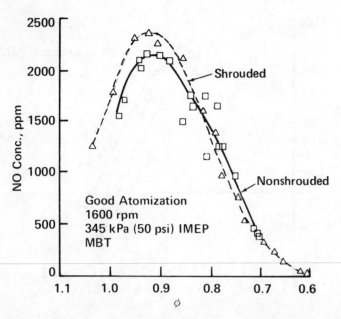

Fig. 4—Effects of combustion duration on NO emissions for good atomization

the effects of cyclic variations in A/F to be more pronounced with bad and wall-wetted atomization and that this would cause larger variations in peak cylinder pressure than for the case of good atomization. However, since this was not the case, we feel that cyclic variations in A/F were not a major factor in the test results.

Since I'm on the subject of A/F, I'd like to make a brief comment on the values of air-fuel ratio reported in the paper. These were computed on the basis of airflow and fuel flow measurements. The air-fuel ratios computed from these measurements were always checked against computed values based on a carbon and oxygen balance of the exhaust gases. The air-fuel ratios determined by these methods agreed within 0.2 of an air-fuel ratio.

Mr. Bishop has asked us to comment on the validity of estimating the percent fuel vaporization based on mixture temperature measurement. First of all, accurate determination of mixture temperatures with thermocouple probes is very difficult because of the presence of fuel droplets. Indicated temperature may differ significantly from the actual temperature because of droplet impingement and vaporization from the probe. Secondly, the mixture temperature probe was located several in upstream of the point where the fuel was introduced for wall-wetted atomization. Obviously, no conclusions can be made about the mixture temperature for this case. Therefore, for these two reasons, we do not agree with Mr. Bishop's estimate of percent vaporization. Also, it has been our experience that most of the available techniques for monitoring vaporization level in an engine appear to be very inaccurate because it is difficult to account for all of the fuel, both in liquid and vapor phases.

Let me conclude my remarks by pointing out that there is little doubt that fuel atomization affects combustion duration as indicated by MBT and peak cylinder pressure measurements. We are convinced that this is caused by some degree of stratification within the combustion chamber resulting from the degree of fuel atomization. Stratification not only resulted in a change in combustion duration, but this idea is also consistent with our emissions results and the extended lean limit. I think in the final analysis, Mr. Bishop reluctantly agrees with our conclusions when he closes his discussion with the following comment, and I quote: ". . . the evidence that flame speed is significantly affected by fuel mixture preparation as presented in this paper appears to be irrefutable and their hypoth that this effect results from stratification appears plausible".

Hybrid Vehicle for Fuel Economy*

L. E. Unnewehr, J. E. Auiler, L. R. Foote
D. F. Moyer and H. L. Stadler
Research Staff, Ford Motor Co.

A HEAT ENGINE/ELECTRIC drive train has been evaluated as a means of improving the fuel economy of various types of automotive vehicles. Computer simulation studies and dynamometer tests on a prototype system indicate that improvements in CVS-Hot fuel economy (miles/gallon) of from 30% to 100% can be realized with this system in a vehicle of identical weight and performance characteristics. Preliminary test data also indicates that these fuel economies may be realizable while meeting the 1975/76 Federal Emission Standards (1.5HC, 15CO, 3.1NO_x) with the use of external emissions controls such as catalytic converters. Although similar in configuration to a standard parallel hybrid drive train, the control strategies and energy flow of this system are considerably different from any known hybrid drives. This system does not appear to be of equal merit for all classes of vehicles, but gives the greatest fuel economy improvements when applied to delivery vans, buses, and large passenger cars. There are certain drawbacks to this particular hybrid system, principally in increased initial cost as compared to conventional systems, but this cost differential may be reduced as improved electrical components are developed and as automotive production and marketing techniques are applied to the electrical components. Other potential limitations of this hybrid system are reduced driving range at very low speeds and reduced capability to supply vehicle auxiliaries at standstill. In general, the replacement of a conventional drive train by this particular hybrid train will not increase the vehicle curb weight.

From almost the beginning of the Automotive Age, various combinations of drive systems have been tried in order to achieve vehicle performance characteristics superior to those that can be obtained using a single type of drive. These efforts have been made in the name of many worthwhile goals, such as increased vehicle acceleration capability, audible noise reduction, operation of an engine or turbine at optimum efficiency, reduction of noxious emissions, and improved fuel economy. These efforts have so far not led to any commercial

---

*Paper 760121 presented at the Automotive Engineering Congress and Exposition, Detroit, 1976.

ABSTRACT

A heat engine/electric hybrid drive train is proposed as a means for improving CVS-Hot fuel economy by an estimated 30% to 100% in various types of automotive vehicles. This drive train, classified as a parallel hybrid, has been analyzed by means of computer simulation studies to evaluate its fuel economy, performance, and emissions characteristics, and has been compared with existing internal combustion engine drive trains and other types of hybrid drives. A prototype system has been assembled and evaluated on a dynamometer test stand and has corroborated the computer analysis and predictions. Problems and limitations of this system are discussed.

applications, although several experimental hybrid buses and rapid transit vehicles are being evaluated at the present time (1,2,3). For private vehicle applications, hybrid drive systems have generally been found to offer insufficient improvement in meeting one or more of the goals stated above to justify the added cost and complexity compared to a singular drive system, particularly compared to the conventional Otto cycle internal combustion engine drive system. Two extensive EPA-sponsored studies of heat engine/electrical hybrid systems have been published (4,5) and generally concur in this conclusion, as does the more recent JPL Report.(6)

It is therefore with some trepidation that the subject of this paper, a heat engine/electric hybrid drive system, is proposed as a viable drive train for modern automotive vehicles of many varieties. However, this proposition has been developed - and to large extent, confirmed - on premises somewhat different from those upon which the EPA studies were based:

1. The critical fuel situation in the U.S. and most Western countries has placed increased emphasis on improved fuel economy for all types of vehicles since the initiation of the EPA studies of Reference 3 and 4. Recent large increases in gasoline prices have led to the conclusion that a sizable increase in initial vehicle cost (resulting from the use of a hybrid drivetrain) can be justified if a sufficient improvement in vehicle fuel economy is realized.
2. Studies performed during the development of this system have shown that the relative size and power rating of the hybrid drive train components with respect to the vehicle weight and performance rating have an important influence on vehicle fuel economy. Hybrid drive trains may not improve fuel economy for vehicles of every size, weight, and application category. Stated in another way, hybrid drive trains are not "scalable" as a function of vehicle size or weight as are singular drive trains.
3. The modus operandi or control philosophy of a hybrid can have a profound influence on both fuel economy and emissions. Past hybrid developments have tended to use the heat engine primarily as a battery charger; the subject hybrid reverses this philosophy and makes minimum use of the electric system.

It is hoped that the validity of these principles will be amplified by subsequent sections of this paper.

SYSTEM DESCRIPTION

A block diagram of the system illustrating functional performance and energy flow paths is shown in Figure 1. This drive system is intended to replace the engine-transmission system in conventional vehicles with the result of increasing the vehicle CVS-Hot fuel economy (miles/gallon) from 30% to 100% at 1975/76 Federal emission levels using the CVS-Hot cycle while maintaining approximately equivalent accelerating, braking, and passing characteristics. The hybrid-electric system consists of the following major components:

1. A different internal combusion engine, considerably smaller in displacement, and, hence, horsepower capability, than the engine in the original drive train.
2. An electric motor/generator (one unit) which may be on a common shaft with the engine output shaft or connected to the engine output shaft by means of a gear, belt, or chain system. The motor/generator may be of the DC commutator, DC homopolar, synchronous, or induction types.
3. A means of controlling power flow between the motor/generator and battery. This may be an electronic controller using power thyristors or transistors, contactor controller using battery switching techniques, or similar devices. The controller must be capable of two-way power flow and should have high energy efficiency.
4. An energy storage device. This may be any device capable of handling the high bursts of power required by the drive train during acceleration and braking and of supplying the energy needs for low-speed driving and the operation of vehicle auxiliaries at low speeds and standstill. At the present time, batteries are the most practical energy storage device, with the nickel-cadmium battery having almost ideal characteristics for this application but suffering a cost penalty. Flywheels, fuel cells in combination with batteries, closed loop cryogenic expander systems, are other possibilities.
5. A differential and a drive shaft. In general, it is desired to use the original drive shaft and differential of the vehicle.

The system can be classified as a

parallel hybrid with engine on-off control, and bears some similarity in configuration with two other recent hybrid developments. (9),(10)

In addition to these major power components, other components required by the hybrid drive train include: control circuitry for the proper operation of the power controller; modified engine throttle and carburetor; sensors for converting vehicle speed, battery voltage and charge level, component temperatures, etc., to electrical signals suitable for use in control and protection systems; protection systems for both engine and electrical system emission controls; and an overall vehicle control system.

Two modifications of the above system (Figure 1) have capabilities for improved system performance but usually add some cost penalties:

1. The use of an automatically-controlled decoupler to permit the engine to be detached from the electrical motor drive shaft when the vehicle is operating in an all-electric drive mode or in a braking mode. It has been shown that the use of such a clutch will result in a further improvement in fuel economy (see Figure 5).
2. The use of an electrically-controlled gear changing system. This will often result in a reduce electrical system weight and an improved electrical system efficiency.

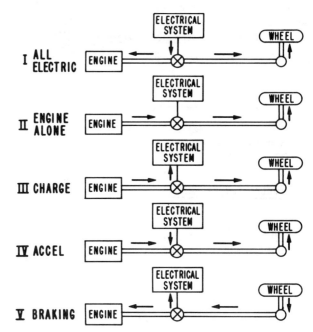

Fig.2 -Five hybrid modes of operation

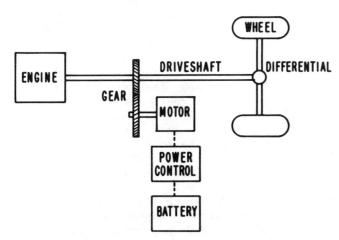

Fig.1 -Ford parallel hybrid

SYSTEM OPERATION

The system has six modes of operation. The first five modes are shown in Figure 2. Mode I is all electric at speeds below 10 to 15 MPH. In Mode II the engine is the primary source of propulsion and there is no energy in or out of the electrical system. Mode III is the battery charging mode. The engine still drives the rear wheels; however, excess energy is used to charge the battery. When acceleration demands exceed the power input of the engine, the motor provides the needed additional power. This is shown as Mode IV. Mode V is regenerative breaking. The deceleration energy of the vehicle is used to charge the battery. Fuel is shut off to the engine during the all electrical mode and during braking. The battery state of charge is maintained between fairly narrow limits by the control system around a state of charge of about 75% of full charge. This strategy prevents deep discharge cycles on the battery. The sixth mode is at vehicle standstill, during which condition both the engine and electrical motor are inoperative or "dead". Required vehicle auxiliaries are supplied electrically at standstill.

The objective of this system is to provide an increase in fuel economy over a conventional automotive drive system while maintaining equivalent acceleration performance. Comparisons between the hybrid system and conventional systems have been stressed in all studies. The manner in which this comparison is viewed from an overall systems standpoint is important in understanding the significance of this particular hybrid configuration and its operation.

Figures 3 and 4 show that the fuel economy for both a conventional and hybrid system can be expressed as follows:

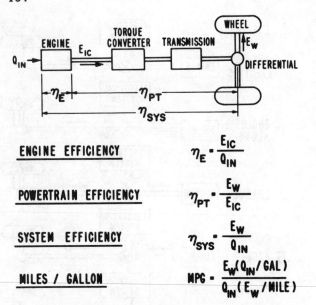

Fig.3 – Average fuel economy for a conventional vehicle in terms of system efficiencies

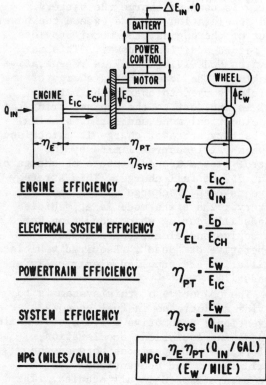

Fig.4 – Average fuel economy for a hybrid vehicle in terms of system efficiencies and energies

$$MPG = \frac{\eta_E \, \eta_{PT} \, (Q/Gal)}{(E_W/Mile)}$$

where η_E is the average engine brake termal efficiency, η_{PT} is the average transmission efficiency, (Q/Gal) is the energy content per gallon gasoline consumed and (E_W/Mile) is the total energy requirement at the drive wheels per mile necessary to accelerate the vehicle and to overcome vehicle friction and aerodynamic drag. The quantities in this expression represent average values over a prescribed driving cycle. It should be noted that the average powertrain efficiency is defined as the ratio of total positive engine shaft work to total positive energy requirement at the drive wheels. Stated in another way, this represents the fraction of total engine work used to propel the vehicle. For the hybrid drive train state of charge is assumed to be the same at the beginning and end of the drive cycle, thus the net energy input to the transmission from the battery is zero.

The task facing the hybrid system can now be clearly seen. In order to provide an increase in fuel economy over a conventional system the quantity $\eta_E \, \eta_{PT} / (E_W/Mile)$ must be increased. The present hybrid system will be described in terms of how it strives to maintain high average engine efficiency, high average transmission efficiency and low work requirements at the drive wheels while maintaining the equivalent acceleration performance of the conventional system it replaces.

A. High Average Engine Efficiency
1. Small engine – The engine used in the conventional system is replaced by a much smaller engine in the hybrid system. The smaller engine operates at higher load factors, resulting in increased efficiencies. The hybrid engine is sized to meet vehicle cruise requirements up to a specified road speed. This enables the vehicle to be propelled by the engine alone for extended cruise periods. This corresponds to Mode II in Figure 2.
2. Fuel off during idle and deceleration – Approximately 20% of the CVS-H fuel consumption is used during idle and braked deceleration for the conventional vehicles with automatic transmission considered in this study. Elimination of idle and braked deceleration fuel flow in the hybrid configuration results in significant improvements in average engine efficiency.
3. Fuel off during low speed operation – Since the engine is

geared directly to the drive wheels the fuel is shut off at low vehicle speeds and the vehicle is propelled by the electrical system. This corresponds to Mode I in Figure 2. The fuel savings must be weighed against the electrical energy dissipated that must be replaced by charging the battery later in the driving cycle. Since this charging is done at a higher engine efficiency, this mode has a positive effect on the average engine efficiency. However, this charging has an adverse effect on the average transmission efficiency since a lower fraction of the engine work shows up as useful work at the drive wheels. The total gasoline used to replace the battery energy expended during this mode can actually exceed the amount of gasoline used in a conventional vehicle in accelerating up to the corresponding vehicle speed. The energy requirements of this mode can be substantially improved by lowering the work required to motor the engine by opening the throttle, collapsing the valves or by de-clutching the engine. Other approaches include gear changes or use of motors with better low-speed efficiencies.

4. Charging the battery at high-engine efficiency - When the battery requires charging from the engine as represented by Mode III in Figure 2, the basic strategy is to provide the charging energy at the most efficient engine operating point. This contributes to a high overall engine energy efficiency; however, this effect must be weighed against the effect on transmission efficiency since the optimum engine efficiency will not in general correspond to the most efficient charging torque level for the electrical system. Additional trade-offs appear when the effect of engine torque on emissions is discussed in a later section.

5. Accelerate at high-engine efficiency - When the vehicle acceleration demands exceed the power capacity of the engine, the electrical system is used to provide the extra needed power. This is described as Mode IV in Figure 2. In general the engine torque level at which the electrical system is called upon corresponds to a high-engine efficiency point. The effect on transmission efficiency must also be considered since a lower engine torque requires more electrical energy.

B. Transmission Efficiency - The transmission in a hybrid drive train is the portion of the system that transmits useful work from the engine to the drive wheels. Since all the energy needed to propel the vehicle ultimately comes from the engine (assuming the battery ends the drive cycle at the same state of charge) the basic objective of the transmission is to minimize the amount of engine energy used for other purposes. This is achieved as follows:

1. Engine geared directly to rear wheels for primary source of propulsion - When the electrical system is not in use, the energy from the engine is transmitted directly to the rear wheels through the differential. This is Mode II in Figure 2. The instantaneous transmission efficiency during this mode is essentially equal to the differential efficiency. The engine is sized to provide sufficient torque in this mode for extended high-speed cruise.

2. Use of electrical system only when needed - To keep the use of the electrical system to a minimum, the motor is used only when needed. The two modes requiring the motor are the all electric mode at low speed (Mode I) and during heavy accelerations (Mode IV).

3. Use of regenerative braking - During braking the kinetic energy of the vehicle is used to charge the battery. This is described as Mode V in Figure 2. This has a substantial effect on transmission efficiency by reducing the charge energy required from the engine.

C. Drive Wheel Energy - In converting a conventional vehicle to a hybrid configuration the total energy requirements at the drive wheel must also be considered in assessing the potential fuel economy gains. The primary factors that could reduce fuel economy are an increase in the vehicle weight and an increase in the rotational inertia due to higher rotational speeds of the engine and motor. System weights will vary considerably with the vehicle acceleration requirements. For the hybrid configurations considered in this study small weight savings were realized. These differences were generally not enough to change the inertial weight class of the vehicle and were not considered in the fuel economy projections. The effects of increased rotational inertias were also seen to be minimal for the configurations investigated.

METHOD OF ANALYSIS

A computer program was developed to simulate all elements of the drive train for the six basic modes of operation over an arbitrary drive cycle. The required power at the drive wheel is computed from the drive cycle data, the vehicle friction, aerodynamic drag, inertial acceleration and rotational inertias. The corresponding power levels are computed throughout the drivetrain based on rotational speeds and torques and component performance characteristics.

Motor/generator and controller efficiencies are computed from efficiency tables in terms of torque and RPM. The efficiency tables used for the D.C. system are based on experimental data from reference (7). Similar tables for a brushless synchronous motor system are based on experimental data from reference (8). Battery efficiency is computer from equivalent circuit models for specific battery types as described in Reference (16).

The engine is sized to provide sufficient power for extended cruise without the electrical system. Fuel flows are computed in terms of engine speed and torque. In general, automatic calibration fuel island data is used with simulated exhaust system, fan on, alternator operated at one-half charge and power steering pump loaded. Engine motoring torque is computed as a function of engine RPM from experimental data.

Axle ratio between the engine and drive wheels and gear ratio between the motor and engine are varied in the analysis until a suitable compromise is reached between fuel economy, top speed, acceleration, maintaining battery charge and, in some cases, emissions.

Comparisons with conventional drivetrains are made by applying the same basic technique of starting at the rear wheels and describing each element individually. Transmission efficiencies are computed for each gear from output speed. Automatic transmission shift schedules are determined from driveshaft RPM and manifold vacuum. Manifold vacuum must be implied from engine torque which cannot be computed until the proper gear is determined. The engine torque and transmission shift schedule must, therefore, be matched iteratively.

The approach is similar to techniques described in Reference (11) for conventional vehicles and in Reference (16) for electric vehicles.

DYNAMOMETER TESTS

Early in the course of the computer simulation and other analytical studies of the hybrid concept, the need for some experimental evidence to support the computer predictions of fuel economy and performance was recognized. Also, emission measurements and engine strategy for emission control were required. The first step in such experimental evaluations has been the testing of an engine-electric drivetrain with a dynamometer and inertia wheel as loading devices. Ultimate evaluation of any alternate engine or other drivetrain component must of necessity by made through a long series of vehicular tests under typical or prescribed driving conditions. However, for systems so far removed from conventional automotive practice as a hybrid drivetrain, dynamometer testing appears essential before vehicular testing is initiated. The principal goals of the hybrid dynamometer tests were:

1. To test the computer predictions of fuel economy, performance, and emissions using a production engine.
2. To establish that the fuel economy improvement is attainable at acceptable emission levels. This required that near optimum engine strategy regarding spark, air-fuel ratio, and exhaust gas recirculation be developed. This was done by dividing the speed torque plane in a grid pattern, studying each area in the grid and summing the total for hybrid operation. This process is called engine mapping in subsequent discussions.
3. To determine that the on-off fuel control required by the hybrid was practical at acceptable performance, emissions and cost. This was determined using a carburetor and minor modifications.
4. To determine that the selected battery was adequate.
5. To determine that the engine is

basically suited to the unique or unusual operations in this concept, such as:
a. Motoring the engine between 0 and 800 RPM as required by the direct coupling to the wheels. Normally an engine is cranked and immediately accelerated to an idle speed of 700 RPM or more.
b. Operation at high torque most of the time.
c. Higher than normal total use and long duration of high torque at high speed.

The experimental hybrid drivetrain was configured as in the block diagram of Figure 1 with two exceptions: The electric motor was on a common shaft with the engine, and the driveshaft was directly coupled to a dynamometer and inertia wheel to simulate the vehicle road, aerodynamic, and inertial loads. The principal components used were:
1. Engine: Ford 2.3L, 4-cylinder, '74 production engine, modified for fuel off operation.
2. Motor: Westinghouse, 40HP, 240 V., 1750 RPM industrial shunt motor; blower cooled.
3. Controller: SCR chopper for motor armature control during motoring and regenerative braking (designed and assembled at Ford); separate power supply for field control.
4. Battery: 140 cells connected in series of Marathon, type 20D120, NiCd; auxiliary forced-air cooling to maintain cells at approximately 20 C; plus required monitoring equipment.
5. Loading Device: Absorption dynamometer of 150 lb-ft^2 inertia and a flywheel of 360 lb-ft^2 inertia.

The combined inertias of the rotating members of the experimental system are equivalent to a vehicle of 7500 lb. inertia weight based upon an engine RPM/vehicle MPH (N/V) ratio of 53.5. Conventional gas analysis equipment was used to measure emissions under conditions of steady state engine operation. Measurements of exhaust CO, CO_2, HC, O_2 and NO_x and intake CO_2 were made. Fuel flow was measured by weight.

Since the hybrid application requires operating an engine under conditions considerably different from those associated with conventional vehicles, preliminary evaluation and modification of the 2.3L engine was necessary:
1. The engine was modified to permit fuel to be turned off during deceleration and at speeds below 15 MPH. This was accomplished by means of a small solenoid valve to block fuel flow in the idle jet, removal of the throttle stop to permit full closure of the throttle plate, a means of admitting air below the throttle, and PCV modification.
2. A sequence control was required for minimum emissions and quality performance during engine fuel turn-on and turn-off. For example, during turn-off, the following sequence was used: (a) close throttle, idle solenoid, and PCV valve, (b) open by-pass air valve around throttle to permit air without fuel into intake manifold, (c) turn-off ignition, with elapsed time between these events.
3. Removal of some engine auxiliaries; for example, the engine alternator is not required in a hybrid drive; air conditioner was not used. The power steering pump was connected and driven.
4. Low-speed engine friction: In a conventional vehicle, the engine is operated below the idle speed (about 800 RPM) for only a few seconds during start-up. In the hybrid, much longer operation may be required. The low-speed friction torques of the 2.3L engine were measured.
5. Low-speed lubrication was evaluated.
6. The EGR valve and plumbing were enlarged to permit large EGR flow at wide-open throttle operation.

Another interesting problem for which there was almost no precedent was the measurement of HC emissions during the frequent engine off/on transitions that the engine passes through during a typical driving cycle. Since CVS equipment for this measurement was not available a technique using diluted samples from the engine-off period was developed and considered to give reasonable accuracy. This method was used to predict the emissions discussed in later sections of this paper.

The resulting experimental system proved to be very "driveable" with smooth transitions between the various operating modes. The system was "driven" through several of the standard test driving cycles with ease and accuracy after a few learning cycles by the operator.

In order to experimentally verify the calculated values of fuel economy that had been obtained from the various computer simulations described above, several dynamic runs over both CVS-H and SAE (17) driving cycles were performed on the experimental hybrid system mounted on a dynamometer test stand. The SAE driving cycle is a simplified version of the CVS-H cycle developed mainly for the electric vehicle tests. Many comparisons of the two driving cycles have shown that both result in approximately the same fuel consumption for both ICE and electric vehicles. Since the "driving" of an experimental drivetrain on a dynamometer test stand over the SAE cycle is much simpler than over the CVS-H cycle, and since the control of the system was not fully automated but required considerable manual control, the SAE cycle was chosen as the means for comparing calculated with measured fuel economy of the hybrid drivetrain. It was found that after only a few tries, manual control was able to follow the required speed and acceleration variations specified by the SAE cycle almost perfectly. The actual efficiencies of the components in the electric branch of the hybrid and the actual road load simulated by the dynamometer were fed into the computer model to obtain the calculated fuel economy. The engine throttle positions were likewise made to correspond between the measure and calculated test runs. The results are summarized below:

TABLE I
COMPARISON OF MEASURED AND CALCULATED
DYNAMIC FUEL ECONOMY OF HYBRID

Simulated vehicle inertia weight	7500 lbs
Length of test run	3 SAE cycles (3 miles)
Calculated fuel economy	15.2 mpg
Measured fuel economy	15.8 mpg

FUEL ECONOMY STUDIES WITH AUTOMATIC ENGINE CALIBRATIONS

A variety of studies was conducted by applying the computer program to hybrid and conventional versions of the same vehicle using fuel island data for stock engines with automatic calibrations. The hybrid electrical systems were sized to provide approximately equivalent acceleration performance. The results of these studies are summarized in Figure 5. The purpose of this section is to discuss the reasons for the fuel economy improvement resulting from a hybrid system and to discuss the effects of fundamental system changes on fuel economy.

A. Reasons for Fuel Economy Improvement Resulting from a Hybrid System - The Econoline Van and the Mark IV configurations received the most emphasis in these studies. Figures 6 and 7 present summaries of comparisons made between typical hybrid and conventional versions of the Econoline Van and Mark IV, respectively. The computations were done for the CVS-H drive cycle and both comparisons are based on equivalent acceleration performance between the respective hybrid and conventional configurations. Both hybrid systems represent typical configurations with automatic engine calibrations, DC motor and controller and normal idle throttle engine motoring friction during fuel off modes.

In Figure 6 a 4500 lb. conventional van with 300 CID engine

Vehicle	Hybrid Power Train		Calculated Fuel Economy[c] (MPG)		% Improvement Hybrid/ICE
	Engine	Motor	Hybrid	ICE	
Van[a]	1.1 with closed throttle	DC	18.5	14.4[d]	28
Van	1.1 with wide-open throttle	DC	18.8	14.4	31
Van	1.1 , valves closed	DC	19.5	14.4	35
Van	1.1 , clutch	DC	20.0	14.4	39
Van	2.3 diesel	DC	22.0	14.4	53
Van	2.3 diesel, clutch	DC	23.7	14.4	65
Van	1.1 , clutch	Disc[e]	23.8	14.4	65
Van	2.3 diesel, clutch	Disc	28.4	14.4	97
Mark IV[b]	2.3 (ICE), clutch	DC	18.7	10.5[f]	75
Mark IV	2.3 (ICE), clutch	Disc[e]	21.9	10.5	109

(a) 4500 lb. Inertia Wt.
(b) 5500 lb. Inertia Wt.
(c) All fuel economy calculations based upon vehicle driving the Federal CVS-H cycle; no net change in battery state-of-charge.
(d) Calculated for unemissionized, 240 CID engine. Calculations based upon the 1975, emissionized 300 CID engine used on 1975 vehicles resulted in a fuel economy of 13.4 MPG.
(e) Axial air-gap reluctance motor developed by Ford. (See References (8) and (16)).
(f) Calculated for 1974 460 CID engine with automatic calibration.

Fig.5 -Calculated fuel economy comparisons

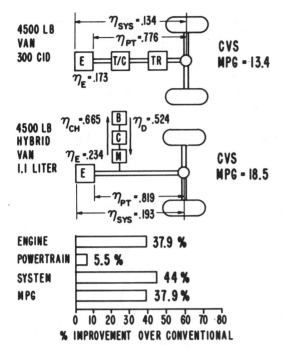

Fig. 6 –CVS-H fuel economy and efficiency comparison between hybrid and conventional econoline van

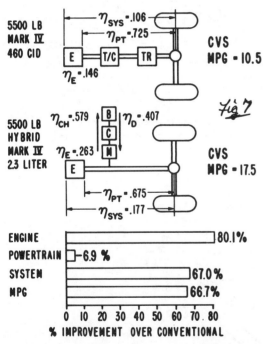

Fig. 7 –CVS-H fuel economy and efficiency comparison between a hybrid and conventional Mark IV

and automatic transmission is compared to a 4500 lb. hybrid van with a 1.1 liter engine, DC motor with 130 ft. lb. peak torque and 45 KW of NiCd batteries. Acceleration and battery charging are both done at wide-open throttle and fuel is shut off at engine speeds below 1000 RPM and during braking. Gearing between the engine and rear wheels gives a ratio of engine RPM to vehicle speed in MPH (N/V) of 100, while the ratio of electric motor RPM to engine RPM is 1.75. The weight summary of this substitution is shown in Table II. The performance predictions (acceleration) for this same vehicle are given in Table III.

In Figure 7 a 5500 lb. conventional 1974 Mark IV with 460 CID engine and automatic transmission is compared to a 5500 lb. hybrid Mark IV with a 2.3 liter engine, DC motor with 260 ft. lb. peak torque and 80 KW of NiCd batteries. Accelerations are done at wide-open throttle, while battery charging is done at optimum fuel consumption. Fuel is shut off at

TABLE II
VEHICLE WEIGHT EXCHANGE

Production Systems - 1975 Nantucket	Curb Weight (lbs)
Delete:	
. 300 CID Engine	631
. C-4 Automatic Transmission	155
. Exhaust System	56
. Fuel System (22 gal. base tank)	29
. Battery and Alternator	54
	925 lbs.
Add:	
. 1.1L Engine	243
. Exhaust System	25
. Fuel System (13.3 gal. base tank)	18
. Motor	120
. 2-spd. Trans. (Provision -- not included in fuel economy)	80
. Controller	70
. Battery and Cooling (Ni-Cad System)	170
. 12V Inverter	5
	731 lbs.

NOTE: Structural and other small component changes may alter this weight comparison.

engine speeds below 800 RPM and during braking. Engine RPM to vehicle MPH is 58.66, and electric motor RPM to engine RPM is 2.27.

The comparisons shown in Figures 6 and 7 illustrate the following important characteristics regarding fuel economy comparisons

TABLE III

Performance (g)	TED(h) (sec)	0-10 sec (ft)	0-60 mph (sec)	25-60 mph (sec)
1.1L Hybrid Wide Open Throttle(b)(e)	13.3	358	19.7	15.1
1966 240 CID - E-100 (b)	12.5	385	17.1	12.9
Memo: 1974 240 CID E-200 (c)	13.3	364	19.0	14.4

(b) <u>Non-emissionized</u>.
(c) <u>Emissionized</u>.
(e) CVS-CH at 4500 lb. test weight, 25 HP motor, a 50 KW battery, and maximum speed of 81 mph.
(g) Computer projections developed by Powertrain Research, PP&R, Ford Motor Company.
(h) "Time Exposed to Danger"; this is the time required to gain 150 ft. on a vehicle traveling at 55 mph.

between hybrid and conventional vehicles over the CVS-H drive cycle:
1. For both the van and Mark IV configurations the improvement in the hybrid fuel economy is approximately equal to the improvement in overall engine efficiency.
2. Average hybrid transmission efficiencies are comparable in magnitude to the transmission efficiencies of the conventional system with automatic transmission and torque converter.
3. The effect of higher rotating inertias on the required work at the drive wheels for the hybrid configurations is not a significant factor in the fuel

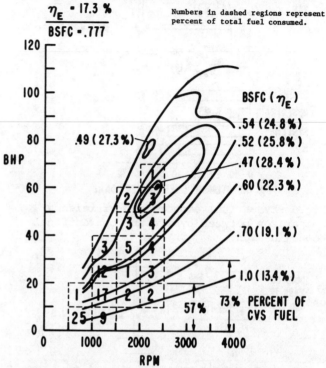

Fig. 8 -CVS-H fuel economy utilization for 4500 lb conventional econoline van

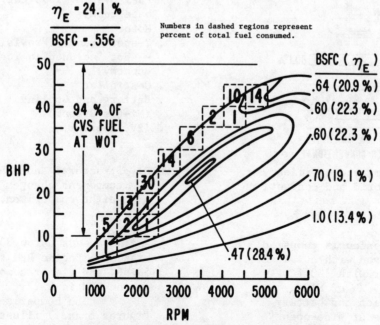

Fig. 9 -CVS-H fuel utilization for a 4500 lb hybrid econoline van (N/V=85)

economy, resulting in fuel economy penalties of less than 2%. Another important feature of these comparisons is the much greater fuel economy improvement shown for the hybrid Mark IV (66.7%) over that shown for the van (37.9%). Fuel economy is stated in miles per gallon. The reasons for this difference are shown in Figures 8 through 11 which show distributions of fuel utilization over the CVS driving cycle for the conventional and hybrid versions of the van and the Mark IV. The percentages of total fuel consumed at various engine operating points are indicated by the numbers enclosed by the dashed square regions. This information is superimposed on the engine fuel island curves which show contours of constant engine efficiency and brake specific fuel consumption in terms of engine RPM and brake horsepower for an automatic engine calibration. The conventional van does not offer as much improvement potential. In addition, it was necessary to charge at wide open throttle with the hybrid van, while in the case of the Mark IV charging was done at optimum fuel consumption resulting in a higher average engine efficiency. The

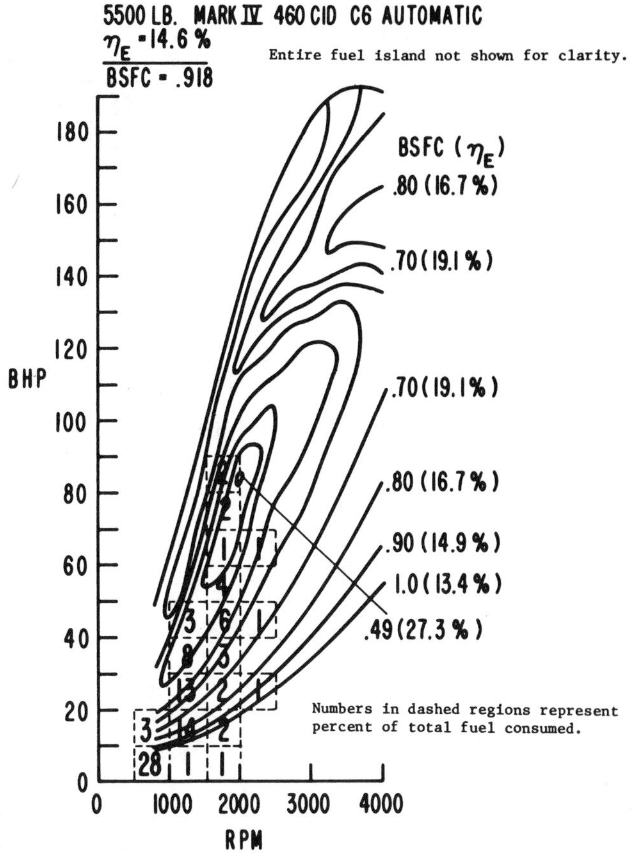

Fig.10 -CVS-H fuel utilization for a 5500 lb conventional Mark IV

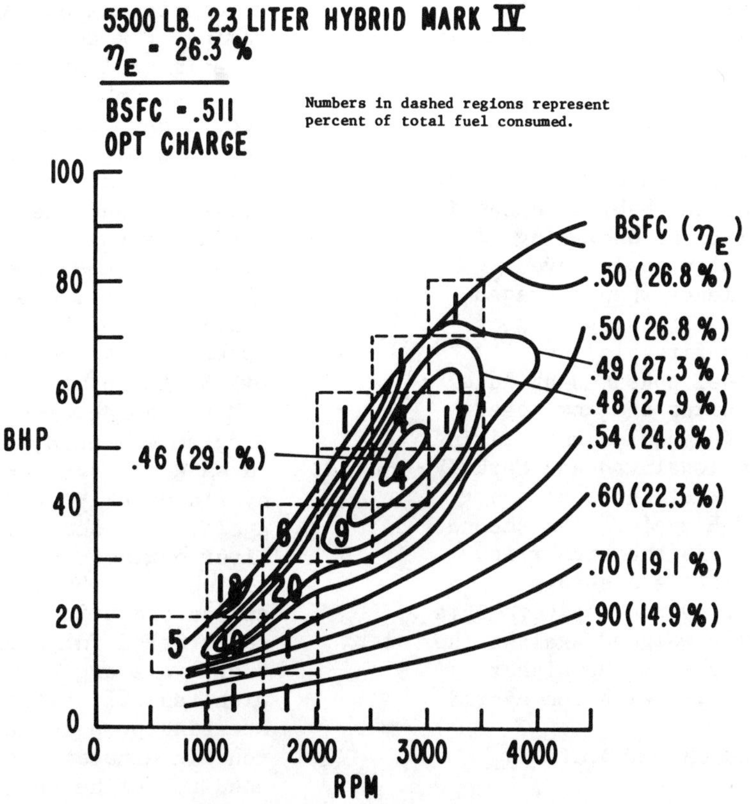

Fig.11 -CVS-H fuel utilization for a 5500 lb hybrid Mark IV (optimum fuel charge)

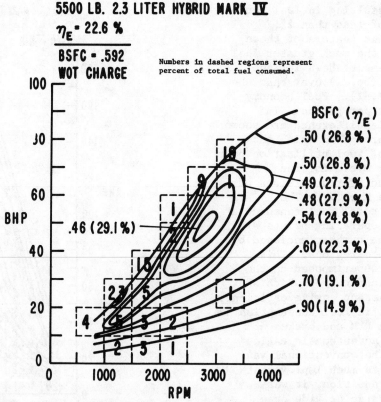

Fig.12 - CVS-H fuel utilization for a 5500 lb hybrid Mark IV (WOT charge)

difference in engine efficiency due to charging strategy is also shown in Figure 12 which shows the reduction in hybrid Mark IV engine efficiency of approximately 14% that would result from charging at wide-open throttle.

It is important to understand the reasons why the average engine efficiency is improved with the hybrid configuration. The key point is that the hybrid engine is operated at more efficient operating points. This results in an improved overall engine efficiency when averaged over the drive cycle. This improvement has two sources. The first is the elimination of all fuel consumed at idle, during braking and during the low speed all-electric mode. The equivalent driving modes for the conventional van and Mark IV account for 25% to 30% of the fuel consumed for the CVS-H cycle. The second source of improvement is the higher load factors and wider throttle openings required by a smaller hybrid engine. This gain must be carefully weighed against the higher frictional losses at the higher engine speeds encountered with the hybrid.

EMISSION AND FUEL ECONOMY PREDICTIONS

The studies described in the previous section indicate that a substantial improvement in fuel economy can be achieved from a hybrid with a conventional automatic engine calibration for spark timing, EGR and air fuel. The experimental program described previously was undertaken to map the emissions and fuel consumption of a 2.3 liter engine. One objective of this program was to provide data for use in computing hybrid emissions and fuel economy over the CVS-H cycle for the Mark IV configuration previously described in Figure 7.

The first step in computing the hybrid emissions was to divide the engine speed torque region into a grid composed of 10 ft. lb. torque increments and 300 RPM speed increments. The hybrid computer program was used to define the time spent in each cell as the engine was used in a hybrid Mark IV driven over the CVS-H cycle. The next step was to map the 2.3 liter engine over the entire speed torque region from 950 to 3350 RPM with the engine operating with a 1974 production calibration for spark timing, EGR and air fuel. All engine mapping was performed with fan off, alternator off, power steering pump loaded and with simulated vehicle exhaust system. These data were used to define the initial emissions and fuel distributions for the baseline configuration. The next step was to select regions of high emissions from the automatic calibration results and to perform additional experiments to reduce emissions. These results were then used with the hybrid computer program to obtain revised emissions and fuel economy predictions for the baseline hybrid configuration.

A. Automatic Calibration Results - NO_x, CO, HC and fuel data were obtained from mapping the 2.3 liter engine with automatic calibration. Figure 13 shows the projected engine energy distribution for the hybrid Mark IV over the CVS-H cycle. Energy distributions are shown as percentages of the total positive engine shaft work over the cycle. These energy values are used with the emission and fuel data obtained experimentally to obtain the CVS-H emission and fuel economy projections for the hybrid Mark IV with automatic engine calibration.

The start/stop value for HC results from shutting fuel off on braked decelerations and starting the engine when it reaches 800 RPM.

Figure 13 clearly shows two regions of high energy usage. One is along the line of maximum torque resulting from hard accelerations. The other region is a band of intermediate torques used to charge the battery. The particular strategy assumed for the baseline hybrid used the optimum fuel engine torque at a given speed to charge the battery. Optimum fuel torques were determined from the automatic engine calibration. The regions of high emissions and fuel utilization were seen also to be located in bands of maximum torque and optimum fuel torque; however, the distributions differ markedly from the energy distributions. The effects of power enrichment were readily seen by the high concentrations of CO and HC at maximum torque. The effect on fuel consumption is similar but not as great. Power enrichment had the opposite effect on the NO_x distributions, tending to lower the distributions at maximum torque.

B. Emission Reduction at Selected Operating Points - Having determined from the automatic calibration results that most of the CVS-Hot emissions and fuel are contained in two narrow bands of torque, the next step was to

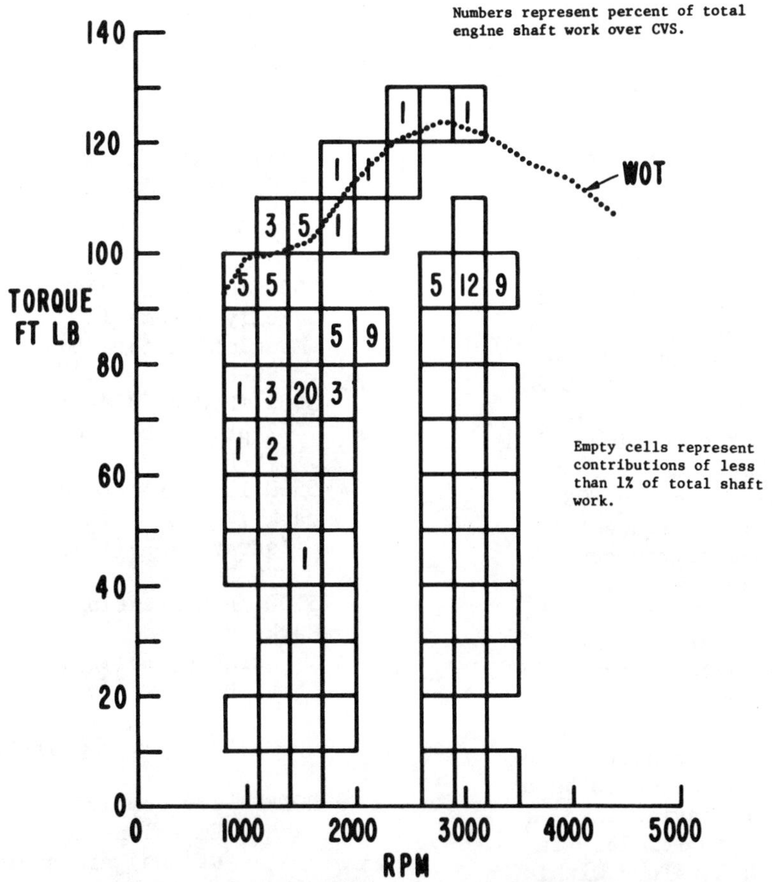

Fig.13 -Distribution of engine energy over the CVS-H cycle for a 5500 lb Mark IV with a 2.3 l engine (optimum fuel charge)

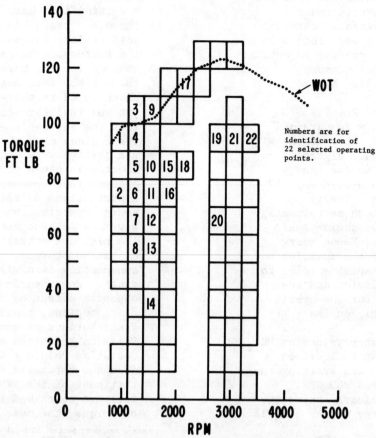

Fig. 14 - 22 operating points selected for additional data (2.3 l engine)

acquire additional data in these regions at lower emission levels. Figure 14 shows the 22 operating points at which additional data was taken.

The general approach was to reduce CO at high torques by eliminating power enrichment and to control NO_x at high torques with EGR and spark retard. Points at or very near maximum torque were obtained with wide-open throttle, optimum or near optimum spark and little or no EGR. At intermediate to high torques the throttle was kept open to reduce pumping work, large EGR rates were maintained, and torque was reduced by retarding the spark. Very little additional data was taken at low torques, since NO_x reduction at intermediate and high torques was considered to be higher priority. An operating point is considered to be at the midpoint of a cell having the dimensions 300 RPM by 10 ft. lb. In arranging the data, all values within the cell were assumed to be at midpoint.

The CVS-H emissions and fuel economy projections corresponding to the lowest measured NO_x at each of the 22 operating points shown in Figure 14

	HC	CO	NO_x	MPG
5500 lb Hybrid Mark IV · Fully automatic engine calibration · CVS-H	1.36	44.6	3.28	17.82
5500 lb Hybrid Mark IV · 22 pt NO_x reduction · CVS-H	1.38	14.5	2.14	18.59
5500 lb Conventional Mark IV · (strategy A)(11); CVS-H	.94	5.98	1.69	10.9
5500 lb Conventional Mark IV · (strategy B)(11); CVS-H	1.30	5.64	6.64	11.7

TABLE IV

NOTE: For comparison, the Interim Federal Emission Control Requirements (49 states) are shown below.

	HC	CO	NO_x
1975/6 CVS-CH Federal Standards	1.5	15	3.1
1977 CVS-CH Federal Standards	1.5	15	2.0

with automatic calibration data assigned to other points is presented below. Also shown for comparison are the fully automatic calibration results from the previous section and computer CVS-H results from Reference (11) for two engine calibration strategies.

The primary emphasis in collecting the data was to adequately describe the regions of high emissions and fuel for the baseline hybrid configuration. The 22 operating points accomplished this purpose. After most of the data had been taken, it was decided that additional information gained by examining emission and fuel trends for different hybrid configurations would be very helpful in evaluating the future potential of hybrids. In order to accurately identify trends, a consistent variation in emissions and fuel with speed and torque was needed. Some additional data was collected and previous data was re-examined. This data revealed consistent trends with torque for NO_x, HC and BSFC at each speed. Considerable scatter was observed for the CO data, which was more sensitive to fluctuations in air fuel ratio. In general, the air fuel ratio varied from 14.0 to 14.6. Initial attempts to describe the entire engine operating region by fitting the data at each speed as a function of torque proved unsuccessful due to lack of consistency with speed variations. The data were again re-examined and some previously discarded data points were included. Plots of emissions and fuel with speed as well as torque were made. As a result of this re-examination, emissions and fuel distributions were approximated over the entire engine operating region. The purpose of these approximations was to provide a reasonable representation of the measured data that clearly shows observed trends with speed and torque. The approximations do not necessarily represent the lowest possible NO_x; however, they do represent projected emissions and fuel values at low NO_x levels with realistic distribution in speed and torque.

C. Emission and Fuel Economy Trends Due to Configuration Changes - The "low NO_x" emission and fuel data obtained experimentally were used to obtain CVS-H fuel and emission projections for various configuration changes.

Figure 15 shows the effects of varying acceleration and charging torque for a 5500 lb. Mark IV. In general, a reduction in either charging or acceleration torque

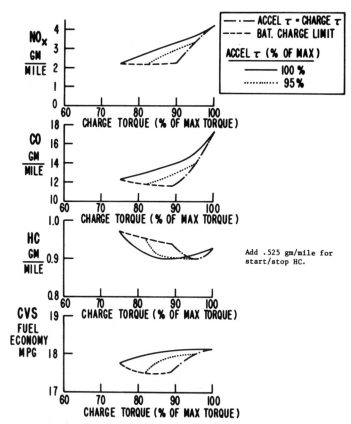

Fig.15 -CVS-H fuel and emissions trade-offs for various charge and acceleration torque strategies for a 5500 lb hybrid Mark IV

results in lower NO_x and CO, lower fuel economy and a tendency toward higher HC. These changes are due to the lower NO_x and CO values and higher BSFC and HC values occurring at lower torques. The battery charge limit clearly shows the limiting torque values needed to keep the battery charged over the CVS-H cycle.

Figure 16 shows the same information for a 4500 lb. vehicle with all other characteristics the same as before. The battery charge limit is shifted toward lower torque values with a significant reduction in NO_x to values approaching 1.0 gram/mile on CVS-H. Substantial reductions are also observed for CO and HC. Substantial increases in fuel economy are observed due to the lower total energy required.

Figure 17 shows the effect of vehicle weight on a configuration having charge and acceleration torque levels of 90% of maximum. The effects of an engine clutch and disc motor are also shown. In general, the effects of a clutch, disc motor and lower inertial weight result in lower emissions and higher fuel economy due to the reduced charging demands on the

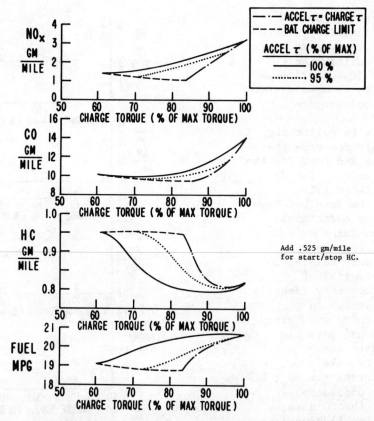

Fig.16 —CVS-H fuel and emissions trade-offs for various charge and acceleration torque strategies for 4500 lb vehicle

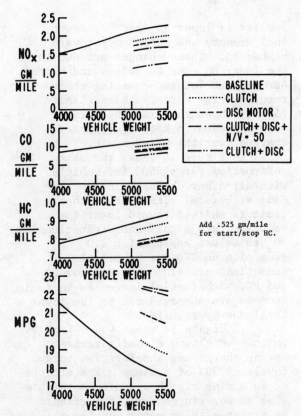

Fig.17 —Effects of vehicle weight, clutch, and disc motor on CVS-H fuel and emissions for a 5500 lb hybrid Mark IV

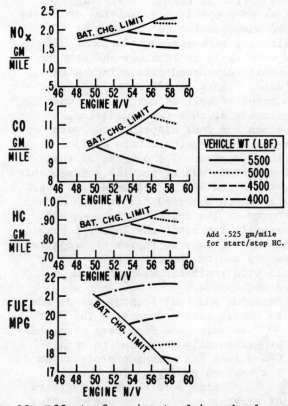

Fig.18 —Effect of engine to drive wheel gearing on CVS-H fuel and emissions for Mark IV with various inertial weights

engine which increased transmission efficiency and reduced high torque operation.

Figure 18 shows the effects of N/V for inertial weights ranging from 4,000 lbs to 5,500 lbs. The ratio of motor speed to engine speed was held constant at 2.27. At the lower weights a reduction in N/V increases NO_x, CO, and HC and lowers fuel economy. The dominant effects are lower engine powers available for charging and lower electrical efficiencies at a given road speed as N/V is reduced. This results in a higher fraction of the engine energy used for high torque charging and a reduction in transmission efficiency. These trends tend to reverse as the weight is increased. At 5,500 lbs. NO_x actually decreases and fuel economy increases as N/V is reduced to the battery charge limit.

SUMMARY AND CONCLUSIONS

An engine/electric parallel hybrid drivetrain has been proposed as a means for improving the fuel economy of vehicles presently powered by conventional ICE drivetrains. The proposed system bears some similarity to several previously studied systems whose evaluation has not appeared very promising. The proposed system is shown to be capable of overcoming many of the deficiencies of the earlier systems through proper matching of hybrid engine to vehicle weight, through use of a single electrical machine for both motoring and regenerative operation, through design of the electrical branch of the hybrid on the basis of short-time power requirements rather than energy requirements, and through maximum exploitation of engine control to achieve both efficient operation and relatively low emission levels. Many analytical studies and corroboration by dynamometer testing have shown that present CVS-H fuel economies (miles/gallon) of existing engine power vehicles can be improved by 30% to 100% while meeting 1975/76 Federal Emission Control requirements with the use of catalytic converters. The percent improvement in fuel economy achievable is largely a function of vehicle weight and performance specification with the larger increases occurring on the larger, high-powered vehicles.

At the present time, nickel cadmium batteries appear to be a feasible choice for the energy storage device in the electrical branch of the hybrid. As a result, substitution of a hybrid drivetrain for a conventional drivetrain would result in a cost penalty on initial vehicle cost.

It is shown that the principal reasons underlying the fuel economy increase realized by this particular hybrid configuration are:

1. The engine used in the hybrid is operated in regions of minimum specific fuel consumption during a much greater portion of its operating time than in conventional drives. The engine is sized more for steady-state (constant speed) driving conditions than for vehicle acceleration requirements. The electrical system serves a function somewhat analogous to that of an infinitely variable transmission and also adds power during vehicle acceleration and stores power during braking.
2. The elimination of the idling condition on the engine. This is a major source of low fuel economy during city driving.
3. The use of regenerative braking.

It should be noted that there are a number of open issues concerning the viability of this hybrid configuration that must be resolved before any thoughts of production can be entertained. Some of these issues, such as the initial cost penalty, meeting more restrictive NO_x standards, low-speed all-electric operation, and obtaining a suitable energy storage device, have been pointed out in the body of this paper. There are other problems which can be solved only through prototype development and lengthy testing of the drivetrain. These include:

1. Drivetrain packaging in a real vehicle.
2. Battery maintenance.
3. Engine lifetime under increased loading (the engine load factor required for the 2.3L hybrid Van operation is .335; for the conventional van, it is .125.)
4. Supplying power to vehicle auxiliaries.
5. Developing the best vehicle control system to achieve driveability comparable to existing vehicles.
6. Driveshaft, differential, and rear wheel performance during regenerative braking.
7. For some applications, such as those requiring driving long distance on upgrades or at very low speeds, larger battery energy capacity than that indicated in this paper is desirable.

ACKNOWLEDGEMENTS

It is a pleasure to thank Mr. Bruce Kopf for economic analysis, Mr. Phil Piatkowski for semiconductor controller development, Mr. Jerry Hough for assistance in the dynamometer experiment, Mr. Ed Peters for performance calculations, and Dr. Robert H. Park of the Battery Development Corporation of New York City, for consultation on NiCd batteries.

REFERENCES

1. "A Flywheel in Your Future", Newsweek, February 11, 1974.
2. "Hybrid Drive with Flywheel Component for Economic and Dynamic Operation", J. Helling, H. Schreck, B. Giera; Paper #7453, Third International Electric Vehicle Symposium, Washington, D.C.; February 9, 1974.
3. "Kritische Betrachtungen zur Einordnung von elektrisch angetriebenen Nutzfahrzuegen in den grosstadtischen Strassenverkehr", H. Albrecht and D. von Scarpattetti; ETZ-A Bd. 94 (1973) h.11; summarized in "Hybrid Bus Responds to Urban Pollution", Product Engineering, pp. 12-13, November 17, 1969.
4. "Current Status of Advanced Alternative Automotive Power Systems and Fuels", Volume IV, Aerospace Report #ATR-74(7325)-2, Vol. IV, April 10, 1974.
5. "Cost and Emission Studies of a Heat Engine/Battery Hybrid Family Car" TRW report #21054-6001-RO-00: April, 1972.
6. "Should We Have a New Engine", JPL Report #SP43-17, June, 1975.
7. Foote, L.R., and Hough, J.F., "An Experimental Battery Powered Ford Cortina Estate Car", SAE paper 700024, presented at the Automotive Engineering Congress, Detroit, Michigan, January 12-16, 1970.
8. Unnewehr, L.E. and Koch, W.H. "An Axial Air Gap Reluctance Motor for Variable Speed Applications", IEEE Transactions on Power Apparatus and Systems, Vol. PAS-93, No. 1, Jan/Feb 1974, pp 367-376.
9. "Vehicle with Hybrid Driving System", Robert Bosch Report K/EVE-2545; February 1, 1973.
10. U.S. Patent #3,791,473; "Hybrid Powertrain", Charles L. Rosen, February 12, 1974.
11. Powertrain Simulation: "A Tool for the Design and Evaluation of Engine Control Strategies in Vehicles", P.N. Blumberg; to be presented at 1976 SAE convention.
12. "Hybrid Electric Propulsion Utilizing Reconnectable Motor Windings in Wheels", E. Reimers; Paper #739116, 8th Intersociety Energy Conv. Conf., Philadelphia, PA; August, 1973.
13. "GM STIR-LEC II", GM Tech Center Publicity Release, May 7, 1969.
14. "The New Cars Get Government Scrutiny", Industrial Research, November, 1969, p.28.
15. "Fundamental Parameters of Vehicle Fuel Economy and Acceleration", D. N. Hwang, SAE Paper #690541, January, 1969.
16. "Electric Vehicle Systems Study", L. R. Foote et al; Ford Scientific Research Staff Report #73-132, October 25, 1973; available upon request; summary published in Electric Vehicle News, Vol. 3, No. 2, May, 1974, p.19.
17. SAE Technical Report J227, March, 1971.

An Analytical Study of the Fuel Economy and Emissions of a Gas Turbine-Electric Hybrid Vehicle*

Sidney G. Liddle
Research Labs., General Motors Corp.

ABSTRACT

A study was made to determine the effect of hybrid operation on the fuel economy and emissions of a vehicle using a gas turbine engine, a continuously variable transmission, and an electric power storage system. Both series and parallel hybrids were considered in a 1815 kg vehicle.

To facilitate this study, a computer program was written which modeled the vehicle and, using experimental data, computed its fuel consumption and emissions over the 1972 FTP driving cycle, starting with a fully warmed up engine.

This study indicates that, under certain conditions, the fuel consumption or emissions of the hybrid vehicle may be reduced as compared to its non-hybrid counterpart, but under other conditions, they may be increased. It is not possible to reduce fuel consumption and all of the emissions simultaneously. The reduction of one pollutant is usually accompanied by an increase in one of the others. The extent of the reduction or increase experienced with hybrid operation depends on the particular type of hybrid and on the engine operating conditions. The following table shows the relative effects of minimizing fuel consumption or any of the various pollutants. The table also shows which type of hybrid is best as well as the operating schedule and fuel control parameter. The fuel consumption and emissions are normalized with respect to the standard vehicle as defined in Assumption 8.

Table 1 - Relative Effects of Minimizing Fuel Consumption

	Min. Fuel Consumption	Min. HC	Min. CO	Min NO_x
Type of Hybrid	Parallel	Parallel	Series	Series
Operating Schedule	1	1	1	3
Fuel Control Parameter	0.14	0.077	0.0	0.077
Fuel Consumption	-5%	-4%	+19%	+94%
Unburned Hydrocarbons	-50%	-75%	-72%	+1025%
Carbon Monoxide	-43%	-21%	-51%	+772%
Oxides of Nitrogen (as NO_2)	+22%	+14%	+70%	-41%

This table indicates that the hybrid offers a small decrease in fuel consumption as compared to the standard car and at the expense of increased NO_x emissions. Minimizing unburned hydrocarbons reduces fuel consumption but at the expense of increased NO_x. Both fuel consumption and NO_x emissions are increased when carbon monoxide is minimized. Very large increases in fuel consumption and in unburned hydrocarbon and carbon monoxide emissions result from minimizing oxides of nitrogen emissions.

It must be concluded that the hybrid vehicle does not automatically guarantee lower fuel consumption or emissions.

*Paper 760122 presented at the Automotive Engineering Congress and Exposition, Detroit, 1976.

INTRODUCTION

In this study only the two basic hybrid systems, series and parallel, are considered. The only heat engine used is a regenerative single-shaft gas turbine and the energy storage device is a battery of the lead-acid type.

In the particular series system shown in Figure 1, all of the power output of the engine, except that used to accelerate the engine itself, is converted into electric power by the alternator. The electric power is reconverted to mechanical power by a DC motor which is connected to the drive wheels. This system constitutes a continuously variable electric transmission. Whenever the power requirement of the vehicle is greater than the power supplied by the engine, additional power is drawn from the battery. When the engine power capacity is greater than the vehicle requirement, the excess is used to recharge the batteries.

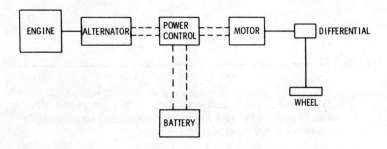

Fig. 1-Schematic of series hybrid

In the particular parallel system shown in Figure 2, only the power required to recharge the batteries is converted to electric power. The majority of the engine power is delivered through a continuously variable mechanical transmission directly to the wheels. When the vehicle requirements exceed the engine capacity, power is drawn from the battery by the DC motor geared to the output of the transmission.

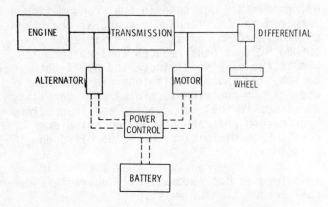

Fig. 2-Schematic of parallel hybrid

It can be seen that the series and parallel systems are very similar in nature, differing in only two ways. The first way is that the series system uses an electric transmission while the parallel system uses a mechanical transmission. The second way is that in the series hybrid the output of the alternator is the sum of the road load power and the battery recharging demands. The engine output is controlled only by the alternator and engine speed-power relationship during steady speed operation can be restricted to a single operating line. In the parallel hybrid, the transmission controls the engine speed as a function of the road load power. The battery recharging power is added onto the road load power but does not affect the engine speed. As a result, the engine speed-power relationship of this parallel hybrid is a fairly broad band depending on the battery condition.

A significant amount of power is required to accelerate a gas turbine engine and during engine transient operation, this power must be accounted for in the calculations. When the engine is decelerated, this power is released and can be used to recharge the batteries if required.

In both of these systems, a DC motor is used with regenerative braking, with the power developed from the decelerating vehicle going to the battery.

A number of studies of hybrid vehicles have been made in the past. Most of these studies (Ref. 1-6) are aimed at system definition and specific components and not at emissions. One study which did attempt to evaluate emissions was made by the Aerospace Corp. (Ref. 7-9). Their estimates of emissions from current gas turbine engines agree reasonably well with the results of this study. These reports by Aerospace Corp. covered a wide variety of heat engines and energy storage devices but concluded that of the heat engines studied only the piston engine and the gas turbine are feasible, and that batteries are the best energy storage device.

The author in an earlier paper (Ref. 10) reported on a combined analytical-experimental program to determine the effect of hybrid operation on a car using a spark ignition piston engine. Both series and parallel hybrids were studied using lead-acid batteries as the energy storage device. The general conclusions were that hybrid operation reduces hydrocarbon emissions but could either increase or decrease carbon monoxide and oxides of nitrogen emissions depending on the type of hybrid and the particular operating conditions. Fuel consumption could also, either increase or decrease depending on the particular hybrid configuration and operating conditions. The standard car used in those comparisons was the same as the hybrid car except that there was no energy storage system.

It has been suggested that a hybrid system might reduce emissions and improve performance and fuel economy by allowing the use of a smaller engine running at its optimum operating condition. It has been further suggested that operation at a fixed speed, fixed power point is the ideal method. The earlier paper by the author indicated that there is no single optimum operating condition and that fixed speed, fixed power operation may, under some conditions, actually increase emissions and fuel consumption over that of the non-hybrid vehicle.

ASSUMPTIONS

For the purposes of this study, the following assumptions have been made.

1. The fuel consumption and emissions are assumed to be continuous functions of engine speed and turbine inlet temperature. The effects of transients, other than as they affect speed and turbine inlet temperature are not considered with the exception of regenerator disk rotation. The effect of the relatively low speed of the regenerator disk on transient operation is included in the analysis because of its effect on burner inlet temperature and, therefore, on emissions.

2. The alternator and DC motor efficiencies are assumed to be functions of speed and output load only.

3. Battery efficiency is assumed to be a constant regardless of battery state-of-charge or its charging rate.

4. Mechanical transmission efficiency is assumed to be only a function of the speed ratio across the transmission.

5. The driving cycle is a hot start cycle. Because of data limitations, cold starts could not be modeled. This would be expected to result in optimistic predictions for hydrocarbon and carbon monoxide emissions as well as fuel economy.

6. The vehicle used in this study is intended for highway as well as for urban use. The engine and other components are sized to handle sustained 70 mph cruising speed power demands.

7. The battery state-of-charge must be the same at the end of the driving cycle as at the beginning.

8. The characteristics of the standard car used in this study are defined to be the same as those of the parallel hybrid car operating on schedule one with a fuel control parameter of 1.

9. The car weighs 1815 kg, has a frontal area of 2.32 sq. metres and a coefficient of drag of 0.50. This vehicle is the same as the family car in the Aerospace Corp. report (Ref. 5).

10. Because of rapid developments in combustion chamber designs and since this study is concerned only with the effect of hybrid operation on fuel consumption and emissions, fuel flow and emissions have been normalized to those of the standard car as defined in Assumption 8 in order to more clearly delineate the trends involved.

11. The energy storage system is assumed not to increase the overall weight of the hybrid car as compared to the non-hybrid car. This would be expected to result in optimistic fuel consumption predictions for the hybrid car.

DISCUSSION

Since a basic assumption of this study is that the car is able to operate on the highway as well as in an urban environment, the engine is sized such that long-distance high-speed cruise (70 mph) and grade climbing requirements are satisfied. This requirement also makes it desirable to vary the engine speed and power to meet the load requirements.

COMPONENTS - The basic engine used in this study is a 67 kW single-shaft gas-turbine engine scaled down and modified from the GT-225 dual-shaft engine. This power rating is based on engine output after accessories. This particular engine uses a centrifugal compressor with a 4.5 to 1 pressure ratio and two axial turbines on a single shaft.

Figure 3 shows the relationship between engine output power, engine speed, turbine inlet temperature and burner inlet temperature.

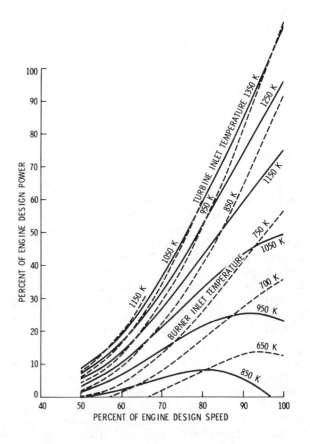

Fig. 3-Burner and turbine inlet temperatures

The turbine inlet temperature and the burner outlet temperature are the same temperature so that the difference between the two sets of curves on the figure is the temperature rise across the burner itself. The lowest temperatures occur in the area of 80-95% engine speed and zero power. The highest temperatures are along the 1350 K turbine inlet temperature line and at low engine speed where the burner inlet temperature can exceed 1150 K.

Emissions data are based on tests conducted in the combustion test rig of the Power Systems Department of the General Motors Research Laboratories. The test data cover a wide range of engine speeds, turbine inlet temperatures and power levels. These data reflect steady-state operation only as no transient data were taken. The burner used is a diffusion-flame can-type combustion chamber. Because of recent advances in burner design, the absolute values of the emissions are not representative of the current state-of-art. However, the general trend of the emissions is quite similar in the more advanced burners, only the quantitative values are different. Since the purpose of this study is to determine the general effect of hybrid operation on emissions and not the absolute emission rates, the emission data have been normalized by referring them to the emissions at the engine design point of 1283 K and 100% engine speed. Similarly, fuel consumption has been normalized to fuel flow at the engine design point.

The normalized emissions are shown in Figures 4, 5 and 6 for unburned hydrocarbons, carbon monoxide and oxides of nitrogen as NO_2, respectively. The peak unburned hydrocarbon emission occurs at 80% engine speed and zero power. This would be expected since this is the region of the engine map where burner temperatures are the lowest and therefore most conducive to producing a high hydrocarbon emission. The same reasoning would hold true for carbon monoxide and Figure 5 bears this out. The highest CO emission is at 82% engine speed and zero horsepower. The lowest hydrocarbon and carbon monoxide emissions should, based on thermodynamic considerations, occur at conditions of high burner inlet and outlet temperatures and both Figures 4 and 5 confirm this hypothesis. Since these temperatures are controlled by the engine cycle rather than the burner configuration, similar trends should occur for other types of burners as well as the diffusion flame burner used in this study.

So far the oxides of nitrogen emission has not been discussed. In Figure 6, the oxides of nitrogen emission (as NO_2) is shown. Whereas the emission rates for hydrocarbons and carbon monoxide are very low at the engine design point, with NO_x it is very high and decreases as power level decreases. A special point to note is that at any particular power level the emission rate is high at low turbine speed and

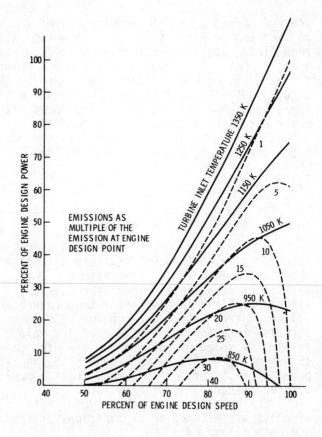

Fig. 4—Unburned hydrocarbon emissions

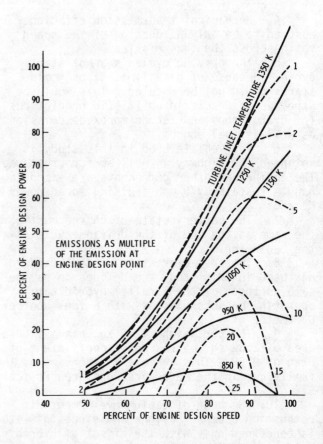

Fig. 5—Carbon monoxide emissions

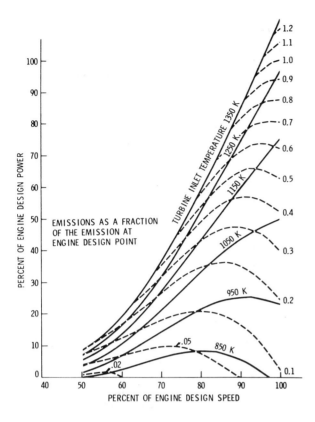

Fig. 6-Oxides of nitrogen emissions

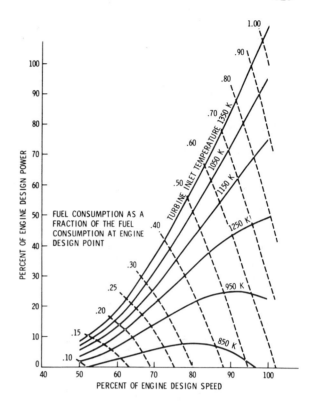

Fig. 7-Fuel consumption

decreases as engine speed increases until a minimum is reached whereupon it begins to rise again. It is therefore possible to establish a line of minimum oxides of nitrogen emission which runs from zero power and 50% engine speed up to the engine design point. This minimum emission line appears to result from two conflicting conditions. One is that at high burner temperatures, the NO_x concentration is high and it decreases as the temperature is lowered and engine speed increased. On the other hand, the mass of exhaust gas increases as engine speed increases. The mass of the emissions is the product of the concentration and the exhaust gas flow rate. These two conditions combine to produce the minimum NO_x emission line. All engines investigated so far, including piston engines, exhibit this minimum oxides of nitrogen line although its location will tend to differ depending on the exact engine configuration. The use of these normalized emissions maps tends to concentrate attention on the effect of hybrid versus non-hybrid operation and not on the mass of the emissions themselves.

The fuel consumption of the gas turbine engine is shown in Figure 7. The notable feature of this figure is the steepness of the constant fuel flow lines. It is evident that at any given power level, an increase in engine speed produces a marked increase in fuel consumption and, therefore, it is desirable to operate at the lowest possible engine speed which, of course, means the highest possible turbine inlet temperature which promotes low unburned hydrocarbons and carbon monoxide but increases NO_x sharply.

The efficiency of the alternator used in this study is assumed to be a function of its speed and load as shown in Figure 8. It has a peak efficiency of 90% at 70% of its design speed and 75% of its design load. The efficiency of the DC motor is also assumed to be a function of its speed and load, and has peak efficiencies of 90.5% at 65% of its design speed and 100% of its load as shown in Figure 9. These efficiencies include controls and power conditioning requirements for an optimized system and should be considered optimistic. The efficiencies of a prototype system would be lower. The same efficiency data are used for both the large alternator and motor used in the series hybrid, and the smaller components used in the parallel hybrid.

The mechanical transmission used in the parallel hybrid is an idealized one in which the efficiency is a function only of the speed ratio across the transmission as shown in Figure 10. Since there are no proven transmissions available for single-shaft gas-turbine engines, this curve is constructed from a series of estimates as to gear losses, bearing losses, churning and spin losses and losses associated with either a torque converter, a toric unit or a slipping clutch. Depending on the exact configuration, the efficiency of a real transmission could easily vary by 10% from the values shown in Figure 10.

The batteries used in this study are

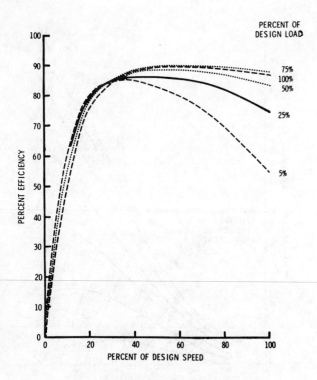

Fig. 8-Alternator efficiency

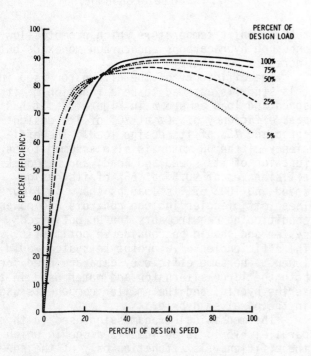

Fig. 9-Motor efficiency

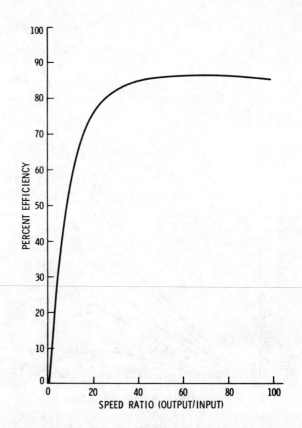

Fig. 10-Mechanical transmission efficiency

assumed to be of the lead-acid type. The overall efficiency of the battery system is assumed to be 75%, with a charging efficiency of 83% and a discharging efficiency of 90%. In this analysis, the batteries are considered strictly as a power storage device with certain capacities and efficiencies. Any other energy storage device such as a flywheel, hydraulic system or other types of batteries which have similar capacities and efficiencies would have the same effect on the overall system. In other words, the type of storage system is not important provided it is capable of absorbing and discharging power as required by the system.

CONTROLS - The accelerator pedal on a conventional spark ignition engine opens the throttle directly. On a gas turbine, it increases fuel flow to the burner to increase the turbine inlet temperature. On a hybrid car, it only signals the driver demand and a separate control system determines how the demand will be met. In the case of an increased demand, the control system adjusts both the power being delivered from the battery to the motor and the fuel being delivered to the engine such that the combined power matches the demand.

In this study, a controlled delay fuel control is used on the engine. A change in accelerator position forces the fuel control to change its output but because of its delay characteristics the full change does not immediately affect the engine. However, if the demand is maintained long enough, such as in climbing a hill, the fuel control will eventually supply sufficient fuel to meet the power demanded. In the meantime, the batteries meet the difference in demand and if sufficient excess power exists after the engine meets the vehicle load, the batteries can be recharged as well. In urban stop-and-go driving the power peaks are reduced and the energy capacity of the batteries are heavily utilized. In operation, if there is a step change in vehicle demand, the fuel control responds by a specific fraction of the difference between the engine output and the vehicle demand during each unit of time. This fraction is referred to in this paper as the fuel control parameter. If it is one, then the fuel control will respond directly to vehicle demand. If the value of the parameter is less than one, the fuel control responds at a reduced rate. If it is zero, then the fuel control does not respond to the change in demand and the engine operates in a fixed speed, fixed power mode.

The motor control is more complicated than the fuel control since it must sense both the demand from the accelerator and the output of the engine and adjust accordingly. The engine output is determined from a combination of engine speed and turbine inlet temperature. The relationship between speed and power is defined as an operating schedule. There are several types of these schedules. The relation of engine output power to engine speed is the engine schedule. In the series hybrid, the relationship between alternator input power and alternator speed is the alternator schedule. The relation of transmission input speed to input power for the parallel hybrid is the transmission schedule. During steady speed operation in the series hybrid the engine schedule and the alternator schedule coincide. The engine output power is controlled by the fuel control while the alternator schedule is controlled by changing the field current as a function of alternator speed which is the same as engine speed. If the engine output is greater than the alternator scheduled power, the excess power accelerates the engine. If the output power is less than the alternator scheduled power, the engine decelerates.

The parallel hybrid is somewhat more complicated. In the series hybrid, both road load power and battery recharging power is produced by the alternator and the engine output is divided into two parts, the alternator demand and the power needed to accelerate the engine. The engine output of the parallel hybrid is divided into three parts. One part of the power goes to the transmission, the second part goes to the alternator to recharge the batteries and the third part is the power required to accelerate the engine. The transmission input power is controlled as a function of input speed which is the same as engine speed. If the batteries are fully charged and the engine is operating at constant speed, then the engine schedule and the transmission schedule coincide. If the batteries are being recharged and/or if the engine is being accelerated or decelerated, then the engine schedule may be above or below the power level defined by the transmission schedule. The engine power varies over a wider range with the parallel hybrid than with the series hybrid and this wider range has an effect on the relative fuel economy and emissions of the two types of vehicles.

COMPUTATIONAL METHOD - In this study the 1972 FTP driving cycle (Ref. 11) is used except that the engine is assumed to be fully warmed up to idle conditions at start. The cold start required by the FTP cycle is not part of this simulation. This driving cycle specifies the vehicle speed as a function of time at one second intervals. The acceleration is calculated by numerical differentiation of the speed.

Knowing the speed and acceleration of the vehicle at any specific instant of time as well as its weight, coefficient of drag, frontal area and rolling resistance, the road load power can be calculated. Using either Figure 1 or 2 as a model and starting at the wheels, the speed, power and acceleration of each component may be calculated in turn. This procedure carried back to the engine results in the engine operating speed, power, emissions and fuel consumption for each integration step. Seven ordinary differential equations for engine acceleration, battery-charge rate, air-flow rate, fuel-flow rate, hydrocarbon emission rate, carbon monoxide emission rate and oxides of nitrogen emission rate are numerically integrated.

At the end of the driving cycle, the total fuel used is divided by the distance covered to calculate the fuel economy. Similarly, total emissions are divided by the distance to obtain the emissions in grams per mile.

RESULTS - In this phase of the overall study, the effects of fuel control parameter and operating schedule on fuel consumption and emissions of the vehicle operated on the FTP driving cycle are considered for both parallel and series hybrid configurations.

The main purpose of the hybrid system is to reduce the range of speeds and power levels which the engine encounters during the driving cycle. The effect of the fuel control parameter is to reduce the rate at which the engine accelerates or decelerates, and hence restricts the range of engine speeds encountered during transient operation. The lower the value of the parameter, the greater the restriction on the range of engine speeds.

Three operating schedules are used in this study and are shown superimposed on the engine map in Figure 11. The same schedules are used for the alternator in the series hybrid and for the transmission in the parallel hybrid. Schedule 1 is a high temperature schedule operating near the temperature limit of the turbine. This schedule is also in the region of minimum fuel consumption as well as minimum emissions of unburned hydrocarbons and carbon monoxide. Schedule 3 approximates the line of minimum NO_x emission discussed earlier. Sche-

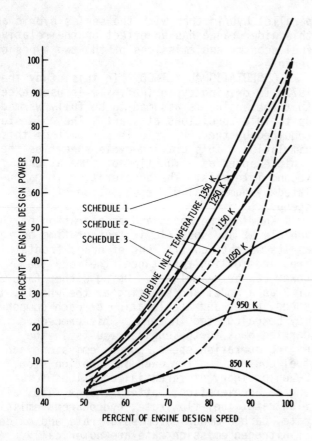

Fig. 11—Engine operating schedules

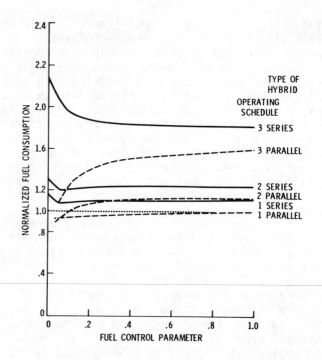

Fig. 12—Effect of fuel control parameter on fuel consumption

dule 2 is a compromise schedule between schedules 1 and 3.

In any non-hybrid gas turbine powered car, it is desirable from a fuel consumption standpoint to operate the turbine at as high a temperature and as low a speed as possible, in effect, along a line such as schedule 1. For the purposes of comparison, the standard car is defined as having a transmission of the same type as the parallel hybrid, operating on schedule 1 with a fuel control parameter value of one. The results for all of the other schedules and parameter values are normalized with respect to this standard car. Reductions in fuel consumption and emissions are characterized by numbers less than one while numbers greater than one indicate that the fuel consumption and emissions are greater for the hybrid than for the standard car.

In Figure 12, the normalized fuel consumption is shown for both of the hybrid types and for all three schedules. For all schedules, the series hybrid has higher fuel consumption than the parallel hybrid. In addition, there are distinct differences in the fuel consumption trends as the fuel control parameter, and hence engine response to accelerator pedal, is lowered. All of the parallel hybrids show a downward trend in fuel consumption as the fuel control parameter is lowered until a value of approximately 0.10 is reached below which the batteries cannot be fully recharged during the driving cycle. In the series hybrid, there is a continuous upward trend for schedule 3 while schedules 1 and 2 show only a slight downward trend until a value of the parameter of 0.075 is reached and is followed by a sharp rise as a value of zero is approached. In the case of the series hybrid where the fixed speed case (fuel control parameter equals zero) can be calculated, the fuel consumption is generally higher than for the non-hybrid version (fuel control parameter equals one). From a fuel consumption standpoint, the parallel hybrid used in this study is definitely superior to the series hybrid. The effect of schedule is also clearly seen in this figure with schedule 1 having the lowest fuel consumption and schedule 3 the highest for both types of hybrids. The hybrid system does not offer a significant savings in fuel economy since the lowest fuel consumption (schedule 1, parameter = 0.077) is only 5% less than the standard car. On the other hand, significant increases in fuel consumption may occur under certain operating conditions.

The effects of hybrid operation on emissions are shown in Figures 13, 14 and 15 for unburned hydrocarbons, carbon monoxide and oxides of nitrogen respectively. Hydrocarbon emission is normally not a problem in gas turbines. For schedule 1, both types of hybrids have the same emissions at a fuel control parameter value of one and both decline as the parameter is reduced, with the series hybrid emission dropping faster than the parallel hybrid emission. Schedule 2 results are similar with the series hybrid emission higher than schedule 1 and higher than the

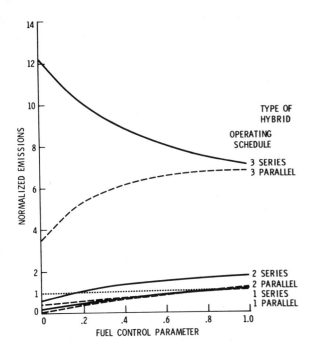

Fig. 13-Effect of fuel control parameter on unburned hydrocarbon emissions

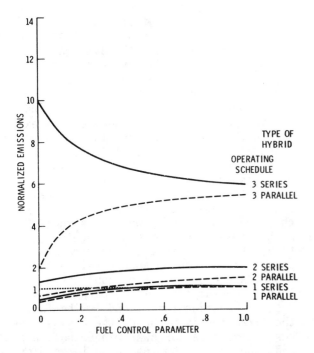

Fig. 14-Effect of fuel control parameter on carbon monoxide emissions

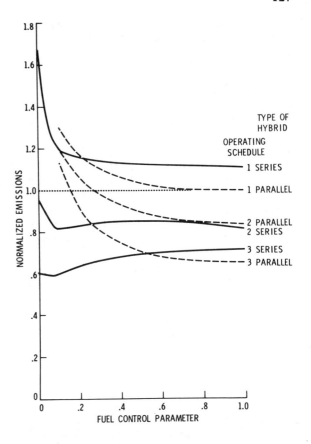

Fig. 15-Effect of fuel control parameter on oxides of nitrogen emissions

standard car. With the parallel system, the emission declines as the fuel control parameter is reduced. The series hybrid has the opposite effect with the hydrocarbon emissions rising as the parameter is reduced.

Carbon monoxide follows the same general trends as hydrocarbons as would be expected since the emission maps, Figures 4 and 5, are qualitatively similar.

The oxides of nitrogen (as NO_2) emission is shown in Figure 15. The parallel hybrid using all of the schedules shows an increase in emissions as the fuel control parameter is decreased to a value of 0.10. When the fuel control parameter has a value of one, there is a marked difference in the NO_x emissions for the three schedules. As the parameter is reduced, the difference is also reduced and all three curves appear to converge.

The series hybrid shows mixed results. Using schedule 1, the NO_x emission rises continuously as the parameter is decreased with the sharpest rise when the parameter has a value less than 0.15. Schedule 2 shows an initial rise in NO_x as the parameter decreases to 0.5, then begins to drop to a minimum near a parameter of 0.10 only to rise sharply as the parameter approaches zero. Schedule 3 shows a continuous decrease in NO_x until the parameter reaches a value of 0.08 followed by a moderate rise. At the minimum value of 0.08 the NO_x emission is more than 40% less than the stan-

parallel hybrid on both schedules. With both hybrids, the emissions decline as the value of the fuel control parameter is reduced. In fact, at low values of the parameter, the parallel hybrid on schedule 2 has a lower emission than either hybrid on schedule 1.

On schedule 3, the minimum NO_x schedule, both the series and parallel hybrids have hydrocarbon emissions, at a fuel control parameter of one, about seven times that of the dard car, but this change is due more to the use of the low NO$_x$ schedule than to the use of the hybrid system. If the effect of fuel control parameter is examined, all of the curves are higher at the very low values of the parameter than at a value of one with the exception of the series hybrid using schedule 3 where a 15% decrease in NO$_x$ is possible.

The emissions from the gas turbine-hybrid vehicle driven over the FTP driving cycle presents two main points to be considered. First, only the series hybrid using schedule 3 is of any value in the reduction of NO$_x$ emissions. Second, this same hybrid and schedule produces the greatest unburned hydrocarbon and carbon monoxide emissions and has the highest fuel consumption as well.

RATIONALE - Having presented the data, an examination of the causes for the effects found is in order. To do this, it is necessary to go back to the emissions maps Figures 4, 5 and 6 and consider what happens during the driving cycle.

If the fuel consumption is directly proportional to the power and the power storage system (batteries and electrical components) is 100% efficient, the hybrid vehicle will have exactly the same fuel consumption as its non-hybrid counterpart. The reason is that the reduction in fuel consumption resulting from using battery power during acceleration will be counterbalanced by increased fuel consumption during battery recharging. If the fuel consumption is directly proportional to power and the power storage system is less than 100% efficient, then the fuel consumption from the hybrid vehicle will be higher than for the non-hybrid vehicle, since more power will be required from the engine to recharge the battery, thus requiring more fuel usage. Hence the potential reduction in fuel consumption using a hybrid vehicle depends on the nature of the variation of fuel consumption with power.

Exactly the same argument can be applied to the emissions from the vehicle and hence the reduction of emissions using a hybrid vehicle will depend on the manner in which the emissions vary with power.

If schedules 1 and 3 are replotted as functions of fuel consumption and power, the result is Figure 16. In this figure, with normalized engine power as the abscissa and normalized fuel consumption as the ordinate, the relationship between power and fuel consumption becomes quite clear. Schedule 3, which produces the lowest NO$_x$ emissions and the highest carbon monoxide emissions, has a fuel consumption well above schedule 1 except at the end points. The reason that schedule 1 on Figure 12 shows very little variation in fuel consumption as the fuel control parameter is varied can be explained using Figure 16 which shows that schedule 1 is very nearly directly proportional to power (dashed line), particu-

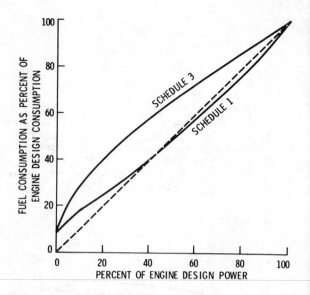

Fig. 16-Fuel consumption for schedules 1 and 3

larly in the range above 20% power. The specific fuel consumption (SFC) at a particular operating point is the slope of the line connecting the operating point to the origin. The dashed line is, therefore, a line of constant SFC. It can be seen that the lowest SFC for schedule 1 occurs at about 60% engine power. It has been found that the average power used during the FTP driving cycle for this car and engine is about 15% of the peak power. During non-hybrid operation the power ranges from zero up to about 85% of the peak power. The series hybrid operates on the schedule line except during the engine acceleration. Since the fuel consumption for schedule 1 is nearly directly proportional to power, the fuel flow for hybrid operation depends on the battery efficiency. As the fuel control parameter is reduced, the engine operation is progressively restricted to the area around 15% power where the SFC is higher than at higher power levels. In addition, the batteries are used more as the parameter is reduced and therefore the battery losses are greater. The result is an increase in fuel consumption as the parameter is reduced. Opposing this trend is the fact that as the parameter is reduced, the engine speed is restricted to a narrower range and the power and, therefore, the fuel used to accelerate the engine is reduced. Since the power involved in accelerating the engine is a significant fraction of total engine power, the reduction of fuel consumption is considerable and almost cancels out the increases due to higher battery losses and the higher SFC. The fuel consumption for schedule 3 is considerably higher than for schedule 1 but is not directly proportional to power, in fact, the SFC of schedule 3 decreases as the power level increases all the way to the 100% power point. As the fuel control parameter is reduced, the operation of the engine is limited to a progressively higher

SFC range and the fuel consumption increases accordingly.

The parallel hybrid reacts differently because the alternator power is taken directly off the engine and the schedule controls the transmission input only. As a result, the engine operates above the schedule curves shown in Figure 11 in a lower specific fuel consumption region when the batteries are being recharged. As the fuel control parameter is reduced, the batteries are used more and alternator pulls more power to recharge them. The engine, therefore, spends more time in the lower specific fuel consumption regions and the driving cycle fuel consumption over all schedules decrease as the fuel control parameter decreases. Differences in the driveline efficiencies also contribute to the lower fuel consumption of the parallel hybrid relative to the series hybrid.

As shown in Figure 17, both unburned hydrocarbons and carbon monoxide emissions follow the same trends as the fuel control parameter changes. For schedule 1, the lowest specific emissions (mass of emissions per unit power) is near 23% of peak power for hydro-

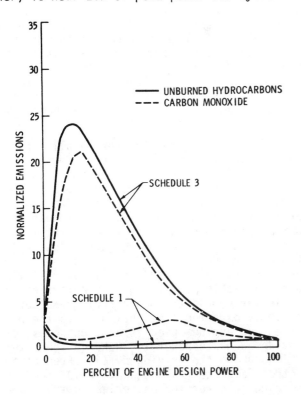

Fig. 17—Hydrocarbon and carbon monoxide emissions for schedules 1 and 3

carbons and near 28% of power for carbon monoxide. As the fuel control parameter is reduced, the engine is forced to operate more of the time in this low emission region with the result that for both hydrocarbons and carbon monoxide, the emissions decrease with fuel control parameter for both series and parallel hybrid vehicles.

Schedule 3 presents a more complicated picture. Peak emissions occur in the 10% to 20% power range. It would, therefore, be expected that emissions would rise as the fuel control parameter is reduced and the engine forced to operate more in this low power region. This is exactly what happens to the series hybrid which follows the schedule closely. The parallel hybrid because of the alternator arrangement tends to operate above the schedule line when the batteries are being charged. The increased use of the batteries as the fuel control parameter decreases and, hence, the increased use of the alternator tends to move the engine operating conditions away from the peak hydrocarbon and carbon monoxide emissions area with the result that the emissions decrease with a reduction in fuel control parameter for the parallel hybrid on this schedule. It should be pointed out that if the alternator was located at the output of the transmission, the fuel consumption and emission would be higher and much closer to the results of the series hybrid. The exact arrangement of the components in a hybrid vehicle can have a marked effect on both its fuel economy and its emissions.

The variation in the NO_x emission with changes in the fuel control parameter differs markedly between the two types of hybrids as can be seen in Figure 15. The emission from the parallel hybrid rise continuously as the fuel control parameter is decreased as does the series hybrid operating on Schedule 1. However, the series hybrid for schedules 2 and 3 show a different behavior. The NO_x emission initially drops as the fuel control parameter is reduced until a value of about 0.08 is reached where the emission begins to rise. The schedule 2 emission is somewhere between schedules 1 and 3 in both values and trends. It starts off with a rise in the NO_x emission as does schedule 1 but reaches a maximum at a fuel control parameter value of 0.6. It then decreases to a minimum at a value of the parameter of 0.10 when it begins to rise again.

To explain this odd behavior, the normalized emissions are plotted as functions of power in Figure 18 for both schedules 1 and 3. The emission of NO_x at zero power is not quite zero but is about 0.01 or 1% of what it is at design power. As a result, the specific emission at zero power is infinite. On schedule 1, the specific emission declines as the power increases until the power reaches approximately 60% of the design power whereupon the specific emission begins to rise again. As the power is increased, the specific emission on schedule 3 drops quickly to a minimum at 16% power but is nearly constant from 8% power to 27% power and rises continuously afterwards as the design power is approached. It should be remembered that the power at the road required to drive a given car over the FTP driving cycle is inde-

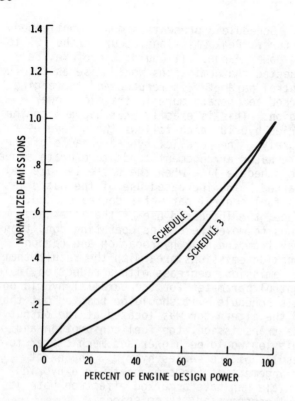

Fig. 18-Oxides of nitrogen emissions for schedules 1 and 3

pendent of the powerplant, so the higher the specific emissions, the higher the absolute emissions. Secondly, the average power required to drive this car over the cycle is about 15% of the design power. Looking at the series hybrid on schedule 1, it is apparent that as the fuel control parameter is reduced, the engine is forced to operate in a narrower and narrower band around the 15% power average. This is where the specific emission is high compared to what it is at higher power levels and therefore the emissions in Figure 15 rise as the parameter is reduced for this hybrid-schedule combination. For the series hybrid using schedule 3, the initial effect of reducing the fuel control parameter is to limit the power range to the area where the minimum specific emission occurs around 16%. The result is a general reduction in emissions with the reduction in fuel control parameter. However, as soon as the power range is restricted to a range below 27% power as it is for very low values of the fuel control parameter, the emission rises because the specific emissions are nearly constant but losses in the energy storage system increase with increased use and the engine has to produce more power to compensate for these losses thus the rise in emissions at the very low end of the fuel control parameter scale.

The parallel hybrid, because of the power split between the alternator and the transmission and the fact that the schedule only controls the power input to the transmission, acts differently from the series hybrid.

Looking first at Schedule 3, the minimum NO_x emission schedule, and considering that if the alternator drew no power then the engine would tend to follow the schedule and the results would be similar to the series hybrid results but, in fact, the alternator does draw power and this power forces the engine to operate above the minimum emission line. As the fuel control parameter is reduced, the batteries are used more and the alternator draws more power from the engine to charge them. The engine operates further from the minimum emission line and the result is a rapid increase in NO_x emissions as the parameter is reduced. The increased battery usage increases the losses in the energy storage system and these must be compensated for. These effects are only partially counterbalanced by the reduction in the specific emission as the power range is reduced to the 16% power point for minimum specific emission.

Much of the above also applies to operation on schedule 1 except that the schedule itself is above the minimum emission line and close to the maximum power limit. The minimum specific emission occurs at 63% power rather than 16% as for schedule 1. Schedule 3 has lower NO_x emission than does schedule 1 at all values of the fuel control parameter but the difference is less at the lower values of the parameter than at the higher values. This is because at the lower values, the engine for both schedules is operating at nearly the same conditions and close to the maximum power limit.

In the arrangement of the parallel hybrid analyzed in this study with the alternator operating off the engine, it is impossible to recharge the batteries completely during the driving cycle if the value of the fuel control parameter is less than 0.10. The reason is that engine speed is controlled by the transmission and when the car is going slow or stopped, the engine speed drops to the idle speed and at this speed there is very little power available to recharge the batteries. In the series hybrid the power from the alternator is simply diverted from the motor to the batteries and engine speed is kept up so as to provide adequate power to recharge the batteries during the driving cycle. Even at fuel control parameter values above 0.10, the charging rate for the batteries must be considerably higher for the parallel hybrid than for the series hybrid. One solution would be to add a controlled differential at the engine output of the parallel hybrid so power could be transmitted to either the transmission or alternator or to proportion it between them at any engine speed. The result would be a system which would be similar to the series hybrid in operation but would still use the mechanical transmission and the smaller electrical components of the parallel hybrid.

Emissions and fuel consumption would be somewhere between those of the present parallel hybrid and the series hybrid.

CONCLUSIONS - Obviously, this study which involves only one engine in two types of hybrids and operates over one driving cycle, the FTP cycle, cannot completely define either the advantages or the disadvantages of the hybrid vehicle concept. However, it does permit certain conclusions to be drawn.

First, the type of hybrid; parallel, series or any other arrangement, has a significant effect on both the fuel economy and the emissions of the vehicle.

Second, the fuel consumption and emissions characteristics of the heat engine used must be known if the optimum system is to be produced. Since the operating schedule of the engine is extremely important to the overall performance of the system, it must be chosen with care. While qualitative predictions of fuel economy and emissions can be made from the fuel consumption-power and the emission-power curves, detailed computer simulation and hardware test data are necessary to design an optimum system.

For the vehicle used in this study, the fuel consumption varied between a reduction of 5% and an increase of 139% as compared to the standard car depending on the type of hybrid and the operating schedule.

The use of these hybrid systems can reduce NO_x emissions by 13% or increase them by 77% depending on the type of hybrid and the engine operating conditions. Generally, unburned hydrocarbons and carbon monoxide emissions are reduced by the use of the hybrid system by as much as 85% but can be increased by 75% under certain conditions.

There is no "optimum" engine operating point for low emissions and high fuel economy. There are, however, compromise conditions which can be chosen depending on the relative importance placed on fuel consumption and on each of the three pollutants.

While the hybrid vehicle has special advantages in some situations, it cannot be considered as a panacea for either fuel economy or emissions problems.

REFERENCES

1. G. H. Gelb, N. A. Richardson, T. C. Wang, and R. S. DeWolf, "Design and Performance Characteristics of a Hybrid Vehicle Power Train," SAE Paper 690169, January 1969.

2. P. D. Agarwal, R. J. Mooney, and R. R. Toepel, "Stir-Lec 1, A Stirling Electric Hybrid Car," SAE Paper 690074, January 1969.

3. D. H. Brown, "Hybrid Battery System," SAE Paper 710236, January 1971

4. G. H. Gelb, N. A. Richardson, T. C. Wang, and B. Berman, "An Electro-Mechanical Transmission for Hybrid Vehicle Power Trains - Design and Dynamometer Testing," SAE Paper 710235, January 1971.

5. R. R. Gilbert, G. E. Heuer, E. H. Jacobsen, E. B. Kuhns, L. J. Lawson, and W. T. Wada, "Flywheel Drive Systems Study, Final Report," Lockheed Missiles and Space Company, Inc., Sunnyvale, California, Report No. LMSC-D246393, July 31, 1972.

6. P. E. Tartaglia, "Achieving High Energy Efficiency for Urban Transportation Through Hydrostatic Power Transmission and Energy Storage," Paper 739114, Eighth Intersociety Energy Conversion Engineering Conference, Philadelphia, PA, August 1973.

7. J. Meltzer, and P. Lapedes, "Hybrid Engine/Electric Systems Study," Volumes I and II. Aerospace Corp., El Segundo, Calif., Report No. TOR-0059(6769-01), June 1971.

8. M. G. Hinton, Jr., T. Iura, W. G. Roessler, and H. T. Sampson, "Exhaust Emissions Characteristics of Hybrid Heat Engine/Electric Vehicles," SAE Paper 710825, October 1971.

9. H. R. Sampson, and H. J. Killian, "Evaluation of Powertrains for Hybrid Heat Engine/Electric Vehicles," SAE Paper 720194, January 1972.

10. S. G. Liddle, "Emissions from Hybrid Vehicles," Paper 739115, Eighth Intersociety Energy Conversion Engineering Conference, Philadelphia, PA, August 1973.

11. Environmental Protection Agency, "Control of Air Pollution from New Motor Vehicles and New Motor Vehicle Engines," Federal Register, Vol. 37, No. 10, January 15, 1972.

The Optimization of Body Details—A Method for Reducing the Areodynamic Drag of Road Vehicles*

W. H. Hucho, L. J. Janssen
and H. J. Emmelmann
Volkswagenwerk AG (Germany)

DRAG AND FUEL ECONOMY - Recently the incentive to reduce the aerodynamic drag of road vehicles has increased again. Where top speed was formerly the motivation for drag reduction, today an improved fuel economy is the target. Prior to the detailed discussion of several possibilities to reduce air drag, its influence on fuel economy in steady state driving as well as under transient conditions should be elaborated.

The engine power required to maintain a constant speed on level road under conditions of no wind is made up, as is well known, of the power needed to overcome internal losses and the power needed to overcome external resistance:

$$P = \frac{1}{\eta_G} \cdot (P_D + P_R)$$

where P_D and P_R are the air drag and

*Paper 760185 presented at the Automotive Engineering Congress and Exposition, Detroit, 1976.

rolling resistance power requirement respectively, and η_G is the efficiency of the drivetrain.

The relationship between air drag and total external resistance to motion of a vehicle is shown in Fig. 1. Above a speed as low as 40 mph the proportion represented by air drag amounts to more than 50 percent of total drag. Depending on the drag coefficient and the weight, the air drag amounts to 80 or 90 percent of total drag at higher speeds.

The momentary fuel consumption is determined from the required engine power and the engine's specific fuel consumption:

$$B = P \cdot b$$

with fuel consumption B (g/h), enginepower P (kW), and specific fuel consumption b (g/kWh).

—ABSTRACT

Different techniques to reduce the aerodynamic drag of cars have been utilized in the past. Although the best results can be achieved with streamlined bodies, these have not generally been acceptable to the buying public. In order to make use of the potential of aerodynamics nevertheless, a procedure called detail optimization has been developed by Volkswagenwerk AG. Representative results achieved with this technique are reported. The step by step drag reduction of several passenger cars is illustrated. The consequent reduction in fuel consumption has been investigated for steady state driving as well as for standarized cycles.

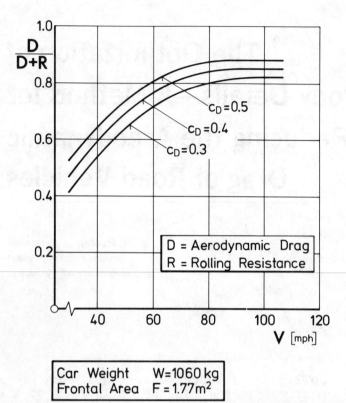

Fig.1 -Ratio of aerodynamic drag to total external Drag (= aerodynamic drag + rolling resistance) of a passenger car

Car Weight   W=1060 kg
Frontal Area F=1.77 m²

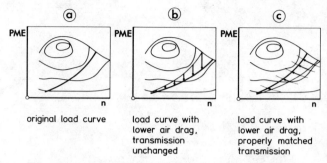

Figs.2A,B&C -Air drag reduction and matching of the transmission to verify minimum fuel consumption (schematic)

This specific fuel consumption can be taken from the engine's SFC-map. The mean effective pressure PME (bar) is obtained from power requirement P, displacement H (liter), and engine speed n (1/min):

$$PME = \frac{P \cdot 1224}{H \cdot n}$$

The engine speed for a given road speed V_F (mph) is derived from the transmission drive ratio i_4 and the dynamic circumference U_{dyn} (meter) of the drive wheels:

$$n = \frac{V_F \cdot 37.5 \cdot i_4}{U_{dyn}}$$

PME and n represent the momentary load point in the SFC-map. A series of these load points at various speeds yields the specific load curve for the vehicle (Fig. 2a).

If the drag coefficient of a vehicle is lowered without alteration of the transmission ratio, the dynamic circumference of the drive wheels and the engine data, the load curve is shifted towards a lower mean effective pressure at the same engine speed. The new load point normally is associated with higher specific fuel consumption (Fig. 2b).

If the new load curve is extended to the intersection with the engine's full power curve, we obtain a new maximum speed point. If the reduction in air drag is sufficient, therefore, the maximum permissible engine speed will be exceeded.

In view of the reduced power requirement, a drop in fuel consumption is achieved, but the effect of this is diminished by the increase in specific fuel consumption.

In order to make use of the full potential of a reduction of air drag, the transmission ratio must be adjusted, as shown in Fig. 2c. The load points will move along hyperbolas of constant power into a region of lower specific fuel consumption.

By means of such a suitable transmission ratio adjustment the new load curve can be brought more or less into alignment with that of the vehicle in its original state, though in this case the same engine speed will be equivalent to higher road speed.

Fig. 3 shows such a modification process in a real engine map of a 55 kW engine. The initial drag coefficient is $c_{D0}=0.5$. The load curves for the reduced drag coefficients $c_{D1} = 0.4$ and $c_{D2} = 0.3$ without gear ratio modification are shown by a broken and a dotted line respectively.

Point ① represents the common top speed load point for the initial drag coefficient and also for the reduced drag coefficients after gear ratio modification, the corresponding load curves lie within

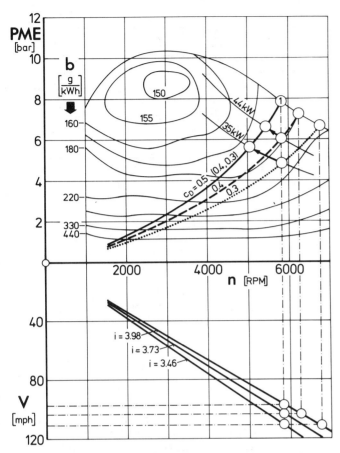

Fig.3 - Road load curves for decreased air drag plotted in the engine map; effect of proper transmission matching on specific fuel consumption

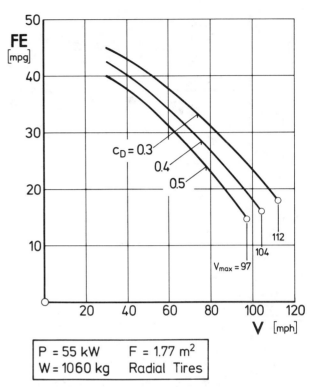

Fig.4 - Effect of air drag reduction on steady state fuel economy; SFC map see Fig. 3

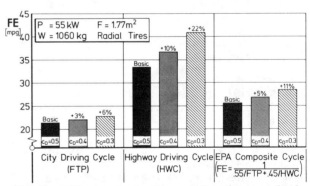

Fig.5 - Effect of air drag reduction on fuel economy in standardized driving cycles

the line width of the initial load curve.

The increase in top speed for the reduced drag coefficients with and without gear ratio modification can be taken from the lower part of Fig.3. The gear ratio modifications do not influence the top speed because the full load curve of this engine under consideration corresponds approximately to the maximum power hyperbola in the required engine speed range.

In Fig.4 the fuel economy is plotted versus speed. The calculations have been made for steady state conditions. If the transmission is carefully matched, the gain in fuel economy is almost twice as high as with the gear ratio unchanged.

In order to get a realistic picture of the advantage of air drag reduction from the average driver's point of view, the fuel economy has been calculated for three different standard cycles taking into account the transients. The results are shown in Fig. 5. As expected, the gain in fuel economy due to an air drag reduction is most striking for the highway cycle. It is still noteworthy for the composite cycle, which is said to be most representative of daily driving in the United States.

PAST AND PRESENT METHODS TO REDUCE THE VEHICLE'S DRAG

HISTORICAL DEVELOPMENT - The development of vehicle areodynamics, considered here only in conjunction with body styling, took place in three main stages, although these cannot be sharply distinguished one from another in terms of time. The major steps in this development process are

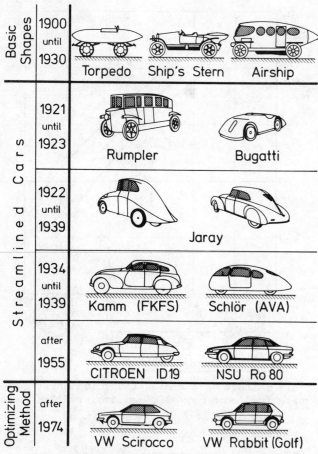

Fig.6 -Historical development of vehicle aerodynamics

illustrated in Fig. 6; the comments are restricted to consideration of passenger cars.

In the first phase, which might be considered as having started about the turn of the century, an attempt was made to adopt directly various streamlined shapes known in other engineering disciplines such as shipbuilding, sailing, or flying. However, these shapes did not succeed in gaining a foothold in automobile engineering.

They were quite unsuitable for road travel as for instance the "airship" outline, or else ineffective as for example the ship's stern. Road speeds which could be attained in those days on the poor roads and with the low engine power available were in any case low, so that air drag had little effect except on speed record attempts and during races.

The second phase was characterized by efforts to apply knowledge of fluid mechanics mostly obtained from aircraft aerodynamics systematically to the automobile and, as development progressed, to adapt such outlines effectively to the practice of automobile engineering. The aim was to build a streamlined car. Aircraft designers in fact produced a whole series of vehicles shapes, but scarcely any of them proved acceptable for road vehicle use, mainly on account of excessive length and a poor relationship between available internal space and overall dimensions. It was not until the work of W.E. Lay, E. Everling, and W. Kamm was published that the facts permitting acceptable road vehicle aerodynamics to be achieved became available for mass-production vehicles. The main result of the work referred to was what we know today as the "Kamm tail". This successfully fulfilled the requirement for a streamlined automobile without an excessively long, flowing tail.

The third phase is still in progress and differs in its approach from the proceding two. No further attempts are being made to develop streamlined shapes and to adapt them subsequently to the demands of automobile engineering. Instead, numerous details of the vehicles overall shape are optimized in order to obtain good airflow characteristics without sacrificing the features of a particular styling. This optimization process is described in detail in the present paper. It has been able to book some notable success, although its limitations can now be discerned more and more.

Fig. 7 represents an attempt to illustrate the trend in the development of automobile air drag. Reliable data are available only for vehicles of fairly recent date. In (1)* a statistical evaluation of European mass-produced vehicles has been undertaken, but this is not broken down into years of manufacture. Similarly comprehensive data are required if one is to be able to reconstruct an accurate picture of the development of drag coefficient. However, such data are not available. Therefore reference is made to result published by MIRA (2) for some older vehicle shapes, and certain older measurements made on scaled models (3) were also taken into account.

---

* Numbers in parentheses designate References at end of paper.

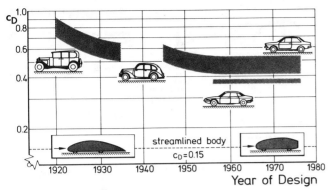

Fig.7 -Progress in drag reduction versus time elapsed

The drop in drag coefficient from $c_D = 0.8$ for cars of the Twenties to an average value of $c_D = 0.46$ for the modern European automobile has occured in two stages. In the first stage, between the two world wars, cars became longer, lower, and smoother in detail while still retaining such dinstinguishing features as free-standing fenders and headlights. In addition to a low drag coefficient of approx. $c_D = 0.55$, frontal area also dropped so that all in all a considerable reduction in air drag was achieved.

The second stage in the reduction in air drag was reached with the introduction of cars with a full-integrated body with its notch-back, fastback, and squareback alternatives. Incorporation of the fenders and headlights into an all-enveloping body shape greatly improved the airflow round the vehicle. Dependent on detail design, drag coefficients of $0.4 < c_D < 0.5$ were now attained. This range of values has not been altered since about 1960. Admittedly, special body shapes have been built to achieve lower c_D values, but for the bulk of European automobiles in the past 15 years no tendency of a systematic reduction in drag coefficients can be detected.

In Fig. 7 the drag coefficient $c_D = 0.15$ is given particular prominence; this value was achieved as early as 1922 by W. Klemperer (3) on a model as shown in the left sketch at the bottom of Fig.7. Other shapes, for example that drawn on the right in Fig. 7, have been used by Volkswagenwerk recently to obtain equally good c_D values. This means that $c_D = 0.15$ can be regarded as the probable optimum value for the automobile even if this figure is not based on any form of natural law. The gap between this value and those achieved by modern mass-produced automobiles is considerable.

THE OPTIMIZING METHOD OF VOLKSWAGENWERK
The optimizing method employed by VW is based on the postulate that the styling concept of a vehicle must be accepted as it stands. Aerodynamic improvements can only be attempted in the form of detail changes. These must be executed in such a way as not to change the vehicle's appearance. An example for this procedure, taking into account the peripheral conditions referred to, is shown in Fig. 8, which summarize the main results of shape optimization on a medium-sized automobile. The left part of the figure shows the shape changes which proved necessary in order to achieve the reductions in drag shown in the diagram on the right. The form modification proposals shown here constitute the optimum values which can be achieved, if we define as an optimum form modification one which produces the full aerodynamic effect with the minimum deviation from the original shape. In order to arrive at these forms it is necessary to vary each individual

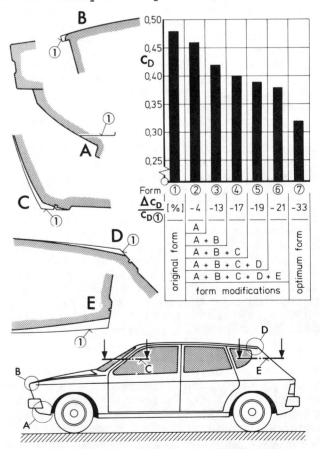

Fig.8 -Step by step drag reduction; an example for the optimization method detail of the form systematically within reasonable limits. A series of examples of how this is done in practice make up the main section of this paper (compare also (4), (5)). From Fig. 8 it will be clear that the range of results which can be obtained through aerodynamic from optimizing is relatively small. On the other hand, the illustrations show that full use can be made of the range, however small, if the detail work is performed with great care. The problems of wind tunnel technique associated with this optimization method are discussed in the final section of this paper.

LIMITATIONS OF THE OPTIMIZING METHOD - It is of course impossible to quote a precise value for the drag coefficient limit which could be achieved by the use of this optimizing method. The value depends to a large extent on the original styling concept. A series of optimizations performed on medium-sized automobiles with non-complex styling has shown that the lower limit for these vehicles lies at approximately $c_D = 0.40$. If it is desired to make use of the potential illustrated in Fig. 7 and to proceed to a further reduction in drag coefficient, the optimizing method will have to be abandoned. Instead, specific low-drag configurations will have to be developed and will be bound to involve a different styling approach. Since the majority of current massproduced automobiles can claim only a drag coefficient far worse than the limit value of $c_D = 0.40$ as achieved by optimizing, it would be appropriate at this time to focus on the potential of the optimizing method.

EXPERIMENTAL PROCEDURE - All the test results quoted in this report were obtained in the Volkswagenwerk's climatic wind tunnel. A drawing of the wind tunnel with its principal dimensions is given in Fig. 9. A detailed description of the wind tunnel including its measuring systems is given in (6); only a few significant features and data are therefore repeated here.

The wind tunnel is of the Göttingen type with closed return duct and open test section. The airstream is restricted at the bottom by the ground plane, which is flush with the nozzle, but has free airstream boundaries at the sides and the top. The nozzles has a contraction ratio of 4:1, with a 7.5 m wide and 5 m high outlet. The

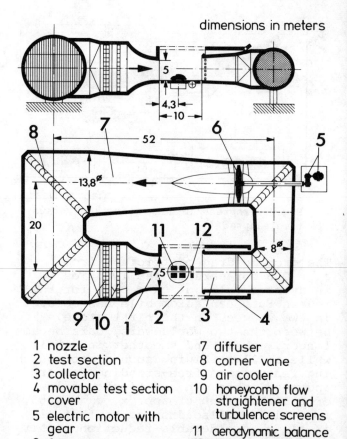

1 nozzle
2 test section
3 collector
4 movable test section cover
5 electric motor with gear
6 fan
7 diffuser
8 corner vane
9 air cooler
10 honeycomb flow straightener and turbulence screens
11 aerodynamic balance
12 roll dynamometer

Fig.9 - Volkswagenwerk climatic wind tunnel

length of the test section is 10 m. A maximum wind speed of 175 km/h (110 mph) is attained. This covers the speed ranges of the majority of standard production automobiles. Wind speed is measured by means of the settling-chamber pressure, being calculated from this with the aid of the nozzle factor. For speeds above 100 km/h, this nozzle factor remains constant. For passenger cars the blockage of the wind tunnel cross section is 5 %. No blockage correction has been applied. The influence of the ground plane boundary layer on the test results can be neglected, as demonstrated in (1).

Forces were measured with a mechanical six component balance, compensated electromechanically with the aid of moving weights. The balance can be rotated through ± 90° around its vertical axis so that cross winds can be simulated. This distance between the balance pivot axis and the nozzle exit is 4.3 m. The pressure gradient in the test section at the location of the vehicle, measured in the empty test section, is equal to zero. The vehicle to be tested rests on four support pads which can be positioned to suit its track and wheelbase. The frictional resistance of these support pads is small enough to be neglected. Their lift can be taken into account by measuring the pressure distribution and deducting their lift from the lift of the vehicle/pad system.

The forces and moments, the reference values, and the coordinate system are shown in Fig. 10. Other items are explained in the text or illustrations as necessary. The measurements listed here were all obtained from full scale models and from actual vehicles. All full scale wind tunnel models possessed a satisfactory imitation of the actual

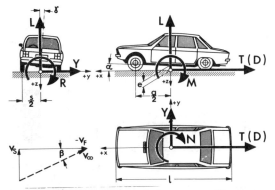

Fig.10 - Definition of forces and moments acting on a vehicle

vehicle's outer bodywork including window recesses, panel and door joint lines, exterior equipment and fittings, and underfloor detailing, as well as the correct cooling air flow pattern.

During force measurements, the vehicles' suspensions were not locked, and the vehicles were able to adopt any attidude corresponding to the lift and pitching moment to which they were exposed as a result of the airstream. This procedure differs from the "classic" wind tunnel technique, which uses rigid models, but has the advantage of corresponding more closely to actual driving practice.

The reference area was taken as the projection of the vehicle along the longitudinal axis, using the shadow process. As shown in (4), the uncertainty in the c_D value is approx. $\Delta c_D = \pm 0.002$ to 0.003, assuming for the drag coefficient and frontal area the following mean values: $c_D = 0.45$; $F = 2\ m^2$.

It is usual to quote only one specific c_D value for a vehicle. However, any automobile in fact possesses a more or less broad c_D range. The parameters influencing the drag are devided in Fig. 11 into those of shape parameters at this stage. The influences of vehicle position (angle of attack, ground clearance) and of cooling airflow have been discussed in detail in (4).

The standard condition for vehicle position was chosen as that obtained at half payload, thus yielding the same test conditions as used for determination of fuel economy and top speed by the German DIN standard test method. All the Information supplied here concerning air drag applies to the condition "with engine cooling air flow". When models were constructed, special care was taken to achieve a correct simulation of the acutal vehicles cooling airflow.

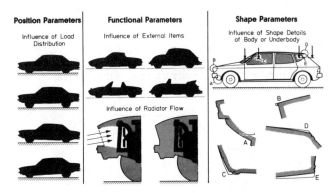

Fig.11 - The three different kinds of parameters influencing the aerodynamic drag

THE OPTIMIZING METHOD

FRONT END DESIGN - The influence of front end shape on air drag has already been the subject of many investigations, for instance (7). However, the margin open to the aerodynamicist is normally far smaller than stated here.

In order to establish what degree of drag reduction can be achieved as a maximum by means of optimum front end design, the airflow around the bow is initially improved by an add-on nose section produced in accordance with pure aerodynamic principles with no consideration given to questions of appearance. It is best to produce this add-on nose section in separate sections, so that the influence of the upper hood shape can be distinguished from the transition to the fender. The drag coefficient obtained with the aid of this auxiliary nose represents the limit value obtainable from the front end dimensions for the given vehicle; this front end form itself is called the optimum one. In Fig. 12 we see an optimum front end of this type. The solid line contour, which splines into the original form in the nose cut-off line region, yields a reduction in drag coefficient of $\Delta c_D = 12\%$. If we depart from the original body, as shown in Fig. 12, it is possible to find forms with the same principal dimensions which yield even better drag coefficient reductions than the so-called "optimum" form. In other words, the term "optimum" is not entirely correct in this context, but considered in terms of actual optimizing practice the discrepancy between this value and the true "optimum" is of minor significance. Working with the chosen "optimum front end" has the additional advantage that other details of the vehicle, for instance the A-pillar or the rear end contour, can be investigated before a satisfactory solution has been found in engineering terms for the front end of the vehicle.

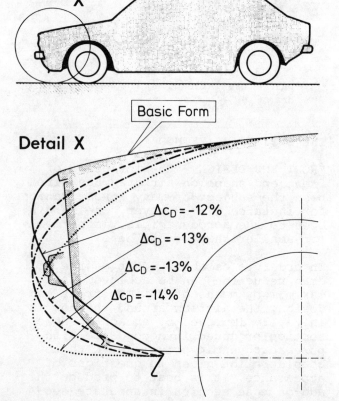

Fig.12 -Low drag front end configurations, leading to the "optimum nose"

Spoiler or air dam development must be carried out together with front end optimizing. The same spoiler arrange-

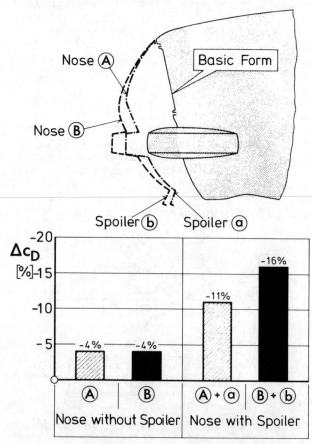

Fig.13 -Matching of the front end shape and the spoiler

ment was arrived at for all the forms shown in Fig. 12. Fig. 13 shows however that the effectiveness of a spoiler can differ greatly when associated with different front end outlines. In this example the add-on nose versions were attached ahead of the basic form, so that the vehicle was increased in length. However, in a later version the optimized nose section can be pushed rearwards parallel to the body axis without influencing the c_D, thus restoring the orginal vehicle length. When noses Ⓐ and Ⓑ were used without spoiler, an identical reduction in drag of 4 % was obtained; attachment of an optimized front end spoiler, however, yielded an 11 % drop in c_D for nose form Ⓐ + ⓐ, but 16 % for form Ⓑ + ⓑ.

This shows that the effect of the front end spoiler is not merely based on a shielding effect involving underbody airflow and thus a reduction in local wind speed along the underside of the vehicle; instead, on certain vehicles there is also an influence consisting of a change in the distribution of over-hood flow and flow past the sides of the vehicles. In this example, the front end of a fastback sedan was optimized. On account of the complex rear end flow pattern, which is strongly three dimensional in nature, this type of tail reacts particularly sensitive to any change in the flow arriving at the tail end of the vehicle.

When optimizing front end spoilers in the wind tunnel, the quality of the airstream close to the ground and the ground plane boundary layer thickness are of particular importance. The displacement thickness of the ground plane boundary layer in the VW climatic wind tunnel is only some 10 % of an automobile's ground clearance, so that even spoiler tests can be carried out unrestictedly, as comparative road tests have shown (1).

The effect of front spoiler on drag and lift for a coupé is shown in Fig. 14.

Tests have been carried out with inclined and vertical spoiler, the latter mounted at three different positions on the front end of the car. Regarding the drag, no definite relationship was detected, whereas spoilers placed in a forward position gave a considerable decrease in lift, corresponding to the length of the spoiler. Spoiler B40 yielded the lowest air drag, the reduction in drag coefficient c_D being 3 % based on the car without spoiler; the front axle lift c_{LF} was reduced by 21 %. For the possibility of drag reduction, the application of a short spoiler (here pattern B40) was recommended as a simple method of obtaining reduced drag. Up to 15 % drag reduction has been achieved, depending on the car under consideration.

The influence of leading edges on drag is becoming more important, since on contemporary passenger cars the general outline of the vehicle front is approximately fixed by the position of the engine and the position of front bumpers, headlights, and traffic lamps to meet the demands of law.

Fig. 15 shows the optimization of the vehicles leading edges. By mounting the "optimumg nose", consisting to the two parts M1 and K1, a decrease in drag of $\Delta c_D = 0.05$ was achieved. Rounding off the vertical leading edges only by means of the ancillary front end section K1 decreased the drag coefficient by $\Delta c_D = 0.015$, i.e. only 30 % of the full drag reduction achieved with the complete optimum nose. The ratio of drag reduction of 70 % and 30 % of both well-rounded bonnet and vertical leading edges of the fenders is certainly not transferable in general. This ratio is influenced by the lateral curvature of the vehicle's front end, position and shape of the front bumper, flow pattern ahead of the radiator, angle of incidence, ground clearance of the vehicle, and inclination of the front bonnet.

The optimum nose was removed, and by a step by step rounding off of both horizontal (m) and vertical (k) edges, an attempt was made to meet the result obtained with the optimum nose. Finally, with the combination M3 and K3, a drag reduction of $\Delta c_D = 0.045$ was achieved. This figure, 90 % of maximum drag reduction, was achieved with a contour still acceptabel to the stylist.

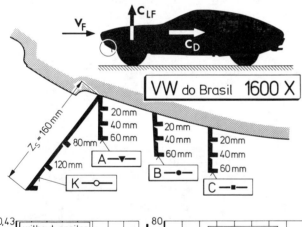

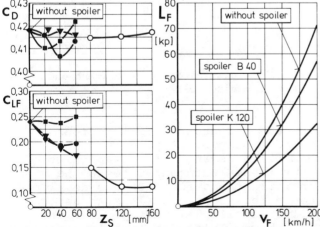

Fig.14 -Systematic front spoiler investigation during the optimization of a coupé

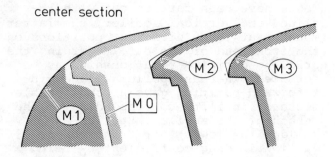

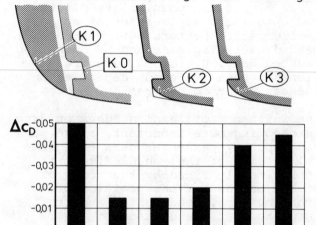

Fig.15 -Optimization of the leading edges of a small car

On vehicles with a short engine hood length, the front edge not only influences the drag but also the pressure at the area forward of windshield, which in turn governs fresh air entry for heating and ventilation (see Fig. 16). Version Ⓐ with fully-separated hood airflow - caused by the sharp-edged front end contour - yields a high drag coefficient of $c_D = 0.48$. Since the separated flow does not reattach on account of the short hood line, only a slight overpressure of $c_p = +0.10$ is obtained at the base of windshield. This pressure value c_p is defined as $cp = \Delta p/q_\infty$ where Δp is contour pressure related at atmosphere and q_∞ is dynamic airflow pressure. If contour Ⓐ were to be used, the low pressure at the inlet grille would mean that the fan would have to run continuously.

In the case of contour Ⓑ, on which the flow around the front edge was improved by an aerodynamically sound molding, the airflow separates from the surface but reattaches at the area of inlet grille. On this version the static pressure at the inlet grille increases to $c_p = +0.30$. The drag coefficient is reduced by the front edge molding slightly.

A rounded-off front edge as in contour Ⓒ yields a flow over the hood free from separation, associated with a drag 15 % lower than that of the sharp-edged front end from Ⓐ, namely $c_D = 0.41$. The pressure of the inlet grille in this case is $c_p = 0.40$, thus ensuring a high fresh air volume and permitting operation without running the blower.

Fig. 17 shows the front end ① for a mediumsized automobile. The drag was

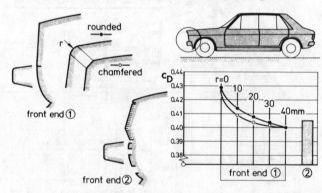

Fig.17 -Optimization of the front end of a medium sized car

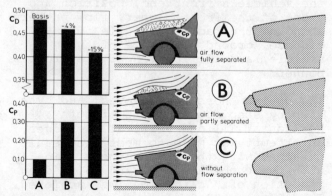

Fig.16 -Influence of front end shape on static pressure at the vent air inlet and on drag

reduced by rounding off the front end edge step by step. Sloping or chamfering is another suitable method. After these fundamental tests the final front end form ② was developed and in this case details such as the grille, cooling air entry, etc. are also shown. This nose form is notable for a slight front end inclination and moderate hood inset. The drag coefficient of this nose form ② is only 1 % above the optimum value obtained in the basic tests with nose form ① using a large radius nose with r = 40 mm.

Fig. 18 indicates possibilities of reducing the drag merely by means of form changes in the front edge region. Once again, the basic form was compared with the optimum add-on nose, form Ⓑ, which yielded a reduction in drag of 9 %. By bringing forward the grille molding and modifying the entire plane of the grille - as illustrated by form Ⓒ - an improvement of up to 5 % was achieved without changing the hood contour. The best results were obtained with a molding located 5 mm ahead of the hood contour. Bringing it forward further did not lead to any additional drop in drag. Since the form Ⓒ variants did not attain the maximum drag reduction revealed by measurements taken with add-on nose Ⓑ, the contour of the hood front edge was also changed, to form Ⓓ. The sketch shows the change in contour compared with the basic form. Despite this geometrical modification, the appearance of the front end was scarcely affected. This modified hood contour, in conjunction with the more projecting grille plane, yielded the highest improvement in drag, namely 9 %.

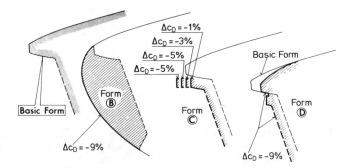

Fig.18 -Optimization of the hood's leading edge and of the position of grille and molding

A further example of optimizing the leading edge of the automobile's body is shown in Fig. 19. The basic from has neither hood molding nor front end spoiler. In form Ⓑ, the hood upper surface airflow has been improved by the addition of an aerodynamically sound molding to the front edge. However, this reduces the drag coefficient by no more than 2 %, although the modification at the same

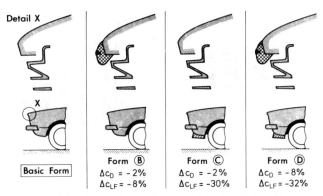

Fig.19 -Matching of an optimized leading edge and a spoiler

time reduces the front axle lift coefficient by 8 %. Form Ⓒ retains the original hood contour but uses an optimum pattern front spoiler. Again a 2 % reduction in drag is achieved, but front axle lift is cut by 30 %. The combination of leading-edge molding as in pattern Ⓑ and spoiler as in form Ⓒ is represented by variant Ⓓ. Here we can see that mere addition of the improvements obtained leads to an incorrect evaluation of the results. Both the individual measures yielded only a 2 % reduction in drag, so that added together a 4 % improvement might be expected; in fact, measurements taken on form Ⓓ showed in 8 % drop in drag. This means that the combination of individual drag reduction measures at the front end of the car must be treated as an entirely separate possible form and that combinations of various measures must be tested again.

DESIGN OF THE A-PILLAR - The influence of the design of the front roof support (usually called the "A-pillar") on aerodynamic drag is dependent to a large extent on position and curvature of the windshield and the design of the vehicle's front. Attention has to be paid to these parts of the body because of manufacturing requirements the obscuring of side windows by water and dirt, wind noise, and pressure conditions at the door gaps.

Fig. 20 shows the influence of A-pillar design on drag for a coupé. The original design ① with prominent rain gutter (drip molding) represents a convenient solution from the manufacturing point of view. Separation of the airflow immediately behind the A-pillar, however, is responsible for undue drag increase and loud wind noise.

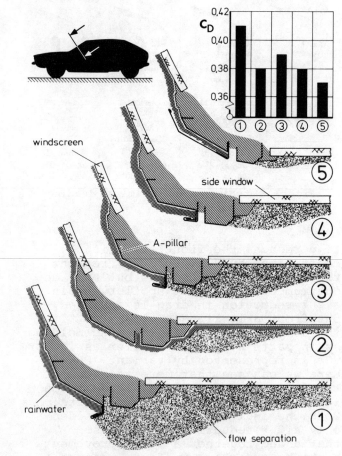

Fig.20 – Design of A-pillar to obtain low drag, low wind noise and proper flow of rain water

The flow of water, also illustrated on Fig. 20, is stopped by the rain gutter, thus preventing the side windows from becoming soiled (if the rain gutter is properly designed). If the lateral curvature of the windshield is small, the water will overflow the rain gutter and spread over the side windows. Under conditions of heavy flow separation, a fine water spray is created, distributing itself over the outside mirror and side windows. This will affect visibility – especially in the outside mirror – and endanger traffic safety.

The A-pillar ② without rain gutter yielded a drag reduction of 7 %. Wind noise was also reduced due to a considerable prevention of flow separation. A disadvantage of A-pillar ② is the fact that rainwater can flow over it undisturbed, thus soiling the side windows.

A-pillar design ③ shows a rain gutter with its flange mounted partially flush to the outer body sheet metal of the pillar. This design, in comparison to form ①, gave a drag reduction of 5 % and a 3 % incrase compared with form ②, which has no rain gutter. Soiling of side windows did not occur with design ③.

Design ④ shows a rain gutter fully incorporated into the A-pillar, thus giving the same drag coefficient as form ②, with no rain gutter. The latter design has the advantage of presenting less difficulty in manufacture and keeps the side windows free of water. No difference in wind noise was found between form ② without a rain gutter and design ④ with the flange incorporated.

A-Pillar ⑤ is characterized by a long watercatching "pocket" and a side window nearly flush with the outer body. The drag coefficient is 3 % less than on design ②, which has a prominent flange. As the side windows are nearly flush with the outer body, the airflow around the A-pillar is nearly undisturbed, resulting in low wind noise.

The extent to which the drag is affected be even quite small changes in front at the A-pillar is shown by the examples in Fig. 21. Basic design Ⓐ does not incorporate any measures

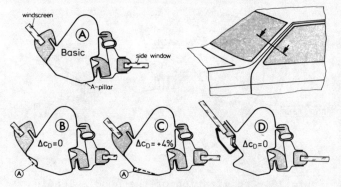

Fig.21 – Modifications on the A-pillar and their influence on drag

to keep the side windows clean. The pillar is aerodynamically good, with the airflow over it almost free from separation effects. In version Ⓑ, a contour change was attempted on the outer face of the A-pillar, in order to prevent rainwater from dropping into the interior of the car when stopped with a door open. On this version of the A-pillar Ⓑ, the water flows from the roof gutter into the recess in the pillar contour and downwards. The recess does not adversely affect the drag coefficient.

In the case of form modification Ⓒ, the same water drain effect at a standstill has been achieved as on version Ⓑ. The drain lip here, however, is produced by an outward extension of the pillar contour, as shown in the sketch in comparison with original contour Ⓐ. This leads to airflow separation at the pillar, which leads to an increase in drag of 4 %.

Whereas pillars Ⓑ and Ⓒ only ensure reliable shedding of rainwater when the car is at a standstill, version Ⓓ is also capable of keeping the side window clear when on the move. The water flowing over the windshield is collected in the drain gutter and is conducted down and also upwards into the roof rain gutter. In this design where the window is fixed by "gluing", the drain channel is formed by the pillar contour and by the shaped trim surrounding the windshield. In order to ensure that this integrated rain channel operates reliable, it is necessary to optimize it for each new vehicle, since the styling of the windshield, its inclination, and the contour of the A-pillar influence the shape of the channel. The drain channel shown as Ⓓ keeps the side windows entirely clear of water and dirt at all speeds, even if a side wind is blowing, and it does not lead to any increase in drag coefficient.

REAR END DESIGN - Whereas in the case of front end design and the results can be transferred with only minor limitations to other vehicles of differnt types, provided that the principal dimensions are similar, any such transfer of optimization results a at the rear end of the car is scarcely possible. The flow pattern at the rear of the vehicle is determined by the flow regime of the front and the windshield and also by the overall dimensions. Furthermore, the three-dimensional nature of the rear end airflow is influenced by details of rear end design.

As an example of rear end optimizing on a notchback car, Fig. 22 shows the influence of the height of the trunk lid on the drag coefficient c_D. The graph shows that a slight lowering (z = -50 mm) or raising to a height of z = 100 mm does not bring any change in drag.

With rear end elevations between 100 and 150mm, the drag coefficient drops from the basic shape with c_D = 0.40 by 8 %, to c_D = 0.37. The contour of this form Ⓐ with 150 mm lid elevation can be seen in the sketch.

Still further elevation of the rear end brings us into the realm of the station wagon, as shown by the contour referred to here as form Ⓑ. This yielded a drag coefficient of c_D = 0.38, in other words 3 % above the optimized notchback form Ⓐ but 5 % below the original basic form.

The test on the influence of trunk lid elevation were conducted without changes to the side panels, so that airflow onto the tail remained unchanged. The influence of narrowing the rear and side panels is shown for the same vehicle as before in Fig. 23. The rear end side panel contraction y was investigated in stages, starting with straight side panels as the basic form. The trunk lid height remained unchanged as the side walls were made narrower. In the region up to y=50 mm (form Ⓐ) a continuous drop in drag was established until the improvement amounted to 5 %, whereas the specific contraction range between 75 and 125 mm brought with it a sudden drop

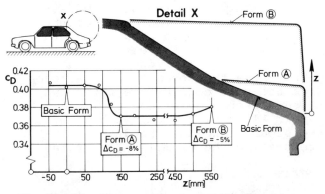

Fig.22 - Air drag versus rear end elevation z

from c_D = 0.43 to c_D = 0.37, a total of 13 % (form Ⓑ). As the side panel was narrwoed down to y = 200 mm on each side of the vehicle (form Ⓒ), the drag remained unchanged.

The following remarks deal with the influence of the rear end inclination angle of fastback and squareback vehicles. A fastback rear end is de-

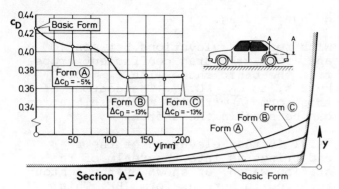

Fig.23 - Air drag versus side panel contraction y

fined as one one which the separation line is located at the base of the sloping rear panel. The rear window, which forms part of the sloping panel, is in a region of attached airflow, and thus remains free form dirt. The square end body, on the other hand possesses a separation at roof level. The entire rear end including the rear window lies in the separated airstream so that dirt is deposited on the glass in the car's wake more or less, depending on the body's rear overhang and the flow pattern in the wake.

In Fig. 24 the influence of the rear end angle of inclination on the drag and the location of separation can be seen. On vehicles with a steep angle tail panel, for instance station wagons with $\varphi > 35°$, the point of separation is at the rear edge of the roof. In the example shown, the drag coefficient is relatively low at $c_D = 0.40$. If φ is reduced, at a given value the separation line moves from the rear edge of the roof down to the lower edge of the inclined rear panel. Along with the downshift of the separation line the drag is increased, in the example illustrated, the increase is of 10 %, to $c_D = 0.44$. The higher drag is attributable to strong trailing vortices with a corresponding rise in lift and thus induced drag.

The transition from square end tail to fastback does not take place suddenly at a specific inclination angle limit, but in a transitional zone shown as a shaded area on the graph. In this transitional zone the separation line oscillates between top and bottom, the degree of fluctuation depending on the edge pattern and the speed. If the angle φ is still further reduced, the drag again drops. At a fastback inclination angle of $\varphi = 23°$ the same drag coefficient of $c_D = 0.40$ is obtained as for the squareback flow pattern.

This angle $\varphi = 23°$ represents the approximate limit of what is accpetable for a sedan, allowing for a reasonable angle of rearward vision. Smaller angles of approx. 15°, have been applied on coupés, and these achieve c_D values up to 15 % lower than those obtained with a squareback.

The relationship between drag and rear end inclination referred to here applies to medium-sized automobiles with an attached upstream flow field. If the flow pattern at the front section of the car is unfavourable, the limit angle is reduced from 30° to approx. 25°; if the front has an excellent flow pattern and no separation occurs at the windshield, the limit angle can be 35°.

The influence of the rear spoiler on drag and rear axle lift for a coupé is shown in Fig. 25. The separation line for all shapes shown if located

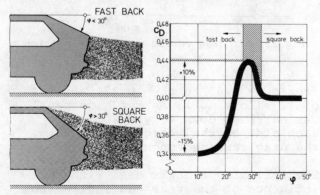

Fig.24 - Fastback and squareback flow regime and the related air drag

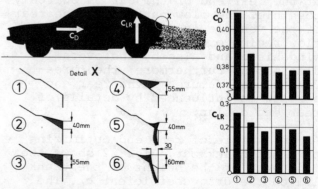

Fig.25 - Systematic rear end spoiler investigation during the optimization of a coupé

on the lower rear edge of the body. Modification ② to ⑥ yielded different improvements in drag and lift. Spoiler ⑤ finally was adopted for this car.

Small changes with the contour of the rear end of the roof as well as with the C-pillar may lead to a considerable drag reduction. Fig. 26 compiles some specific results achieved with a squareback car. The rear end of the roof was rounded off as shown in detail ⓓ, leading to a decrease in drag of 9 %. Building up the C-pillar as shown in section C-C yielded a lower drag too. But most surprising, none of the combinations of roof and side panel modifications led to a drag improvement of more than 9 %.

From Fig. 27 it can be seen, that the drag is not only influenced by the outer contour of the body but also by

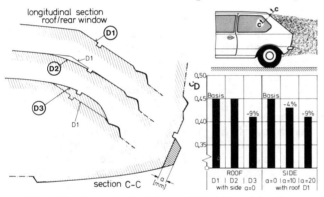

Fig.26 - Drag reduction with the aid of a rounded roof end and an elongated side panel

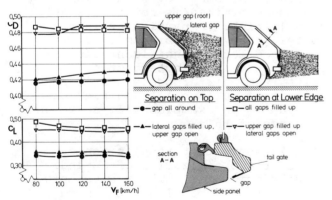

Fig.27 - Effect of sealing the tail gate to body gap on wake pattern, on drag and lift

small gaps of lids. The example shows the effect of the rear door lid gap on the wake and thus on drag.

If the gaps between the rear door and the roof and C-Pillar are open, a low drag coefficient of $c_D = 0.42$ is obtained, together with an airflow separation line at the rear end of the roof. If the two side gaps are sealed there is no change in drag nor in the wake pattern. However, if the roof gap or both roof and side gaps are sealed, the separation line is shifted down to the lower edge of the rear door. The strong trailing vortices associated with this conditions cause high induced drag, which shows up as a 14 % higher drag of $c_D = 0.48$. At the same time, the lift coefficient c_L rises by some 28 %.

Fig. 28 shows the streamlines past the VW SCIROCCO, which was optimized by the method just described, It is clearly seen that there is no separation of the airflow over the entire front part of the car. The deflecting action of the integrated front spoiler is obvious. The point of separation at the rear is not clearly visible. By filling up the separated flow of the wake with smoke, it can be shown that the flow separates at the rear spoiler. The picture was taken with the aid of multiple exposures of a single smoke trail, the height of which was varied in increments.

EXPERIMENTAL PROBLEMS RAISED BY THE OPTIMIZATION METHOD

INTRODUCTORY REMARKS - In this final section the specific demands imposed by the optimizing process on the experimental procedure will be discussed. It is well known that simulating the airflow past a vehicle in a wind tunnel is fraught with a series of shortcomings. The requirements for complete similarity between regular road driving and the wind tunnel test is necessarily disregarded to some extent. This introduces a considerable risk into the adoption of wind tunnel test results for road vehicle use. The wind tunnel correction factors developed in aircraft aerodynamics are not suitable as they stand for application to automobile aerodynamics. These corrections are integral by nature, in that factors such as free stream air speed or angle of incidence can be corrected. In this way the boundary conditions not in accord with actual practice, for example, the finite character of the airstream dimensions can be compensated for. However, the

Fig.28 -Streamlines past the VW SCIROCCO in the Volkswagen full scale wind tunnel

influence of these modified boundary conditions on local airflow zones cannot be dealt with. Unfortunately, the optimizing process is based on just this circumstance, namely that in specific areas use is made of marginal effects. It is impossible to state with certainty that the same effects will persist if the boundary conditions themselves change.

TYPE AND SIZE OF WIND TUNNEL - As we have just implied, the flow pattern past a vehicle in the wind tunnel differs from that on the open road, even if the test section in the tunnel had no ground plane boundary layer. The flow around the vehicle is influenced all the more by the boundary conditions at the airstream's boundaries the closer these boundaries are to the vehicle. In closed test section the streamlines around the vehicle will be pressed together slightly. In an open test section they will diverge more or less. In the first condition, the apparent undisturbed speed will be somewhat lower than the present speed. This deviation can be estimated by means of blockage corrections; for details, see f.e. (8). The main factors in the correction formula are the blockage ratio and the drag of the body in the flow field. However, these corrections can only be applied when they themselves are small in magnitude, in other words at low blockage and low drag.

Very often the demand for low blockage can not be met. Some 5 % of blockage may be regarded as acceptable. In some cases aerodynamic tests are done with a blockage ratio of 20 % or even more. Regardless whether the test section is open or closed, it is doubtful that results achieved under such circumstances can be reproduced on the road.

With the aid of calibration models, empirical corrections can be elaborated for small wind tunnels by comparison with large facilities. Since these corrections are of integral nature, as mentioned above, the risk in transferring results achieved with the optimizing technique remains considerable.

SCALE OF MODELS - Drag experiments and other aerodynamic tests are very often performed on reduced-scale models. Scales of 1:5, 1:4 or 3:8 are used for the purpose. Using models of this size, in the majority of small scale wind tunnels an acceptable blockage ratio can be achieved, and many problems of vehicle aerodynamics can be dealt with successfully.

However, the accuracy with which drag can be determined is not always sufficiently high. In particular when applying the optimizing method described here, significant discrepancies have been noted between results from quarter scale models and those from full scale vehicles or mock-ups. Two examples will serve as evidence of this:

Fig. 29 shows test results achieved during the optimization of an individual form detail. The radius between the front body panel and the side section of a van was increased from its starting point as a sharp corner until further enlargement yielded no additional improvement in drag. The optimum corner radius obtained on the quarter scale model of the van Ⓐ proved to be r/b = 0.06. The same graph, however shows the results ob-

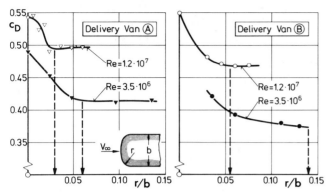

Fig.29 -Influence of Reynolds number on the "optimum" radius of the vertical leading edge, measured on two different vans

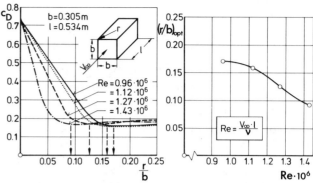

Fig.30 -Influence of Reynolds number on the "optimum" front end radius of a parallele piped, after F.W. Pawlowski

tained using the same method on a full scale vehicle body. Here, the optimum corner radius for the large version with r/b = 0.03 proved to be just half the value obtained from testing the reduced-scale model. Very similar results were obtained when optimizing the same radius on van B, also shown in Fig. 29. Obviously this is due to a Reynolds number effect. The aim of drag reduction using the optimizing method described in this paper is to avoid airflow separation locally. The separation of the turbulent boundary layer, as is well known (9), is influenced by the Reynolds number. As the Reynolds number is increased the separation point moves rearwards on a given body. If the Reynolds number is fairly large, the turbulent boundary layer can withstand a quite sever rise in pressure, that is to say it can flow around relatively sharp corners without separation. This explains why the large version can be made with relatively sharper corners. The optimized radius for the quarter-scale model is no longer the optimum for the full-scale version, and in the case of van B it was so large as to be rejected by the stylists. Subsequent measurements on the full-scale version yielded the results shown in Fig. 29. The correct optimum radius could in this event be realized on the vehicle.

A result published by F.W.Pawlowski (10) as long ago as 1930, and reproduced as Fig. 30, confirms our own conclusions. Similar Reynolds number effects have been observed on other body details. We can therefore state that reliable quantitative results - the only ones which are of use in practice -can only be secured by working with full-scale models.

Often the proposal has been made to increase the air speed in small scale wind tunnels, Thus fulfilling the demand of the Reynolds similarity criterion. However, Fig. 31 illustrates that only a small degree of room to maneuver is available in this respect. Road vehicles are fairly blunt bodies, high excess speed is attained at certain points on the contour. As calcultions and measurements on ellipsoids have shown (see (11), (12)), the critical Mach number - this is the Mach number of the undisturbed flow at which sonic speed is first reached somewhere on the body's contour - is extremely low. Even at a Mach number which is below the critical value there is an increase in drag with the Mach number on account of compressibility. This raise in drag related to the Mach number is likely to be far more difficult to calculate in the case of the complex forms represented by automobile bodies than the influence of the Reynolds number on separation. Accordingly, there are limits to the possibility of working at high air-

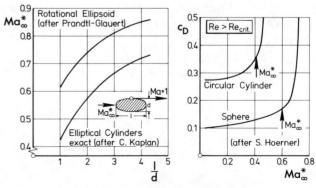

Fig.31 -Critical Mach number and drag coefficient for blunt bodies

flow speeds.

Other proposals aim to increase the effective Reynolds number for scale model testing by artificially raising the turbulence level of the wind tunnel. This procedure has been employed several times for aircraft aerodnamic testing using scale models. However, it still lacks quantitative data to correlate the turbulence level of a wind tunnel with a simulated Reynolds number for arbitrary geometrical shapes. Accordingly, this method cannot be recommended for use in automobile aerodynamics at the moment.

CONCLUSION

With the aid of numerical results it has been shown that the fuel economy of cars can be improved considerably by reducing the aerodynamic drag. This is valid not only for steady state driving under high speed but in realistic cycles as well. It has been pointed out that a large gap exists between the present state of aerodynamic quality of cars and the ultimate air drag. The optimization method, which is presented here in detail, serves to reduce this gap. This method, if applied systematically during the design procedure, can yield a remarkable improvement in the aerodynamic quality of cars. Finally it was mentioned that wind tunnel tests must be carried out with great care to warrant the results achieved with the optimization method.

ACKNOWLEDGEMENT - The authors wish to thank Volkswagenwerk AG for the permission to publish the paper.

References

(1) W.-H. Hucho, L.J. Janssen, G. Schwarz: The Wind Tunnel's Ground Plane Boundary Layer - Its Interference with the Flow Underneath Cars. SAE-Paper 750 066, Detroit 1975

(2) R.G.S. White: A Method of Estimating Automobile Drag. SAE-Paper 690 189, Detroit 1969

(3) W. Klemperer: Luftwiderstandsuntersuchungen an Automobilmodellen. Zeitschrift f. Flugtechnik und Motorluftschiffahrt 13 (1922), S. 201-206

(4) L.J. Janssen, W.-H. Hucho: The Effect of Various Parameters on the Aerodynamic Drag of Passenger Cars. Advances in Road Vehicle Aerodynamics, pp 223-253, Cranfield, UK, 1973

(5) L.J. Janssen, W.-H. Hucho: Aerodynamische Formoptimierung von VW GOLF und VW SCIROCCO. ATZ 77 (1975) S. 309-313

(6) W. Mörchen: The Climatic Wind Tunnel of Volkswagenwerk AG. SAE-Paper 680 120, Detroit 1968

(7) G.W. Carr: Aerodynamic Effects of Modifications to a Typical Car Model. MIRA-Report 1963/4

(8) R.C. Pankhurst, D.W. Holder: Wind Tunnel Technique. Sir Isaac Pitman, Sons, London, 1965

(9) H. Schlichting: Boundary Layer Theorie. 6th Edition, McGraw Hill, New York, 1968

(10) F.W. Pawlowski: Wind Resistance of Automobiles. SAE-Journ. 27 (1930), pp 5-14

(11) H. Schlichting, E. Truckenbrodt: Aerodynamik des Flugzeuges, Bd. 1, Springer-Verlag, Berlin 1962

(12) S. Hoerner: Fluid Dynamic Drag, Published by the Author, Midland Park, N.J. 1965

An Analytical Study of Transmission Modifications as Related to Vehicle Performance and Economy*

Howard E. Chana, William L. Fedewa, and John E. Mahoney
Engineering Staff General Motors Corp.

THE PURPOSE OF THIS PAPER is to show the fuel economy potential of various automotive powertrains. With todays increased emphasis on fuel economy and the rigid emission requirements for the vehicles produced by the automotive industry, new methods have to be used to help guide the selection of the future powertrains to provide the best overall compromise of size, weight, and cost for optimum performance and economy. Time does not allow for all the various powertrain combinations to be designed, fabricated, developed and evaluated with the actual vehicles in which they would be used. There is also difficulty encountered in measuring small changes in fuel-economy and performance with road or chassis dynamometer tests. These factors and the availability of a previously developed vehicle simulation program led to the set-up of an analytical process to evaluate the various combinations to indicate their potential in meeting these goals.

The transmissions used today are primarily of two types, manually shifted type and hydrodynamic drive automatics. They utilize a varied number of geared ratio steps and overall ratio spreads. To determine the maximum potential of these conventional transmissions, various parameters of these designs were evaluated to show any improvements that could be made to further their contribution to the overall performance and economy of a vehicle. For the manual transmission, the effect of the number of geared ratios and the overall ratio spread was examined to find the optimum combination for best vehicle performance and economy. In addition to these two parameters, the automatic transmission also considered the efficiency of the hydraulic torque converter.

After the potential of these optimized conventional drivelines had been determined, an evaluation of the continuously variable transmission was made and compared to determine any advantages that it might have. This comparison was made since several continuously variable transmissions are under development and there is need to establish the benefits of such a transmission.

The information presented here is directionally and qualitatively correct; however, their magnitudes are only representative and

*Paper 770418 presented at the International Automotive Engineering Congress and Exposition, Detroit, 1977.

— ABSTRACT —

A method of vehicle performance measurement has been developed so that selection of optimum fuel-economy-performance trade-offs can be made for a vehicle having various powertrain components. This method was utilized in an analytical study of drivetrain component features such as--overall ratio range, number of ratio steps, locked converters, continuously variable drives, etc. Both manual and automatic type transmissions are considered. Indications are that ratio range is an important consideration in the selection of transmission design parameters and also conventional transmission concepts can be competitive with the more exotic continuously variable type units.

subject to qualifications. The simulation uses only normal operating temperature engine data while the EPA car test requires a cold start. There also is no allowance for any engine adjustments that may be required to meet driveability criteria and emission standards. Since the drivetrain concepts force the engine to operate at different load-speed conditions, emissions will be affected.(1) Only a single set of vehicle parameters other than the powertrain was evaluated. Other vehicle parameter changes could alter the magnitudes of the differences shown in the study.

THE METHOD

The method used to precisely evaluate the effect of drivetrain variations on vehicle performance and economy is called the "Optimum Performance Versus Economy Line" Technique. Such lines are developed and used as follows.

The 0-60 MPH time (or other performance parameter) and composite EPA fuel economy (or other) of a specified vehicle can be computed, obtained by test, or obtained by a combination of both methods. The result is one discrete point on a plot of 0-60 MPH time versus EPA composite MPG as shown in Figure 1. Changing the final drive ratio of the vehicle produces another discrete point. Obtaining the points over a range of final drive ratios produces a curve which depicts the trade-off of performance and economy caused by final drive ratio changes. This is illustrated in Figure 2.

Repeating the above process on the specified vehicle for a range of engine sizes will produce a series of final drive performance-economy trade-off curves. Constructing a tangent line to this series of curves produces a curve which represents the best fuel economy at a given performance level that can be obtained from the specified vehicle through changes of engine size and final drive ratio. This is shown in Figure 3. Such a tangent curve is referred to as an "Optimum Performance vs. Economy" curve. It should be noted that it is not possible to specify an engine final drive combination that will produce a point to the right of the optimum curve. All combinations but the optimum will fall to the left.

Constructing such "Optimum" curves for driveline changes in the specified vehicle provides a means of precisely evaluating the effect of the changes. Figure 4 shows the "Optimum" curves for two different transmissions in the specified vehicle. Note the ease of determining the economy difference at a given performance level.

The "Optimum" lines technique overcomes several difficulties in evaluating drivetrain changes. Other methods often compare the economy gains at different performance levels for the changes. This makes assessments difficult. They may also compare at less than

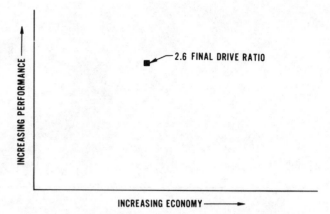

Fig. 1 - Performance versus fuel economy

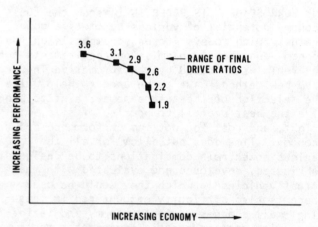

Fig. 2 - Effect of final drive ratio on performance and fuel economy

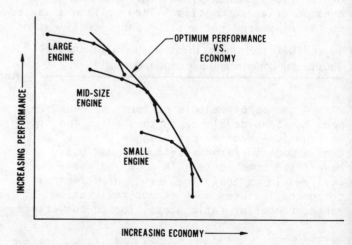

Fig. 3 - Construction of the optimum performance versus fuel economy line

optimum conditions or at different levels of optimization for the changes, both of which may produce less accurate conclusions.

The "Optimum Line" was used in the following analysis of powertrains.

COMPUTER PROGRAM USED FOR ANALYTICAL STUDY

The analytical tool used to evaluate the various parameters in this study was a computer program referred to as GPSIM (2) (A General Purpose Automotive Vehicle Performance and Economy Simulator). The GPSIM computer program is an automotive vehicle simulator used to evaluate the performance and economy of a vehicle by computing the operating conditions of the engine and driveline as the vehicle is directed through a prescribed operating schedule. The program will handle transient and steady state vehicle operating conditions. This provides the capability to simulate any performance criteria or to evaluate fuel economy for many driving schedules.

The data required by the program consists of data tables needed to describe the various components of the vehicle and powertrain and to provide adequate information for the simulated vehicle to operate normally. The data tables used to describe the engine, torque converter and accessory and spin losses are taken from test data measured at stabilized conditions. Information describing other driveline and vehicle parameters such as the transmission, shift schedule, driveline inertias, road load power requirement, etc., are taken from design information and test results. In addition to this data, information for the desired operation of the vehicle must be provided. This information could be a simple wide open throttle performance request or a more complex set of performance instructions to enable a vehicle to follow a desired economy driving schedule.

Data from various existing and experimental vehicles and powertrain components were used for this analytical study. The output data generated for each vehicle powertrain combination was the fuel economy for the EPA Urban and EPA Highway driving cycles and the 0-60 MPH wide open throttle performance time. Fuel consumption for these driving cycles were used to determine the EPA Composite (55%-45%) Fuel Economy. It must be remembered that exact simulation of the EPA Urban cycle cannot be made since only normal operating temperature engine data is used. To obtain the optimum values for fuel economy and performance for each of these combinations, an optimum shift schedule was used. This schedule chooses the best gear of a specific transmission to be used to satisfy a particular vehicle operating condition. For the wide open throttle condition, the ratio allowing the best wheel power is always the operating gear; for part throttle conditions, the ratio allowing the best fuel economy to meet a specific driving schedule requirement becomes the operating gear.

The flexibility and convenience provided by this program made it possible to determine the optimum performance vs. economy lines for the transmissions considered in this study.

TRANSMISSIONS INVESTIGATED AND SIMULATION ASSUMPTIONS

As mentioned earlier, both manual and automatic transmissions were studied. The basic manual was of 4-speed design while the basic automatic had 3-speed Simpson type gearing with a conventional hydraulic torque converter.

For the manual transmission comparison, units having wider ratio range and more speeds were investigated. Three 4-speed units and one 5-speed unit were included in the study and the basic schematic for the transmission is shown in Figure 5. All 4-speed transmissions used similar spin losses relative to input speed. These losses were determined from tests of a transmission having a similar capacity. A percentage loss was also added for each active gear mesh. When the 5-speed unit was investigated, its input shaft losses were adjusted in relation to the number of additional gear tooth meshes required to provide the additional geared

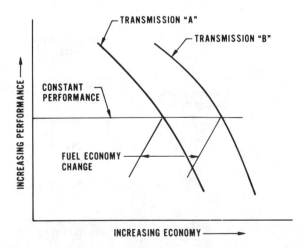

Fig. 4 - Comparison of two transmissions by the optimum performance versus fuel economy method

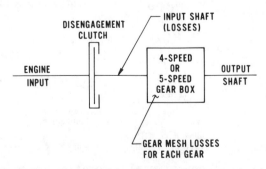

Fig. 5 - Conventional manual transmission schematic

ratio. All transmissions were sized to handle the same engine torque and vehicle weight.

The basic automatic transmission, shown in Figure 6, consisted of a unit designed to meet the same capacity requirements as the manual. The losses associated with the gear box portion of the transmission consisted of the front pump losses running at converter input speed and the gear box losses (including clutch losses) running at gear box input shaft speed. The losses applied to these characteristics were obtained from test data of units having a similar capacity. Once again, a percentage loss was added for each active gear mesh. The torque converter in front of the gear box was of conventional construction and sized to give the same stall speed for all engines used. The basic unit used an open converter; however, in several of the comparisons, lock-up clutch converters were also used. As ratio spread was increased, losses were assumed to stay the same relative to input shaft speed. A 4-speed automatic type transmission was also investigated. This transmission was of conventional construction utilizing a torque converter ahead of the gear box. The torque converter characteristics were similar to those in the 3-speed units. The gear box component changes were of such a nature that the spin losses could be assumed to be similar to that of the 3-speed automatic. Active gear mesh losses were included similar to the other automatic units.

The simulation used to represent a continuously variable transmission was an automatic unit having 12 equally stepped geared ratios. This model, Figure 7, included a torque converter with a lock-up clutch directly behind the engine and ahead of the simulated gear box. The gear box was assigned a constant efficiency for all operating conditions and spin losses and pump losses were assumed to be the same as the basic 3-speed automatic unit. No additional losses for increased pump capacity or pressure requirements were added. It was assumed that this unit would provide the same degree of automatic driving convenience that the conventional automatic transmission provides. To further represent the conditions of a continuously variable transmission, shift time and energy loss during the shifts were assumed to be zero. The converter was locked in all gears except its highest numerical ratio gear where it could be either locked or open. In this highest ratio gear the converter was selected to be locked or open and switched in mode at the appropriate instance to provide either maximum fuel economy or performance depending on the schedule in which it was operating.

Several assumptions were made for the computer simulation. Geared ratio shifts occurred at the optimum schedule to provide maximum fuel economy on the driving schedules and did not have to be in sequence. They were also selected to provide maximum acceleration for performance comparisons. Shift time for manuals was 0.5 seconds per up-shift with a zero energy transfer during the shift. Conventional automatics had a 0.6 second shift time; however, energy transfer during the shift was 80%. As mentioned previously the simulation of the continuously variable transmissions assumed a zero shift time with a 100% energy transfer during the shift.

The various manual transmission combinations evaluated in this study are tabulated in Figure 8 and the automatics and continuously variable units in Figure 9.

RESULTS

MANUAL TRANSMISSIONS — The manual transmissions were evaluated using the technique described in the previous sections. The four manuals simulated in this study have been described in Figure 8. The first manual investigated, labeled "A", which we will use as a baseline, was one having an overall ratio spread of 3.56:1. The "Optimum Performance vs. Economy" curve for this unit is shown in Figure 10. It must be remembered that for a given vehicle, the engine displacement must be adjusted in order to de-

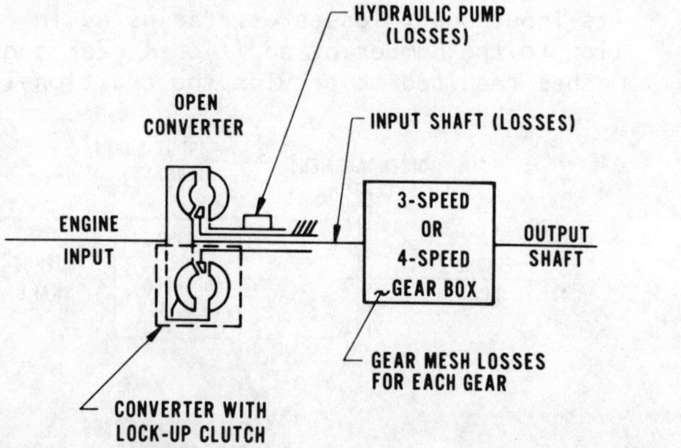

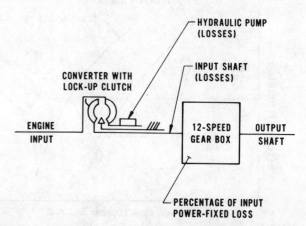

Fig. 6 - Conventional automatic transmission schematic with converter lock-up clutch option

Fig. 7 - Schematic for continuously variable transmission as used in simulations

velop the performance and fuel economy shown by the "Optimum Performance vs. Economy" curve.

Four-speed transmissions having overall ratio spreads of 4.72:1 and 5.94:1 were also investigated. These transmissions are identified as units B & C respectively. "Optimum Performance vs. Economy" curves were determined for these units and are shown in Figure 11. For the purpose of this discussion a performance level of 13.5 seconds to accelerate from 0-60 mph will be assumed as being the vehicle requirement. This requirement was set forth by the Office of Air Programs of the Environmental Protection Agency in 1972 for several powertrain investigations. At this level of performance, unit B will provide an improvement in fuel economy over the base of 3.8% while unit C provides a 4.4% increase.

To further study improvements that could be made in manual type transmissions, a 5-speed unit was also investigated. This unit is identified in Figure 8 as unit D and had an overall ratio range of 4.72:1--the same as 4-speed unit B. The "Optimum Performance vs. Economy" curve for this transmission is shown in Figure 12 and the improvement in fuel economy over the base manual at the required performance level is 4.1% or only 0.3% better than its 4-speed counterpart.

AUTOMATIC & CONTINOUSLY VARIABLE TRANSMISSIONS — The automatic and continuously variable drive transmissions were evaluated in a similar manner as the manuals. The base three-speed (unit E, Figure 9) was operated with an open converter as well as with a lock-up converter--the lock-up clutch being applied in 3rd gear only. The optimum lines for these transmissions are shown in Figure 13 with the improvement at the selected 13.5 second 0-60 MPH performance time being 6.7%. It is not the intent of this paper to imply that a lock-up clutch could be satisfactorily adapted to today's transmissions since the driveability and emission targets must also be met and this study does not take these factors into consideration.

A 4-speed transmission which is designated unit F having an overall ratio range of 4.22:1 was mathematically modeled and operated with a converter lock-up clutch applied in both 3rd and

AUTOMATIC TRANSMISSIONS
Geared Ratios

	3-SPEED	4-SPEED		C.V. SIMULATION	
OVERALL RATIO	2.84	4.22	6.00	4.22	6.00
UNIT IDENTIFICATION	E	F	G	H	I
GEAR NO.					
1	2.84	2.87	3.42	2.87	3.42
2	1.60	1.60	1.78	2.52	2.91
3	1.00	1.00	1.00	2.21	2.47
4		.68	.57	1.94	2.10
5				1.70	1.78
6				1.49	1.52
7				1.31	1.29
8				1.15	1.09
9				1.01	.93
10				.88	.79
11				.78	.67
12				.68	.57

Fig. 9 - Table of automatic transmission gear ratios

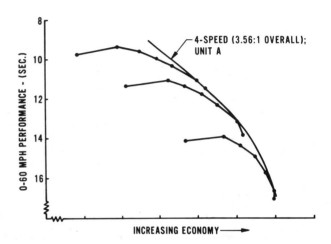

Fig. 10 - Optimum performance versus EPA composite (55%-45%) fuel economy for base manual transmission

MANUAL TRANSMISSIONS
Geared Ratios

OVERALL RATIO	4-SPEED			5-SPEED
	3.56	4.72	5.94	4.72
UNIT IDENTIFICATION	A	B	C	D
GEAR NO.				
1	3.45	3.45	3.45	3.45
2	1.94	1.94	1.79	1.98
3	1.37	1.03	.97	1.42
4	.97	.73	.58	1.02
5	-	-	-	.73

Fig. 8 - Table of manual transmission gear ratios

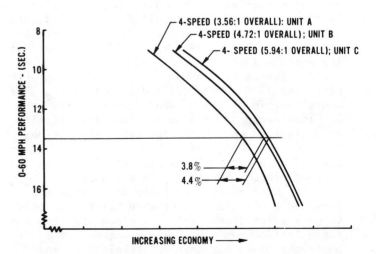

Fig. 11 - Effect of 4-speed manual transmission overall ratio on EPA composite (55%-45%) fuel economy

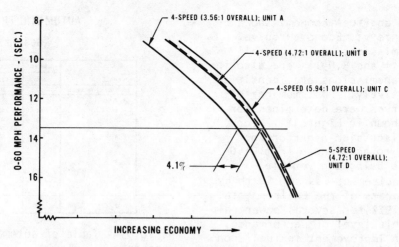

Fig. 12 - Effect of a 5-speed manual transmission on EPA composite (55%-45%) fuel economy

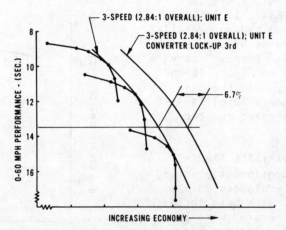

Fig. 13 - Improvement in EPA composite (55%-45%) fuel economy using a converter lock-up clutch in 3rd gear

4th gears. It was also operated with the lock-up clutch applied at the most beneficial time in first gear and fully applied in 2nd, 3rd, and 4th gears. The optimum lines for these units are shown in Figure 14. It should be noted that the economy improvement is 14.3% when the converter lock-up clutch was applied in 3rd and 4th gears and 18.0% when utilized in all gears. The overall ratio of the 4-speed type transmission was then expanded to 6.00:1. This transmission is designated unit G on Figure 9. The optimum line for this unit, also shown on Figure 14, indicated an economy improvement of 19.7% when the lock-up clutch was applied in all gears.

The continuously variable units used the 12-speed simulation described earlier. In order to give these transmissions maximum benefit, the converter lock-up clutch was engaged at the optimum time in 1st gear and remained engaged in the remaining 11 gears at all times. The geared ratio ranges investigated were the same as the 4-speed conventional automatics--that is, 4.22:1 and 6.00:1. These units are identified as H & I in Figure 9. The optimum lines generated by these units when the continuously variable portion of the transmission was assumed to be 95% efficient are shown in Figure 15. Improvement for the narrower ratio unit over the base 3-speed is 18.0% and the wider ratio unit 21.6%. The same data for the 6.00:1 overall ratio range transmission was generated for a unit having 90% efficiency in the continuously variable portion of the gear box and is also shown in Figure 15. Improvement here was only 15.7%.

In order to assure that the 12-speed gear box adequately simulated the continuously variable transmission, a transmission having 20 speeds with equal geared steps and a 6.00:1 overall ratio was developed. The optimum line for this unit indicated an error of only 0.16% for the 12-speed unit. It was, therefore, concluded that the 12-speed units resulted in a satisfactory simulation.

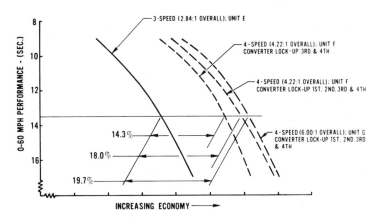

Fig. 14 - EPA composite (55%-45%) fuel economy improvements for various 4-speed automatic transmissions

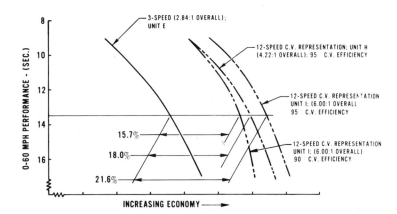

Fig. 15 - EPA composite (55%-45%) fuel economy increases anticipated for continuously variable transmissions having various characteristics

SUMMARY

It is important that drivetrains be evaluated at their optimum performance--fuel economy capability since it is by this means that the true potential of a transmission can be determined. This study is a step in that direction. Erroneous results can be obtained by investigating transmissions in vehicles having other drivetrain parameters such as fixed axle N/V. Estimates of fuel economy benefits of wide ratio, overdrive, and continuously variable units can be forecast unusually high if the comparison is based on only a high axle N/V vehicle. The system established here eliminates this possibility and compares the transmissions on an equal basis.

Figure 16 is a tabulation which compares the type of improvement that was predicted for manual transmissions having wider ratios and increased number of geared speeds (from 4 to 5) over a base manual 4-speed. It should be noted that increased ratio range has almost equal effects whether done in a 4- or 5-speed design configuraiton. Figure 17 is a similar tabulation for the automatic transmissions and the continuously variable units. Here it can be noted that automatic transmissions utilizing conventional design techniques could perform as well as the continuously variable units and that the continuously variable ratio portion of such a transmission must have high operating efficiency to remain competitive. The magnitudes of the changes indicated are subject to qualifications which have been described previously.

Since emissions and driveability have not been considered in this study, the benefits indicated may not be totally achievable; however, the most promising areas of gain are indicated and it remains for the engineer to resolve the problems.

None of these areas can be solved independently since they are all inter-related with the powertrain selection and the final product offered the public must meet the emission requirement levels and fuel economy targets for that model year as well as driveability and acceptability standards.

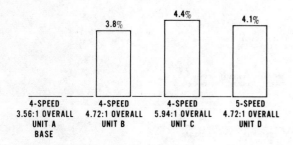

Fig. 16 - Comparison of fuel economy improvements at equal performance level for manual transmissions

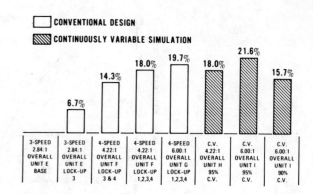

Fig. 17 - Comparison of fuel economy improvements at equal performance level for automatic transmissions

REFERENCES

1. Marks, Craig; Niepoth, George, "Car Design for Economy and Emissions" SAE Paper 750954 presented at Automotive Engineering and Manufacturing Meeting, Detroit, Michigan, October 1975.

2. Waters, William C., "General Purpose Automotive Vehicle Performance and Economy Simulator." SAE Paper 720043 presented at the Automotive Engineering Congress, Detroit, Michigan, January 1972.

An Overall Design Approach to Improving Passenger Car Fuel Economy*

Edward K. Hanson
Buick Motor Div., General Motors Corp.

TODAY, MORE THAN EVER BEFORE, automotive engineers are faced with a myriad of difficult yet exciting product design challenges. Pressures for change are being felt in virtually every aspect of automotive design. Industry is faced with problems of rapidly rising costs for labor and materials. Shortages of raw materials are no longer uncommon, both short and long term, while projections for future raw material usages are being required earlier and for increasingly longer lead times. Federal, state, and local regulations in the areas of ecology, worker safety, and environmental health are challenging engineering management on every product design decision made. Product serviceability, reliability, functionality and insurance costs are becoming more important as criteria by which consumers evaluate potential vehicle purchases. Of course, safety and exhaust emissions are in the forefront of activity in industry, Government, and in the public forum.

But perhaps the most critical challenge facing the automotive industry, and indeed the world's economy in general, is that of managing energy consumption more effectively and conserving world petroleum supplies. Automobile manufacturing and production of raw materials and components used in automobiles combine to make the automotive industry a very energy intensive one, and motor vehicles are currently the largest single users of petroleum at about 33% of total annual consumption.

With respect to petroleum reserves, the situation is critical in both short and long terms:
- Short term, because many nations have become dependent upon uncertain oil imports to a significant extent creating large monetary outflows from their economies, and
- Long term, because a world-wide petroleum shortage is projected to occur in the next century.

Thus, the importance of designing more energy efficient vehicles for both manufacture

*Paper 780132 presented at the Congress and Exposition, Detroit, 1978.

---- ABSTRACT ----

The critical challenge of improving fuel efficiency of new vehicles must be met if the world's finite petroleum supply is to be conserved for future generations. Automotive engineers will require new, more efficient design techniques to successfully meet the challenge. This paper presents the approach used in the design of GM's new 1978 intermediates as an example of how improved efficiency can be achieved. Factors affecting fuel consumption are outlined along with program goal setting. Areas of emphasis including weight reduction, powertrain optimization, and road load horsepower reductions are discussed as well as explanations of new engineering tools used.

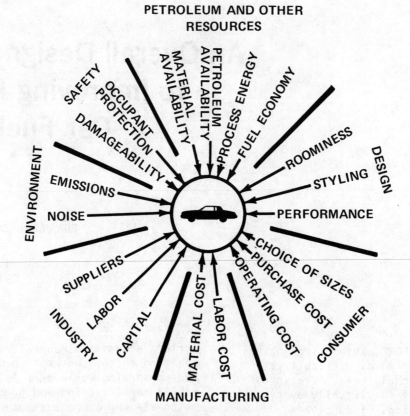

Fig. 1 - Pressures influencing automotive design

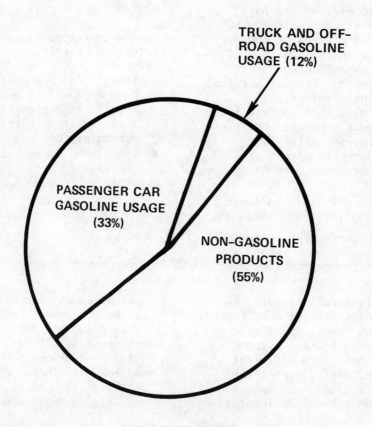

Fig. 2 - 1977 U.S. petroleum usage

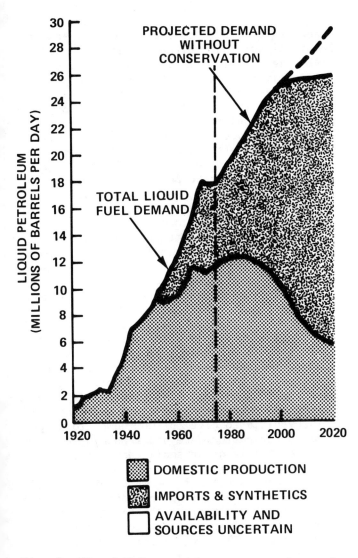

Fig. 3 - Total U.S. petroleum consumption and sources over time (Ref. 1)

and customer operation becomes readily apparent. Collectively, the impact these challenges will make on automobile design over the next decade will be tremendous.

The purpose of this paper is to outline how one such new vehicle design program, the 1978 General Motors intermediates, addressed these challenges, specifically that of improving fuel economy. Major factors affecting fuel economy are discussed as well as the techniques used for the establishment of an overall program fuel economy goal. Specific steps taken to improve fuel economy are described in detail and include design changes related to:
- Weight reduction
- Powertrain optimization
- Road load horsepower reduction

Finally, the author discusses new design approaches, engineering tools, and analysis methods used during the vehicle design process and suggests that these techniques have general application to any new vehicle program. The author does not presume this is the

- Governmental regulations
  - Fuel economy standards
  - Emission standards
  - Inertia weight class definitions
  - Standards affecting vehicle weight, size, and equipment content
- Marketing pressures
  - Manufacturer's fleet sales mix
  - National economy
  - Fuel availability and cost
- Design factors
  - Vehicle weight
  - Powertrain parameters including emission hardware
  - Aerodynamics
  - Misc. chassis items (suspension, brake and wheel bearing adjustment)

Fig. 4 - Factors affecting fuel economy

ultimate design approach, rather it is one which has been successfully applied and may provide the basis for ideas for future vehicle programs utilizing yet further improved design techniques.

FACTORS AFFECTING FUEL ECONOMY

The first step toward designing vehicles for improved fuel economy is to identify all factors which influence fuel consumption. These can be generally categorized into three classes; governmental regulations, marketing pressures, and design factors.

GOVERNMENTAL REGULATIONS - The Energy Policy and Conservation Act (*) regulates manufacturer's sales weighted fleet minimum fuel economies beginning in model year 1978 and extending through 1985. Certainly this is the primary Governmental influence currently affecting individual vehicle fuel economies.

The Clean Air Act Amendments of 1977 (**) is another piece of legislation critical to fuel economy since it defines maximum exhaust emission limits. (Figure A in the Appendix.)

While it is not within the scope of this paper to discuss the relationship between emissions and fuel economy, it certainly can be said that the two do, in fact influence each other. Therefore, assumptions as to exhaust emission levels must be made when evaluating new vehicle design scenarios for improved economy.

(*) Pub. L. 94-163, 89 stat. 871 - Title 3 of this act amends the Motor Vehicle Information and Cost Savings Act, 15USC 1901.
(**) Pub. L. 95-95, Clean Air Act Amendments of 1977, 42 USC 7401.

Other Governmental legislation, in effect or pending must also be considered. Examples include the revision of the EPA dynamometer horsepower setting procedure for 1979 and the redefinition of EPA inertia weight/test class limits and option content for 1980 (***). All Federal and state legislation proposals which deal even indirectly with fuel economy, emissions, inertia classes, vehicle size definition, equipment content, vehicle weight, etc., must be monitored closely and continuously to assess their potential impact on fuel economy. This is essential if vehicle family fuel economy design goals are to be meaningful.

MARKETING PRESSURES - Vehicle manufacturers historically estimate the sales penetrations of their various product lines with respect to total Corporate and industry sales. In planning for fuel economy improvements, the best possible sales penetration projections are needed between both the operating vehicle divisions of a given manufacturer and the car lines offered by particular divisions of that manufacturer. Sales penetration projections are also essential for the distribution of engines available within a given car line, manual versus automatic transmission sales, 49 state versus 50 state sales, and option penetrations within engine families and model series.

Consumers' purchase preferences for vehicles within a given product line-up can usually be expected to remain relatively constant with time <u>providing</u> that the image of the vehicle line remains constant and <u>also providing</u> that external pressures such as the general state of the economy, fuel costs, fuel availability and competitive vehicle offerings do not change significantly. In recent years, however, these external pressures have not remained stable and as a result have compounded the difficulties faced by marketing analysts in projecting future sales mixes.
Nevertheless, <u>the importance of having the most accurate possible sales estimates cannot be overemphasized</u> in planning flexibility into future vehicle fuel economy improvement programs.

DESIGN FACTORS - The third major area affecting fuel consumption includes a wide variety of design factors. Vehicle curb weight becomes a key element in determining the EPA inertia weight class into which the new vehicle is to be classified. There are both direct and indirect effects of reducing vehicle weight. First, fuel consumption is directly proportional to vehicle weight based on actual test experience with a wide variety of production vehicles, so that as weight is reduced, fuel consumption is reduced. Secondly, specific weight reductions may have compounding effects on other component designs such as fuel tank capacity, engine displacement, brake performance, bumper requirements, etc., which can result in yet further mass reductions and corresponding increases in fuel economy.

Powertrain parameters are also key design variables. Examples include engine efficiency differences, displacements, and carburetion systems. Included with engines are emission hardware effects. Fuel economy can be significantly affected by the use of various emission devices necessary to enable an engine to comply with emission standards. It is quite probable that these emission hardware requirements will vary depending on the weight of the vehicle to which the engine is applied. Also of great importance is the proper sizing of the powerplant to the vehicle to give optimum power to weight ratio for performance and economy. Engines either too large or small can be inefficient.

Another significant group of design influences involve items which affect road load horsepower. Since no mechanical device is perfectly efficient, there will be inevitable energy losses associated with its operation. The obvious goal, however, is to minimize these losses in order to improve efficiency. In automobiles, these losses are generally attributable to the following.

- Engine thermal inefficiencies
- Friction losses - engine mechanical, transmission hydraulic, transmission mechanical, propshaft/axle mechanical, wheel bearing mechanical, suspension alignment, tire to road surface, etc.
- Aerodynamic drag losses
- Engine accessory loads - electrical, air conditioning, vacuum losses, etc.

In any program to improve fuel economy, each of these areas must be examined in great detail to develop improved component designs which will reduce these losses.

OVERALL PROGRAM GOAL SETTING

After identifying all possible factors which can affect fuel economy, the next principle step is the establishment of vehicle design goals. The primary objectives which must first be defined are vehicle sales weighted family fuel economy (EPA composite) and vehicle weight (EPA inertia test weight). Of course, other basic marketing goals must

---

(***) 40 CFR Part 861 - Control of Air Pollution from Motor Vehicles, and 40 CFR Part 600 - Fuel Economy Emission Testing and Other Procedures for 1978 and Later Model Year Automobiles, 42 FR 45641, 9-12-77.

also be initially established including vehicle image, market segment, price range, and tooling/investment program limitations. These goals can be easily expanded into more detailed objectives later. It is not the intention of this paper to discuss how marketing objectives are established, rather to concentrate on the engineering objectives of vehicle fuel economy, packaging, etc.; so let it be assumed for the balance of the discussion that the marketing goals have already been defined.

Fuel economy and weight goals can be determined concurrently; however, each influences the other, and further analysis iterations must be performed as changes are made. To affect EPA reported fuel economy, weight reductions must be large enough to cause the new vehicle to be reclassified into lower inertia or test weight classes as defined by the EPA or large enough to have an impact on reducing dynamometer set horsepower as determined by the alternate coastdown test method. The greater the reduction, the greater the fuel economy gain. Current automotive design experience shows that an average reduction of one inertia weight class is typical for a vehicle line undergoing a complete redesign.

A reduction of this magnitude would typically represent a vehicle family change of 500 to 700 lbs although specific models could be affected to more widely varying degrees. Reductions of this size would most likely involve vehicle resizing, material substitutions, and more efficient (state of the art) component designs.

In 1980, the inertia weight classes are redefined, and narrower band width test weight classes are inserted within the inertia weight classes. This change must be comprehended in any future vehicle analyses.

Determining a vehicle line fuel economy goal, as the next step, involves accumulating the following information:

(1) Total Corporate sales weighted fuel economy requirements for the model year in question (Figure 6).

(2) Projected vehicle line sales for the entire Corporate vehicle line-up upon which the Corporate sales weighted fuel economy Federal requirement is based. This sales breakdown need not be in great detail as to models, engines, etc., but should include projections for vehicle sales by inertia weight class.

(3) A relationship between inertia weight classes and fuel consumption of vehicles within those classes.

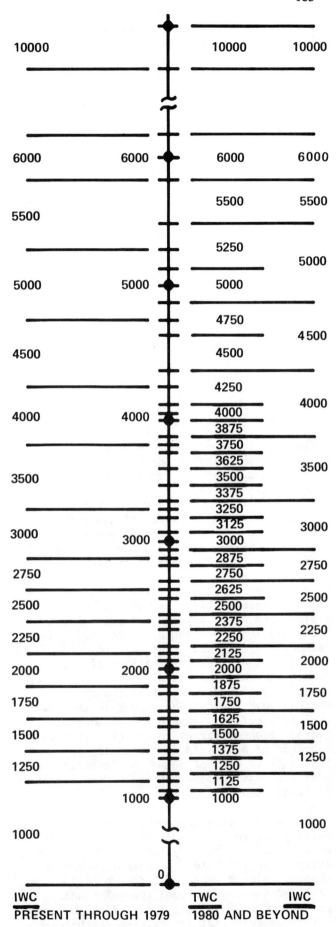

Fig. 5 - Inertia weight/test weight class definition

Item #1 is straightforward, and item #2 can be developed by forward planning analysts. Item #3 requires a bit more analysis. The approach is to compare EPA composite fuel consumption to EPA inertia class weight for the total spread of current production Corporate vehicle families. The general relationship which results will be that <u>fuel consumption is directly proportional to inertia weight</u>. For goal setting purposes, this approximation is entirely adequate; however, it should be recognized that vehicles with widely different acceleration performance characteristics will vary the relationship somewhat. If significant volumes of new, more fuel efficient "hardware" are planned for <u>all</u> model lines in the new model year in question, the analysis should be revised to reflect these improved projections.

Otherwise current production data is quite satisfactory. The fuel consumption relationship to inertia weight can then be easily transposed to an expression for the fuel economy contribution or "responsibility" of a given vehicle line to the Corporate average as follows.

> Specific vehicle family fuel economy responsibility (mpg) = Corporate average inertia class weight (lbs) x Corporate fuel economy target (mpg)/vehicle family average inertia class weight (lbs).

Derivation of this relationship is developed in Figure B in the Appendix. Solving the equation will yield fuel economy responsibilities for each of the vehicle lines offered. Since many of the vehicle lines may be carryover for a particular year, their computed responsibilities must be tested against projected actual fuel economies. Recalculating a new vehicle line fuel economy responsibility based on actual projected economies and sales penetrations of the remaining vehicle lines will quickly indicate the amount of additional fuel economy "technology" improvement needed in the new vehicle program to enable the Corporate average economy target to be met. This value can then be evaluated for technological feasibility. If not deemed achieveable, then modifications to other carryover vehicle lines are indicated.

This analysis approach, therefore, can quickly set the basic design parameters of weight and fuel economy for the new vehicle design, can indicate the general amount of economy related improved hardware technology required above and beyond the gains anticipated from vehicle weight reduction, and can point out potential inconsistencies between the mandated manufacturer's total sales

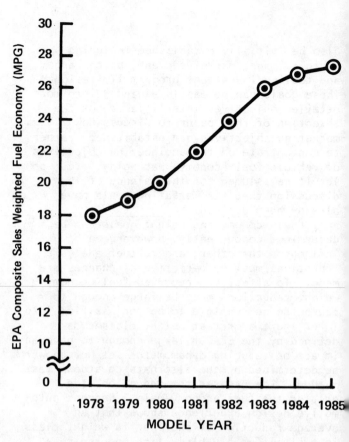

Fig. 6 - Manufacturer's fleet fuel economy requirements (Energy Policy and Conservation Act 94-163)

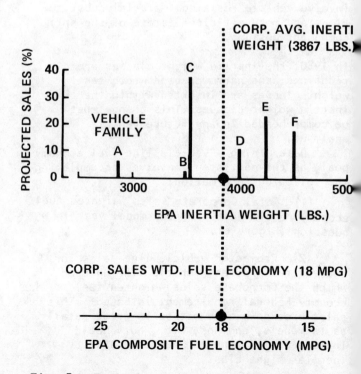

Fig. 7 - Typical example for establishing vehicle family fuel economy responsibilities

weighted economy and individual car line contributions. The analysis should be updated as required to keep it current with changes in estimated sales penetrations, weight, and fuel economy projections. Note that this phase of goal setting has not placed any restrictions on engine size, transmission type, emission effects, etc.
These factors must also be considered, but are better handled using special design sensitivity analyses to be discussed later in the paper It might be of interest to note that the 1978 GM intermediate vehicles had resulting goals established of 3500 lb inertia weight class and 20.0 mpg EPA composite sales weighted family fuel economy as a result of using this technique.

WEIGHT REDUCTION APPROACH

After establishment of the total vehicle inertia class weight targets, goals must be ultimately set for the systems, subsystems, and components. The first step in this process is to develop the basic packaging of occupants and luggage (number of passengers, head/leg/hip/shoulder rooms, chair heights, seat back angles, trunk volume, etc.). Once the people and utility package "shell" has been defined, closer examination can be made of the rest of the package for minimum space and weight. Factors to be considered in this step are drivetrain configuration, powerplant sizes (usually a variety must be accommodated), tire sizes, suspension types, front and rear tread widths, and structure type. This is typically an iterative process to optimize the size, type and arrangement of components.

After the minimum outline of the automobile has been established, previously determined guidelines for weight reduction per unit vehicle length and width can be applied to project the total weight savings effects of these changes compared to the past model vehicle design. Examples for typical weight changes per unit change of overall dimension for intermediate size body/frame cars are:
- 11.4 lbs per inch width
- 6.0 lbs per inch length (passenger compartment)
- 3.5 lbs per inch length (front and rear bumper overhang).

Design engineers are then asked to furnish accurate estimates of new component weights, and as these are received, a weight model for the new vehicle is assembled. This then represents the first complete vehicle weight estimate from which can begin the task of searching for further reductions.

An important concept found useful in determining the impact of component weight estimate fluctuations is that of weight compounding. In its simplest form, weight compounding is a description of the iteration which takes place between components or systems when the weight of one component is changed. For example, as car weight increases, tires, wheels, suspensions, brakes, steering, and structure may all be affected to the point that their weights would also increase to provide the same level of performance, function, and durability. This also works in reverse for weight reductions. Execution of the concept requires categorizing parts into "functional groupings". Influence coefficients generated from previous vehicle design analyses can then be applied to parts within a given functional group to enable "total vehicle" effects of individual component weight changes to be ascertained. The concept will be discussed in more detail later in the paper under "Tools and Methods Used".

Development of design weight targets for vehicle systems (UPC or Uniform Parts Classification Groups) also proved to be a valuable aid in controlling weight. Targets should only be detailed to the point that a

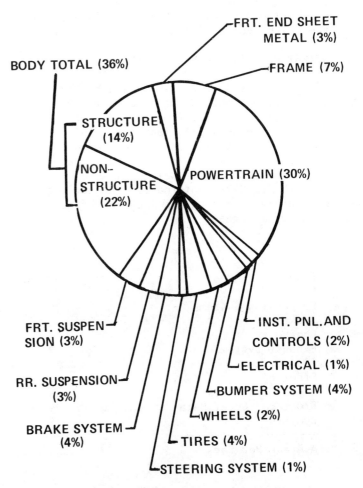

Fig. 8 - Typical optimized weight target distribution for a body/frame vehicle

single target exists for the collection of parts for which an individual design engineer has responsibility. Narrower target increments are unnecessary, and wider increments result in confusion between designers. Targets can be based on many factors, some of which are:

- Proportionality of weight reduction per functional group equal to the proportion of that group's projected weight to the total automobile
- Proportionality established by influence coefficients described previously
- Weight analyses of similar construction, size and/or weight cars
- Packaging size
- "Probability of success" projections of future component weight reduction programs not currently considered part of the official vehicle design definition.

As the design process continues, weight estimates must be closely and continuously monitored since significant weight reductions are difficult to achieve in terms of cost, tooling, lead times, etc. As design pressures rise in these other areas, management must maintain an effective program monitoring discipline which will insure that the weight goals will ultimately be met.

WEIGHT REDUCTION ACHIEVEMENTS

The 1978 GM intermediates achieved weight reductions compared to their 1977 counterparts ranging from averages of 530 lbs on coupes, 615 lbs on special coupes, and 1004 lbs on station wagons. As a result, all models were reduced at least one EPA inertia weight class from their 1977 counterparts. Reductions were achieved through engineering efforts in three basic areas.

Body	140 LBS.
Frame	85
Suspension Systems	36
Brakes	40
Engine	73
Exhaust System	17
Fuel System	27
Steering	15
Wheels and Tires	43
Front End Sheet Metal	79
Bumpers	75
Average Per Car	630 LBS.

Fig. 9 - Typical 1978 GM intermediate weight reduction by major area

- Vehicle repackaging and exterior size reduction 24%
- Material substitutions 20%
- Improved component and system design efficiencies 56%

  100%

Major factors and design features which contributed to these improvements are next discussed.

VEHICLE SIZE REDUCTION - To put the vehicle family exterior size reduction in perspective, the following dimensions describe the general reductions compared to 1977 comparable models.

- Tread 2.5 to 3.4 inches narrower
- Wheelbase 4 to 8 inches shorter
- Overall length 9 to 18 inches shorter
- Overall width 5.5 inches narrower
- Overall height 0.8 inches higher to 1.2 inches lower

Using the packaging techniques described previously, the occupant and luggage packaging was actually improved over 1977. Applying the weight change per unit dimensional change factors presented earlier, the repackaging of the exterior shell alone resulted in the following weight reductions.

- Wheelbase reduction 48 lb savings
- Width reduction 68 lb savings
- Front & rear overhang reductions 34 lb savings
  Total 150 lb savings

MATERIAL SUBSTITUTIONS - Accomplishments in this area were basically realized through the use of more high strength steels, aluminum, and plastics. Compared to 1977 there was a 3.8% shift (by weight) of vehicle materials composition primarily toward the lighter materials mentioned. Table C in the Appendix provides a more detailed comparison of the proportions of different materials to total vehicle weight and also the changes from 1977. Table D outlines several major examples of material substitutions employed.

IMPROVED COMPONENT/SYSTEM DESIGN EFFICIENCIES - Improvements in components and systems provided over half of the total weight reduction ultimately achieved. Weight compounding effects, described previously, were responsible for much of this reduction. As an example, the exterior size of the vehicle allowed narrower bumpers to be used for a typical savings of only 9 lbs. But as the weight of the entire vehicle was reduced, the bumper structural loads were also lowered proportionally resulting in an eventual typical bumper system savings of 67 lbs, over 7 times the magnitude of the original "size savings" alone.

Other examples of where weight compounding contributed significantly are shown in

Table E in the Appendix. Note that virtually all chassis, powertrain, fuel, and structural systems were affected. Since these systems represent over 80% of vehicle curb weight, compounding reductions in these areas also accounted for a high percentage of total weight reduction.

Weight compounding, however, did not comprise the total extent of improved design efficiency. Further design refinements were made with the intent of weight reduction and product improvement. Examples are outlined in Table F of the Appendix.

POWERTRAIN COMPONENT OPTIMIZATION

The next major area in which engineering emphasis was concentrated to improve fuel economy was that of optimizing the powerplant size to the vehicle. Several techniques were used to finally arrive at the production engine line-up. First, a computer simulation of a typical new vehicle and its powertrain was used to screen a variety of engine displacments to determine their impact on the new vehicle's performance and economy. Results were consolidated into power to weight ratios as functions of performance.

This information was then compared to a large group of actual test data obtained on current production vehicles to determine limits of performance acceptable to customers. After determination of these limits, the analysis was repeated to obtain a better projection of optimum fuel economy at limits of acceptable performance.

It was recognized from the outset of the powertrain analysis that a variety of powertrains would be required because of marketing needs, production capacity constraints, and a variety of vehicle weights and design limitations. When these factors were then considered, an even more accurate analysis of vehicle family sales weighted fuel economy resulted, and it became apparent that a wider variety of smaller displacement engines was desirable.

There are 11 engines in the 1978 GM intermediate vehicle line which cover a total of 18 applications. Both V6's and V8's are available in a wide variety of displacements and carburetion, each one carefully tailored for optimum economy and performance.

It should be noted that five of these engines are new or revised designs, and that four of them deal with smaller displacement 6 cylinder configurations, emphasizing the conclusion discussed previously that more small displacement engines were indicated from vehicle analysis.

A turbocharged 6 cylinder engine was also introduced in the 1978 GM intermediate vehicle

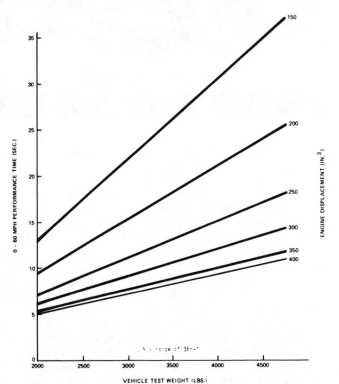

Fig. 10 - Engine displacement versus performance and vehicle weight

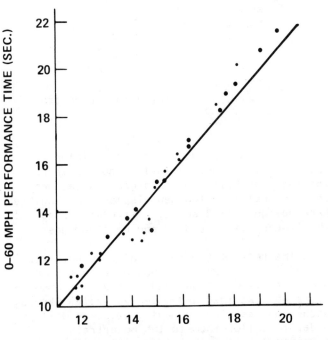

Fig. 11 - Vehicle weight per unit engine displacement versus performance

	49 State	50 State	Alt.
*3.2L-V6 (2 BBL)	X		
*3.3L-V6 (2 BBL)	X		
3.8L-V6 (2 BBL)	X	X	X
*3.8L-V6 (2 BBL Turbo)	X		
*3.8L-V6 (4 BBL Turbo)	X	X	X
4.3L-V8 (2 BBL)	X	X	
4.9L-V8 (2 BBL)	X		
4.9L-V8 (4 BBL)	X		
5.0L-V8 (2 BBL)	X	X	X
*5.0L-V8 (4 BBL)	X		
5.7L-V8 (4 BBL)			X

11 Engines - 18 Applications
* New or revised engines

Fig. 12 - 1978 GM intermediate engine availability

Fig. 13 - Buick's 3.8 L turbocharged V6

- Transmissions
  - Manual — 3, 4, & 5 Speed
  - Automatic — 3 Speed
- Axles
  - Ratios
    * 2.28:1
    2.41:1
    2.56:1
    2.73:1
    * 2.93:1
    3.23:1

*New for 1978

- Tires
  - New Metric Sizes
    P185/75R14
    P195/75R14
    P205/70R14
    P205/75R14

Fig. 14 - 1978 GM intermediate powertrain components

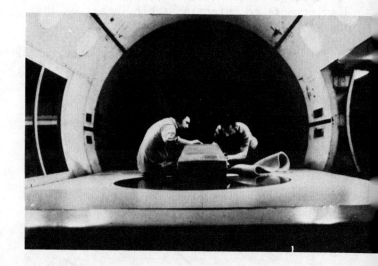

Fig. 15 - Wind tunnel testing

line-up as another approach to obtaining optimum (small engine) fuel economy with performance capability of a larger displacement V8. In addition, two new displacement engines were designed (3.2 and 3.3 litre) to round out the line-up of standard engine offerings.

The rest of the powertrain was not overlooked in the search for better economy. A wide variety of manual and automatic transmissions was examined and made available. Careful attention was given to axle ratios as well in order to tailor them to the powertrain characteristics for optimum performance and economy. This was accomplished with a variety of techniques including computer drivetrain simulations and on-the-road testing. The outcome was a family of six ratios, two of which were new to the vehicle family. One of the ratios (2.28:1) is lower than ratios previously used in the intermediate car line, and was added strictly to optimize fuel economy.

Engine related accessories were also examined closely to improve efficiency. A new type of air conditioning compressor was introduced on selected models with an approximate system operating efficiency improvement of 50% over the old system. Another example of improvement is the automatic leveling control option where a vacuum pump was replaced with an electrically driven compressor thus relieving the engine of an external vacuum load and potential leak points -- a small item but one which shows the extent to which designers went in exploring for ways to reduce powertrain losses.

REDUCED ROAD LOAD HORSEPOWER

The final area in which emphasis was concentrated for improving fuel economy was that of finding ways to reduce total vehicle road load horsepower requirements. Previously described vehicle performance/economy computer simulations were again used extensively to evaluate the effects of aerodynamics and vehicle mechanical friction losses on fuel economy.

A graph (item G in Appendix) was prepared for use by body and sheet metal design engineers showing the relationship between aerodynamic drag coefficient reductions, weight reductions, and fuel economy; and was used as guidance during body surface development along with actual test data taken from wind tunnel tests to stress the importance of minimizing aerodynamic drag.

Much wind tunnel work was done during all stages of the design process beginning with 1/4 scale model evaluations of early body shape concepts. Tests were later expanded into analysis of divisional styling specifics; and tests were performed on more detailed scale models, full size fiberglass models, and finally on real prototype vehicles. At each step, styling changes were tested for their effects on fuel economy by equating the drag coefficient change to an equivalent weight change. Drag was also reduced by the fact that the new vehicles' frontal areas are smaller than their predecessors (approximately a 9% reduction). Tires also received a great deal of design attention in that they are the second largest single contributor to on-road rolling resistance after aerodynamic effects. At low speeds they are, in fact, the largest contributor. Efforts toward achieving tire constructions with less rolling resistance resulted in an eventual reduction of about 4%. This contributed significantly toward fuel economy gains.

Thus, when the effects of weight reduction, powertrain optimization, and rolling resistance reduction were eventually combined, the resulting new vehicle sales weighted family fuel economy was improved 2.6 mpg or 15% compared to the 1977 comparable models.

A final analysis of the complete vehicle shows that the fuel economy gain can be divided into the following proportions.

- Weight reduction   41%
- Powertrain component optimization   47%
- Reduced road load horsepower   12%
                                       100%

Conservative estimates indicate that an improvement in fuel efficiency of this magnitude will potentially result in a national gasoline consumption reduction of 245 million gallons for the 1978 model year or about 0.3% of all gasoline consumed yearly by private motor vehicles in the U.S. (based on 1977 projections).

TOOLS AND METHODS USED

Many design tools, methods, and procedures were used in redesigning the 1978 GM intermediate vehicles to achieve the fuel economy improvements discussed previously. While some of these techniques were not entirely new, their application to automotive design has been fairly recent and their usefulness substantial. The intent of this section is, therefore, to provide an overview of the principle techniques used and discuss their relative importance in the overall design scheme.

WEIGHT COMPOUNDING - One of the first steps in understanding weight compounding influences, or how "weight begets weight" is to analyze the distribution of weight among vehicle functional subsystems. An analysis by functional subsystems is necessary because the design engineer's decision making process is based on vehicle functional characteristics. Functional groupings of components are similar to the traditional UPC (Uniform Parts Classification) breakdown with a few notable exceptions. Table H in the Appendix depicts a typical vehicle's functional grouping of components. Once the functional groups have been established, a series of weight influence coefficients (one for each functional group) can be generated by plotting functional group weight versus total vehicle weight for a number of typical vehicles. The result will usually be a linear function ('X' lbs. of functional system weight increase per lb. of total vehicle increase). Exceptions to this can be found but are insignificant to the accuracy of the analysis. Having established these coefficients, the weight analyst can provide extremely early guidance to the design engineer as to the effects potential design changes will have on overall weight. Once these influence coefficients are established, functional group weight targets can be set. Having targets set by group instead of by component provides the program manager with overall vehicle weight control while giving the design engineer flexibility to develop an optimum system design. Maintenance of management discipline on weight control throughout all phases of the program is essential if the end product is to meet its overall weight objective. While many techniques exist to effectively monitor weight, one of the more useful methods was found to be frequent and regular reports on EPA inertia class and gross vehicle weight <u>reserves</u> compared to target. Costs of weight reduction proposals were also found to be effectively controlled by the establishment of a cost/weight rating system (cost per lb. saved) where specific changes could be compared to vehicle average values.

FINITE ELEMENT ANALYSES - Many technical papers have been written on this subject, thus the author does not intend to repeat the principles behind finite element analysis; but rather to emphasize the importance this tool played in the overall design process. Finite element analysis was used extensively from the outset of the program and continuing through production. Three basic types of models were employed including complete vehicle, system, and component models.

Complete vehicle models were generated very early to provide guidance as to the most suitable type of structure system to achieve specific objectives in the areas of noise, vibration, harshness, and ride (NVHR). Output from these early models also gave estimates of body, frame, and sheet metal weights as needed to maintain acceptable structure. Total vehicle barrier impact simulation models were also constructed to predict vehicle crush characteristics of various structure proposals and engine sizes (Table I in Appendix). As the program design progressed, these models were continually refined and updated to permit structural engineers to closely monitor the overall effects of numerous design developments. As hardware became available, early test vehicles were instrumented to substantiate the models' output.

In the process of establishing correlation, several state-of-the-art improvements evolved. First, refinements were made in modeling techniques for suspensions and complex structural members. One particularly noteworthy improvement was modeling of shock absorbers. Next, a better understanding was attained as to the correlation of subjective evaluations of real vehicle/road situations to computer output. Finite element analysis output tends to be voluminous, and a knowledge of the key elements which correlate most closely with subjective ride evaluations speeds the interpretation of results tremendously. Another finite element innovation used extensively was the technique of stress shading. This tool provides the design engineer with a pictorial representation of stress gradients in complex components such as frames for various structural vibration modes, static, and dynamic loading conditions. This approach was used extensively to guide designers in the determination of potential structural durability problem areas prior to actual testing as well as indicating where increased metal gage, flanges, gussets, braces, etc., could be used to best advantage and with minimum added weight.

Finally, the library of road surface models used as excitation input to the structure models was refined and expanded to cover an even wider range of roads which customers experience.

System models were also used extensively. Examples included models of body, frame, front sheet metal, front and rear suspensions, bumper systems, tires, powerplants and mountings, body/sheet metal mountings, and body cavity acoustical models.

These tools were used to optimize systems designs and reduce weight. Many additional component models were constructed to permit detailed analysis of parts for stress, stiffness, and material distribution characteristics, all of which led to better utilization of materials.

An interesting sideline application of the finite element technique was the creation of a luggage loading program used to optimize the placement of luggage within the trunk for better utilization of space and without the need for extensive "manual" iteration using real vehicles.

3/8 SCALE PLASTIC STRUCTURAL MODELS - Plastic scale models were used extensively in conjunction with finite element analyses to insure an efficient structural design. While both finite element and plastic models are powerful structural tools, plastic models have several characteristics which make their use advantageous. First, they are visual and provide the design engineer with a quick grasp of structural mode shapes (bending,

Fig. 16 - Finite element model evaluation

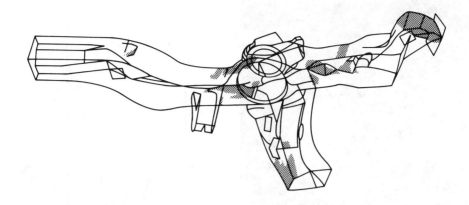

Fig. 17 - Example of stress shading analysis

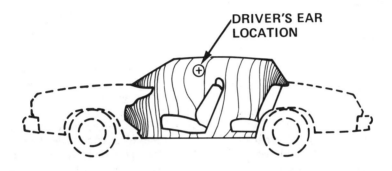

Fig. 18 - Example of acoustical cavity model (3 dimensional)

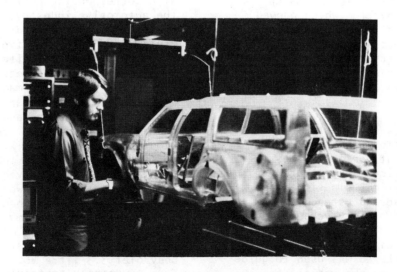

Fig. 19 - Plastic structural model evaluation

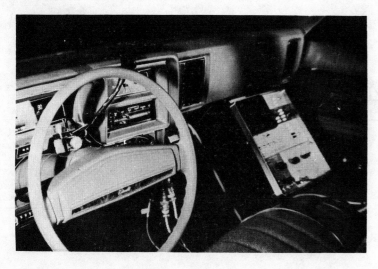

Fig. 20 - Typical installation of real time structural analyzer

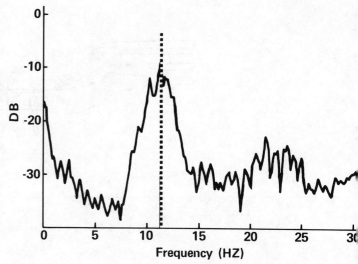

Fig. 21 - Example of structural real time analyzer output

weak points, etc.). Quite often, experimentation with added gussets, braces, ribs, flanges, etc., can be modeled and their effects determined more quickly than with finite element analysis. Plastic models proved especially valuable in body design where it was impractical to develop the same level of detail in the finite element analysis (for example, small formations such as ribs in the underbody). Throughout the program, results from both finite element and plastic models were compared to provide full benefit and utilization of these tools.

VEHICLE "ON-BOARD" STRUCTURAL INSTRUMENTATION - Another innovation which proved to be extremely valuable to the program was the development of a special compact, lightweight, easy to use instrumentation package which could be quickly placed in a test vehicle to analyze particular structural problems.

Essentially, it consists of a fast Forrier real time analyzer coupled with an accelerometer to measure input at various points on the vehicle. Accelerations are measured over a given period of time, classified as to frequency, summed, sorted, stored, and fed back as a plot of frequency versus acceleration. This device proved invaluable in identifying structural problem frequency modes. Information gathered was used to guide finite element analysis of the vehicle on the computer at specific problem frequencies, thus reducing problem identification time. Since it is highly portable, it was used by development engineers to augment their subjective interpretations of problems on real road surfaces.

ENGINE SIZE OPTIMIZATION ANALYSIS - Several techniques were employed to provide early guidance as to engine sizes considered for the 1978 GM intermediates. Maximizing fuel economy was obviously the primary goal; however, the powertrain engineers recognized from the outset that acceleration performance was also a very critical consideration. The approach used was to identify limits of acceptable performance, and then optimize both economy and performance within these limits.

Establishing performance limits involved extensive surveys of current power (or displacement) to weight relationships along with corresponding performance times across the automotive industry. Additionally, data collected on customer driving habits was reviewed to determine typical field performance requirements. After careful consideration of all data, targets were established.

Once these performance limits were set, a linear optimization approach was used to maximize fuel economy and performance. The general model for this technique is shown in Figure J in the Appendix. Note that performance is defined as the ratio of engine displacement (CID) to vehicle weight. Results from this analysis showed a need for smaller displacement engines than were currently available; thus programs were begun which led to the development of 2 new engine displacements, the Buick 3.2 litre V6 and the Chevrolet 3.3 litre V6.

POWERTRAIN PERFORMANCE AND ECONOMY SIMULATIONS - Extensive use was made of vehicle computer simulations for evaluating powertrain characteristics. This technique was used to supplement the engine displacement optimization analysis discussed previously and to provide more detailed information as to the effects of specific engine characteristics, manual versus automatic transmissions (gear ratios and torque converter sizes), axle ratios, and tire sizes (Fig. K in Appendix). Perhaps the most useful aspect of the tool was a sensitivity analysis performed to determine the effects of weight and aerodynamic drag coefficient changes on EPA fuel economy.

WIND TUNNEL TESTING - Much emphasis was placed on designing GM's new intermediate vehicles for reduced aerodynamic drag. 1/4 scale testing was done as soon as basic body shapes began to be proposed, and evaluations were made on all body styles. The next level of tunnel testing involved evaluation of divisional front sheet metal styling differences with numerous styling refinements being tested. Fiberglass full size models were tested with detailed undercarriages to validate the clay scale model results. Finally, actual prototypes were tested and compared to the previous year's models. Wind tunnel results were used to provide design guidance in areas other than drag coefficient. For example, data obtained was used for wind gust sensitivity studies, engine cooling development, wind noise, and side glass and backlight dirt contamination investigations.

ENERGY BALANCE CONSIDERATIONS - Energy balance analyses are becoming increasingly more important as an added consideration in new vehicle designs, especially where material substitutions are considered. Basically, an energy balance is the relationship of the energy required to produce the raw materials and manufacture a component in comparison to its anticipated fuel savings contribution over the life of the vehicle. The obvious objective is to use this approach to select component designs having net energy savings. In cases of material substitutions where the proposed materials are more energy intensive to manufacture or have petroleum stocks as part of their composition, it quite often can be shown that there is a net energy savings in the long term. An example of this technique as used in the 1978 GM intermediates was the bumper system.

CUSTOMER CLINICS - Customer clinics, properly structured, can provide invaluable feedback to the engineer. In the 1978 GM intermediate program, three types of clinics were utilized; ride quality, packaging, and styling. Clinics are usually set up through an independent agency to maintain the autonomy of the sponsor and assure that the results are unbiased. Much can be learned in this manner in terms of anticipating future customer reaction to a new product, an aspect especially important in new vehicle designs where traditional customer concepts of size and weight must change.

Fig. 22 - Customer clinics provide valuable feedback

MISC. TRADITIONAL DESIGN TECHNIQUES - Many other design techniques were used but were more along the lines of traditional approaches such as competitive product evaluations, worldwide state-of-the-art design surveys, and tradeoff studies. However, the items previously mentioned in detail were of paramount importance to the success of the 1978 GM vehicles. Finally, the importance of the automotive design engineer cannot be overemphasized for he or she is, after all, the key ingredient which must take all of the analysis output, however simple or complex, and through hard work turn it into a finished product which meets all of management's objectives.

SUMMARY

The mandate of improving passenger car fuel economy has become, and will continue to be, one of the most critical challenges facing automotive engineers in the coming years. This paper has provided an overview of how this challenge was addressed by GM engineers in the execution of the new 1978 intermediate vehicles.

Governmental regulations, marketing pressures, and design factors, all of which can affect fuel economy, were outlined. The processes were discussed by which vehicle EPA fleet fuel economy and inertia weight class goals were established. Improvements in fuel consumption have been attributed to 3 categories, and each area was discussed in detail.
1. Weight reduction
2. Powertrain component optimization
3. Reduced road load horsepower

Weight was reduced by application of vehicle repackaging and exterior size reduction, material substitutions, and improved component and system design efficiencies which resulted in an average (sales weighted) reduction of 630 lbs. or at least one inertia weight class compared to 1977 models.

Powertrains were optimized by a thorough analysis of sizing the engine and all other drivetrain components to the vehicle. As a result, 5 new engine designs evolved, primarily in smaller displacements. Engine related accessories such as air conditioning were also improved for increased efficiency.

Road load horsepower requirements were minimized through careful attention to aerodynamic drag effects as well as rolling resistance reductions in drivetrain components, primarily tires.

The combined effect of all these improvements was an increase of 2.6 mpg (EPA composite schedule) or 15% over comparable 1977 models. Conservative estimates place this fleet fuel savings at 245 million gallons for the 1978 model year.

Finally, the principle engineering techniques utilized in the program to achieve improved economy were discussed with particular emphasis on new methods and analysis tools employed. Included in this discussion were the following.

1. Weight compounding technique
2. Finite element analyses
3. 3/8 scale plastic structural models
4. Vehicle "on-board" structural instrumentation
5. Engine size optimization analysis
6. Powertrain performance and economy simulations
7. Wind tunnel testing
8. Energy balance considerations
9. Customer clinics
10. Miscellaneous traditional design techniques

In conclusion, the 1978 GM intermediates provide significant fuel savings over their previous counterparts and are indicative of the inevitable future trend in automotive design. The author does not presume to suggest that the approach used in this program is the best possible, rather is one which has by experience been successfully applied and may offer ideas for future application by other automotive engineers.

REFERENCES

1. Elliot L. Richardson, Chairman, Energy Resources Council, "The Request by the Federal Task Force on Motor Vehicle Goals Beyond 1980 - Volume 1 - Executive Summary", September 2, 1976

2. R. M. Ormiston, Standard Oil Co. and G. G. Pollock, Chevron Research Co., "Gasoline Supply for the Car Fleet of Tomorrow". S.A.E. Paper No. 770670 presented at the West Coast Meeting, August 8 - 11, 1977.

3. Dr. Craig Marks, General Motors Engineering Staff, "Which Way to Achieve Better Fuel Economy?". Seminar proceedings at California Institute of Technology, December 3, 1973.

APPENDIX

FIGURE A

1979-1984 PASSENGER CAR EMISSION STANDARDS

(CALIFORNIA STANDARDS 1980 AND LATER SUBJECT TO WAIVER APPROVAL)

	49 STATE	CALIFORNIA		ALTITUDE
1979	1.5-15.0-2.0	.41-9.0-1.5		NO SPECIFIC REQUIREMENT
1980	.41-7.0-2.0	.41-9.0-1.0		
1981	.41-3.4-1.0	.41-3.4-1.0	.41-7.0-.7	TO BE DETERMINED BY EPA
		OR		
1982		.41-7.0-.4	.41-7.0-.7	
1983		.41-7.0-.4		
1984				ALTITUDE COMPLIANCE TO 49 STATE STANDARDS WITHOUT SPECIFIC CALIBRATIONS

FIGURE B

DERIVATION OF THE VEHICLE FAMILY FUEL ECONOMY RESPONSIBILITY EQUATION

The equation for computing vehicle family responsibility is:

$$\text{Vehicle Family Responsibility (MPG)} = \frac{\text{Corp Avg Inertia Class Wt (Lbs)} \times \text{Corp fuel economy target (MPG)}}{\text{Vehicle Family Average Inertia Class Weight (Lbs)}}$$

Basic assumption: Fuel consumption is directly proportional to weight.

Given the basic assumption, the following equation (1) can be written:

$$\text{Vehicle Family Fuel Consumption (FC) Responsibility} = \frac{\text{Avg Vehicle Family Weight}}{\text{Avg Corporate Weight}} \times \text{Corporate FC Target} \qquad (1)$$

Let,

$$\frac{1}{\text{Vehicle Family Responsibility (MPG)}} = \text{Vehicle Family FC Responsibility}$$

And,

$$\frac{1}{\text{Corporate Fuel Economy Target (MPG)}} = \text{Corporate FC Target}$$

Substituting into (1) gives:

$$\frac{1}{\text{Vehicle Family Responsibility (MPG)}} = \frac{\text{Average Vehicle Family Weight}}{\text{Average Corporate Weight}} \times \frac{1}{\text{Corporate Fuel Economy Target (MPG)}}$$

Taking the reciprocal gives:

$$\text{Vehicle Family Responsibility (MPG)} = \frac{\text{Average Corporate Weight}}{\text{Average Vehicle Family Weight}} \times \text{Corporate Fuel Economy Target (MPG)}$$

Which is the vehicle family fuel economy responsibility equation expressed in MPG.

FIGURE C

MATERIAL DISTRIBUTION IN GM'S 1978 INTERMEDIATES

MATERIAL	PROPORTION OF VEHICLE WEIGHT	CHANGE FROM 1977
STEEL	61.7%	-1.5%
IRON	14.9	-1.7
PLASTICS	5.3	+1.4
ALUMINUM	3.3	+1.7
GLASS	3.3	+0.3
RUBBER	2.7	-0.1
LUBRICANTS	1.2	-
FABRIC	1.1	+0.2
COPPER	0.8	+0.1
LEAD	0.8	+0.1
ZINC	0.6	-0.3
PAPER	0.2	-
NICKEL	0.1	-
MISCELLANEOUS	4.0	-0.2
	100.0%	3.8% SHIFT

FIGURE D

EXAMPLES OF MAJOR WEIGHT REDUCTION ITEMS

- MATERIAL SUBSTITUTIONS (SELECTED MODELS)

  - ALUMINUM STAMPINGS (HOODS, DECK LIDS, RADIATOR SUPPORT, AND BUMPER REINFORCEMENTS)

  - ALUMINUM CASTINGS (ENGINE INTAKE MANIFOLDS, REAR BRAKE DRUMS, AND BRAKE MASTER CYLINDER)

  - POLYPROPYLENE FRONT WHEELHOUSES

  - URETHANE (SOFT) BUMPERS

  - HIGH STRENGTH STEEL STAMPINGS (DOOR BEAMS, BRAKE PEDAL BRACKET)

FIGURE E
EXAMPLES OF MAJOR WEIGHT REDUCTION ITEMS

- VEHICLE DOWNSIZING - WEIGHT COMPOUNDING EFFECTS

    - ENGINE DISPLACEMENT
    - TRANSMISSIONS
    - AXLE SIZE
    - BRAKE SIZE
    - TIRES AND WHEELS
    - FUEL TANK VOLUME
    - BUMPERS
    - BODY STRUCTURE
    - FRAME
    - FRONT SHEET METAL

FIGURE F
EXAMPLES OF MAJOR WEIGHT REDUCTION ITEMS

- MORE EFFICIENT COMPONENT DESIGNS

    - MODULAR A/C DESIGN
    - A/C 'R4' COMPRESSOR
    - COMPACT SPARE TIRE
    - EVEN FIRING V6 ENGINE DESIGN
    - FIXED REAR DOOR GLASS
    - 'HALO' ROOF STRUCTURE
    - FRAME
    - POWER STEERING GEAR
    - REAR BRAKE CONFIGURATION
    - EXHAUST SYSTEM

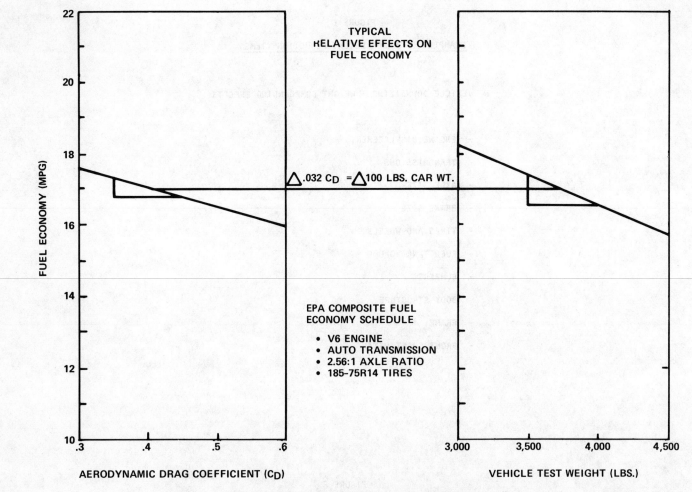

GRAPH G — RELATIONSHIP OF AERO DRAG COEFFICIENT AND WEIGHT TO FUEL ECONOMY

FIGURE H

FUNCTIONAL GROUPS FOR DETERMINING WEIGHT INFLUENCE COEFFICIENTS

GROUP

Powertrain
    Engine
    Starting System
    Transmission
    Driveline
    Fuel System
    Exhaust System
    Cooling System

Structure
    Body Shell
    Sheet Metal
    Frame

Front Suspension

Rear Suspension

Brake System
    Front Brakes
    Rear Brakes
    Apply System

Steering System

Tires

Wheels

Electrical System

Instrument Panel & Controls

GROUP

Bumper Systems
    Front
    Rear

Miscellaneous Functional
and Appearance Items
    Glass
    Doors and Hardware
    Deck Lid and Hardware
    Hood and Hardware
    Seats
    Insulation
    Trim
    Ornamentation
    Heating and Ventilating
    Exterior Lighting
    Wiper and Washer
    Other Functional Items

Passengers

Luggage

Options Normally Installed
    Specific Heavy Options
    Other Options

Low Volume Options
    Optional Engines
    Specific Heavy Options
    Other Options

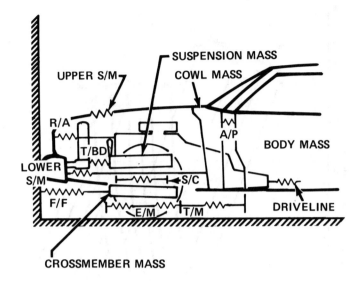

Table I — DYNAMIC FRONT CRUSH SIMULATION MODEL

$$\text{Maximize} \quad \sum_{j=1}^{n} \sum_{i=1}^{m_j} P_{ij} X_{ij}$$

Such that the following are satisfied:

1) $\sum_{j=1}^{n} \sum_{i=1}^{m_j} X_{ij}/FE_{ij} \leq$ (vehicle sales)/(fuel economy target)

2) $\sum_{i=1}^{m_j} X_{ij} \leq$ (Production capacity for engine j) $\quad i=1,2,\ldots,m_j$

3) Σ (engine sales projections by displacements considered) $\leq$ (Total combined production capacity for these engines)

4) Σ (weight class car sales) = projected weight class car sales

5) (engine CID/weight class) car sales = projected (engine CID/weight class) car sales

Where:

n = number of engine/weight classes

m_j = number of weight classes for engine j

X_{ij} = sales for j^{th} engine in i^{th} weight class

P_{ij} = (engine CID)/(Vehicle weight class) for j^{th} engine in i^{th} weight class

FE_{ij} = fuel economy for j^{th} engine in i^{th} weight class

Fig. J - Optimization model for maximizing fuel economy and performance

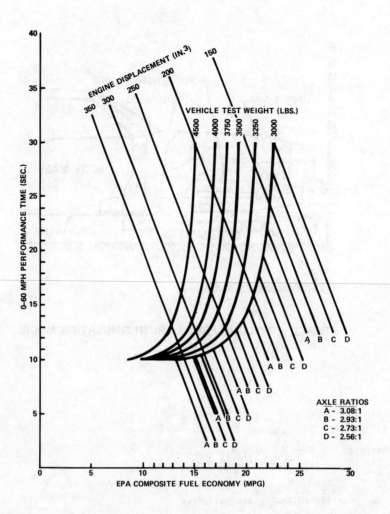

GRAPH K — ENGINE SELECTION CRITERIA

Engine Air Control—Basis of a Vehicular Systems Control Hierarchy*

Donald L. Stivender
Engine Research Dept.
General Motors Research Lab.
Warren, MI

THE CLASSICAL (AND MOST SUCCESSFUL) method of metering liquid or vaporized fuel to the air inducted by spark ignition engines remains that introduced in the 19th century, namely carburetion [1]*. Yet, even in its infancy, the carburetor's detractors appeared: "In this connection, it may be well to say that there appears to be opportunity for great improvements in the way of introducing a measured quantity of liquid gasoline into a measured quantity of air and of having little more than enough of the explosive mixture on hand at any one time than is necessary for the charging of the cylinder at the next inspiration." [2]

The mixture ratios employed were generally those designed to fulfill the requirement to "furnish a completely combustible and uniform mixture under all conditions of demand, by the attainment of perfectly regular and complete ignition, . . ." [3] These minimum criteria led to the general use of fuel-air mixture ratios with an excess of fuel over that which is chemically correct. Nicolaus August Otto determined in the year 1860 that ". . . within the narrow range of combustibility a mixture too rich would ignite easily, but the shock of the explosion would be too strong for the engine; a mixture too lean would burn more gently, but ignition was difficult." [4] This led Otto to search for a method of stratified charging by introducing near the piston both combustion products and air as diluents in order to effect both a readily ignitable mixture and a controlled slow burn. This was his Silent Otto engine of 1876 [5]. He also patented a more overtly stratified charge engine in 1877 [4,6].

Later Otto proposed the rudiments of a homogeneous charge, weak mixture, turbulent

---

*Numbers in brackets designate References at end of paper.

*Paper 780346 presented at the Congress and Exposition, Detroit, 1978.

ABSTRACT

Recent improvements in digital microprocessor hardware have given impetus to synthesizing a consistent set of central-processor engine-system control laws. As an approach to the problem, algorithms for control of airflow, EGR and spark advance were postulated, considering interactions of engine torque, fuel consumption, exhaust emissions, cold-starting and driveability. Development of an analog, real-time driver/vehicle model provided appropriate transient vehicle loads to the experimental engine/transmission/digital controller implementation throughout cold-start vehicular driving cycles. A Transient System Optimization procedure applied continuously over the federal urban driving schedule, including cold start, validated the postulated control laws.

flame propagation theory in which combustion proceeded by means of eddies which ". . . could be in irregular swirls, . . ., but the fuel-air particles would still have to be distributed in such a way that the combustion was propagated from one fuel-air particle to another through an inert medium that did not participate in the combustion and had the function of softening the shock or slowing the burning." [4] Thus were born our contemporary innovation of Exhaust Gas Recirculation (EGR) and Sir Harry Ricardo's further development of the stratified charge engine in 1918 [7].

Spark advance controls also were early in their introduction by being ". . . automatically maintained in some machines. . . ." [3] However, both spark advance and EGR control schemes have generally remained autonomous, i.e., independent of the air-fuel metering system. An early, classical experiment by Draper, et al [8,9] characterized such autonomous control subsystems as "interconnecting regulators." These approaches were shown to depend upon ". . . designing regulators that incorporate the functional relationships required to match input adjustments to each other and to the output for optimum performance . . . ." [9] Limitations of these approaches were attributed to "complex interactions of the many variables." [9]

In order to overcome these complexities, Draper, et al., disclosed an "Optimizing Control" scheme as related to internal combustion engine control. The principles of operation were characterized as encompassing a ". . . closed-chain feedback arrangement that forces the relationships among input adjustment settings to depend directly on the system output without the use of detailed assumptions as to performance characteristics." [9] The advantages of this approach were described as: "Beyond the assumption that some optimum performance condition exists, no knowledge either of the form of the input-output relationship or of quantitative data to describe this function for the installed system is required for the controller design." [9] They utilized this optimizing control on a CFR engine, using constant speed and fuel flow, an output-controlled airflow servo motor and feedback of engine torque (dynamometer armature current), which was the parameter to be optimized. Although this engine torque optimizing controller was slow and exhibited a limit cycle oscillation at the optimum point, it served to demonstrate clearly the principles of optimal control applied to an internal combustion engine.

More contemporary applications of these optimizing control and engine air control principles have been described by Schweitzer and Woods [10-15]. Their work has included a feedback of engine acceleration (a measure of the magnitude and polarity of engine torque) in applications of output control ranging from spark advance to engine airflow. Throughout his work, Schweitzer has adopted the Draper, et al., constraint enumerated above [9], i.e., that some optimum performance condition exists, and it is manifested in maximum engine torque.

In addition to just the minimization of fuel consumption, broader criteria include minimum release of unburned hydrocarbon (HC), carbon monoxide (CO), and oxides of nitrogen (NO_x), minimum fuel consumption, minimum fuel octane quality requirement, maximum cyclic combustion stability, minimum exhaust temperature, rapid transient torque response and maximum torque stability. Therefore it can be argued that optimal control based on maximizing engine torque output is not sufficient. Without the necessary further sophistication of the optimal control approach in hand, the industry has taken recourse to the "interconnecting regulators" approach, as evidenced by virtually all of our current automobile population.

However there does appear to be a middle ground between these diverse philosophies, which in part mitigates the "complex interactions of the many variables" limitation delineated by Draper, et al., if a systems approach is used. It is with our contemporary (and presumably improved) knowledge of spark ignition engines - turbulent combustion phenomena, diluent effects on combustion and spark advance requirements, and the interaction of stoichiometry, diluent and spark advance on combustion stability, thermal efficiency, output torque, and HC, CO, and NO_x emissions - that we herein endeavor to provide a rationale for an integrated, hierarchical engine systems control philosophy.

ENGINE TORQUE/EFFICIENCY CONSIDERATIONS

A common method of characterizing dynamometer engine torque and efficiency relationships over the load range is by means of constant throttle brake specific fuel consumption (BSFC) fishhooks, as shown in Figure 1. These data are for a constant engine speed of 1200 RPM, at six constant throttle settings from wide open throttle to light load. Engine torque is represented by its proportional analog, brake mean effective pressure (BMEP), and efficiency by the brake specific fuel consumption (BSFC). These representative data (extracted from Reference 16) were obtained using a 6.56 ℓ (400 CID) modified 1968 Pontiac engine with a nominal compression ratio (CR) of 10:1 and operated at minimum best torque (MBT) spark timing.

Each constant throttle (essentially constant airflow) parametric fishhook was generated by reducing fuel flow from about 13:1 air-fuel ratio (A/F) to nearly the lean combustion stability limit. The resulting Oxygen Balance [17] A/F are indicated as a parameter on the Figure. It can be noted that, for A/F lower

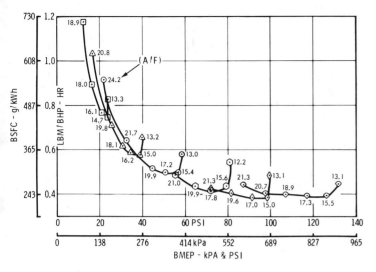

Fig. 1 - Conventional engine BSFC fishooks, 1200 rpm

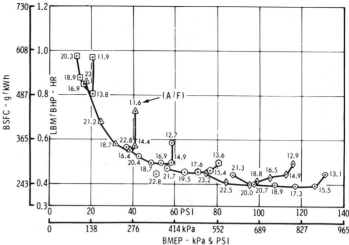

Fig. 2 - Intake valve throttled engine BSFC fishooks, 1200 rpm

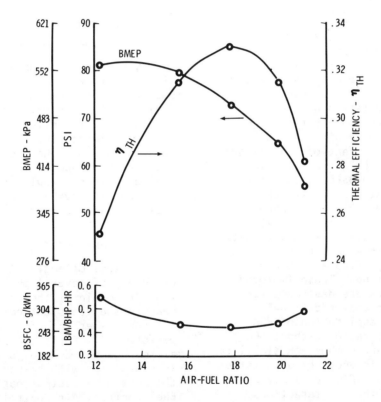

Fig. 3 - BMEP, BSFC, and brake thermal efficiency relationships from Fig. 1

(richer) than about 15:1, the engine torque (or BMEP) is essentially independent of A/F (and fuel rate). Conversely, at lean mixture ratios above 15:1, the BMEP is strongly dependent upon A/F. The BSFC data must increase without bound as the BMEP approaches zero as the definition of BSFC (g/kW h) contains BMEP implicitly in the denominator; and due to friction losses, the fuel flow cannot approach zero in proportion to the BMEP.

The effect of A/F, independent of load, is also apparent on the Figure. At light load, say 30 BMEP, it can be seen that the fishhooks do not coincide - the leaner (higher airflow) fishhook showing significantly higher fuel consumption at about 22 A/F, compared to that at about 18 A/F. It can be appreciated that the minimum BSFC over the entire load range is defined by an envelope of all possible throttle settings which is tangent to the six fishhooks.

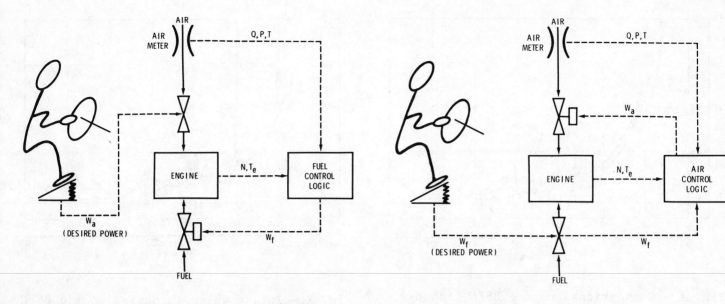

Fig. 4 - Engine Fuel Control (EFC)

Fig. 5 - Engine Air Control (EAC)

Hence, the A/F for minimum BSFC are determined at the points of tangency. At the light loads particularly, these A/F are leaner than those at the minimum BSFC points of each fishhook. It is this observation that underlies several of Schweitzer's papers [12-14]. These fuel consumption-load characteristics generally apply at all engine speeds and are typical of conventional spark ignition engines.

An engine which is tolerant of weak (lean) mixtures is characterized by a reduction in, or absence of, the increase in BSFC at lean mixtures. Figure 2 illustrates BSFC data obtained for this same 6.56 ℓ (400 CID) engine when operating with Intake Valve Throttling [16]. It can be noted that the weak-mixture BSFC data lie almost coincident with a best BSFC envelope, in contrast to the data of Figure 1 obtained with conventional throttling. These improved efficiency characteristics are desirable for weak-mixture operation in that additional latitude is provided for application of EGR and/or very lean air-fuel ratios without the usual penalties in BSFC and combustion stability.

These torque-A/F-efficiency characteristics of the conventional engine (Figure 1) may be recast better to portray their interactions. Figure 3 shows one fishhook of Figure 1 plotted on BMEP and BSFC vs. A/F bases. In addition, since thermal efficiency (η_{TH}) is inversely proportional to BSFC, the corresponding brake thermal efficiency has been also plotted. Note that the engine torque (or BMEP) decreases from a maximum near 13:1 A/F as fuel flow is decreased at constant airflow and RPM. Simultaneously, the thermal efficiency increases to a maximum at about 18:1 A/F. At constant speed

$$\eta_{TH} = K \cdot \frac{BMEP}{W_f} \tag{1}$$

where K = constant

W_f = fuel flow, g/s.

Hence, although the engine torque decreases with fuel flow at constant airflow, the torque per unit fuel flow increases to a maximum as fuel flow is reduced. Had the fishhook been carried out at constant fuel flow with the parameter increasing airflow, the thermal efficiency peak would have occurred at an even leaner A/F, since the engine BMEP, hence mechanical efficiency, would have increased with airflow and A/F, rather than having decreased with fuel flow as observed for the conventional fishhooks.

FUEL/AIR MANAGEMENT CONTROL

Control of engine A/F may be effected using any of several philosophies. The operating A/F domain of the engine dictates in large measure the transient compensation required in the controller in order to optimize the plant performance. For purposes of discussion, fuel-air management control strategies may be broadly classified with respect to the identity of the controller output parameters, i.e., either Fuel Control, Air Control, or both.

Separately, Figures 4 and 5 simply depict the general logic flow for Engine Fuel Control (EFC) and Engine Air Control (EAC), respectively. Figure 4 depicts, for the more common fuel control mode, the operator input to the system to be the modulation of engine airflow, W_a. In this mass measurement example, this airflow modulation is sensed by an airmeter as a change in volume flow (Q) at a meter ambient absolute temperature (T_a) and absolute pressure (P_a). The fuel control logic assimilates these signals and computes an air mass flow (W_a), using the perfect gas relationship $W_a = KQP_a/T_a$, where K is a constant.

From a memory internal to the logic, a programmed A/F is computed as one of a number of possible functions of W_a, P_a, T_a, Q, engine speed (N) and engine temperature (T_e). As shown, the EFC output is fuel flow (W_f), or fuel flow per revolution (W_f/N) (when injected at a periodic rate proportional to N). This fuel flow then enters the engine and results in a torque response. It can be appreciated that this EFC mode of control as depicted approximates the functional aspects of the vast majority of today's spark ignition engine fuel-air management schemes, i.e., venturi carburetion, air-valve carburetion, and a plethora of electronic fuel injection (EFI) concepts.

In all of these EFC systems, operator modulation of engine airflow results in a response in the controller output variable - fuel flow. The fuel subsequently enters the engine passages, is admitted to the combustion spaces, and ultimately results in a torque response. It should be observed here that response lags in the air sensing elements and, in larger measure, response and transport lags of the fuel circuits all contribute to a total effective time lag observable between the operator's input airflow modulation and the resulting change in engine torque when the perturbation in engine fuel rate reaches the combustion spaces. This fuel lag and attendant weak mixture generally require compensation, such as afforded by an accelerator pump (carburetor) or additional fuel injections (EFI), in order to mitigate the emission and driveability consequences otherwise resultant.

In contrast, Engine Air Control (EAC) does not directly involve control of engine fuel flow rate [8-11,15,18]. Instead, as Figure 5 depicts, the EAC controller utilizes direct modulation of engine fuel flow rate as input by the operator; simultaneously monitors P_a, T_a, Q, N, and T_e; ascribes an A/F as an *a posteriori* function of the inputs; and provides an output to servo engine airflow to the appropriate value. As shown in the Figure, airflow is controlled in a closed loop, feedback circuit; whereas the bulk of EFC applications have encompassed open loop control of fuel flow. With EAC, the response and transport lags of the "slower" fuel circuit occur simultaneously with those of the airflow circuit. Hence the EAC system tends to be self-compensating in this regard. As will be developed, this feature is particularly important with respect to transient engine operation in the lean A/F region - especially during cold engine operation.

Although applicable to any lean engine, the EAC philosophy grew out of the unique requirements of the Intake Valve Throttled (IVT) engine [16,18-21]. Due to the absence of intake manifold vacuum in this engine, carburetion was extremely difficult, and conventional speed-density fuel injection strategy could not be employed. Also, due to the mechanical forces

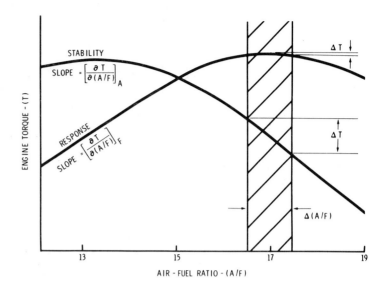

Fig. 6 - Generalized EFC engine torque characteristics - controller input (RESPONSE) and output (STABILITY)

required by the IVT valve lift mechanism, an air controller servo system was required. Hence, with both a fuel controller and an air controller required for an IVT application, it became expedient to combine these features into one system which would combine the separate controllers for airflow and fuel flow, as well as generate more suitable process parameters (described below) for control of necessary attendant systems which have commonly utilized engine manifold vacuum as a control variable (such as A/F, spark advance, EGR, and transmission modulation).

In order to illustrate further the salient distinctions between EAC and EFC in lean combustion engines, the engine torque-A/F relationship data of Figure 3 have been generalized in Figure 6, as applied to the commonly employed EFC mode of operation. The curve labeled STABILITY is the classical fishhook torque characteristic resulting from varying fuel flow at constant "throttle" or airflow. Maximum torque occurs at an A/F of about 13, which is commonly labeled the "best power mixture" for the engine. Note that the slope of this characteristic curve is equal to the partial derivative of torque with respect to A/F at constant airflow, $[\partial T/\partial (A/F)]_A$. For a fixed operator input (airflow in this fuel control example), any perturbations, dither, or instability in the fuel control output (fuel flow) results in an engine torque characteristic which follows this STABILITY curve.

For reasons both of fuel economy and engine-out exhaust emissions, the spark ignition engine is desirably operated in the lean region (such as the indicated region encompassing 17:1 A/F). Note that in this lean region engine torque stability is significantly more sensitive to variations in fuel control output than at 13

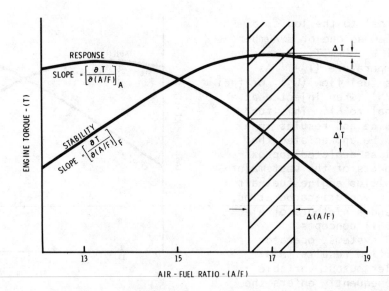

Fig. 7 - Generalized EAC engine torque
characteristics - controller input (RESPONSE)
and output (STABILITY)

A/F. This torque sensitivity is shown as the ΔT corresponding to the A/F dither band shown. Hence, under fuel control, engine torque stability (driveability) is highly sensitive to control system output stability at lean mixtures.

Engine torque response to operator input (airflow) is illustrated in the RESPONSE curve of Figure 6. In this case, again using fuel control, the input variable is airflow and, at constant fuel flow, response is characterized by $[\partial T/\partial(A/F)]_F$. Note that in the lean A/F region, response to the input airflow perturbation is a minimum, depicted by a small ΔT for the A/F band. Hence arises the "hole" or "tip-in" driveability problem observed frequently with lean engines under EFC control.

EAC mode of operation, as it influences engine torque, is shown in Figure 7. Although the partial derivative characteristics for the engine are not altered, the EAC mode interchanges the RESPONSE and STABILITY characteristics (as compared to EFC). The torque stability (ΔT) with respect to EAC output dither (airflow) is a minimum, and actually passes through zero at some lean (best economy) A/F. This characteristic makes EAC a method for application of scheduled airflow dither (together with a suitable torque response sensor) to find best torque and fuel economy.

The engine torque response (to the EAC operator input variable W_f) is shown on the $[\partial T/\partial(A/F)]_A$ curve labeled RESPONSE. For the lean A/F band of interest, it can be seen that the engine torque response is relatively large. Hence the application of EAC principles to lean, economical spark ignition engines should result in both improved engine torque stability and better response as compared to conventional fuel control methods (carburetion, EFI, etc.).

It can be observed that a fuel dither approach is not feasible using only EFC, as best torque is found along the $[\partial T/\partial(A/F)]_A$ operating line at rich mixtures, such that the result of dithering fuel flow would be minimization of engine airflow at maximum torque and about 13:1 A/F.

Experimental evidence of the transport and response time phase distinctions between EFC and EAC fuel-air management during engine transient operation was obtained using an IBM System/7 minicomputer as an engine controller. The controller processed the required logic shown in Figures 4 and 5 for EFC and EAC operation, respectively. Sensors were read and output variables were updated at a frequency of 100 Hz. An experimental 3.28 ℓ (200 CID) four-cylinder IVT engine [16,18-19] was operated on a programmable dynamometer vehicle model (to be described), employing a vehicle mass of 1134 Kg (2500 lb). Figure 8 shows the "vehicle" (dynamometer) speed (MPH) and Carbon Balance engine exhaust A/F [17] signatures during transient maneuvers while employing a digital EFC algorithm. An arbitrary step increase, followed by a step decrease, in the engine airflow resulted in a vehicle acceleration and deceleration.

It can be seen that the A/F has a lean excursion at commencement of acceleration and the typical rich period at the onset of deceleration. These A/F effects occur due to the fuel control input function being step changes in airflow, followed by airmeter lags, subsequent fuel control computations, step changes in fuel flow signal to the injectors, and finally the fuel transport delays between injectors and the engine cylinders. Again, this observed transient performance is repre-

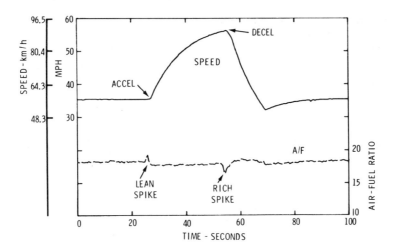

Fig. 8 - Uncompensated EFC A/F and vehicle speed transient response signatures

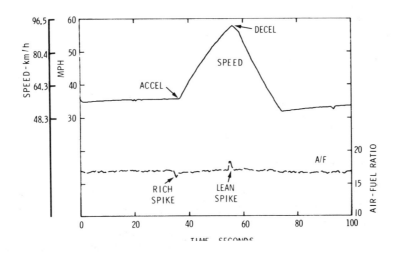

Fig. 9 - Uncompensated EAC A/F and vehicle speed transient response signatures

sentative of the nature of engine applications of fuel control (carburetion, EFI, etc.). From the above discussion of the engine torque response to A/F under EFC (Figure 6), it can be appreciated that, without compensation, the lean A/F excursion at acceleration results in a low instantaneous torque response. This corresponds to the "hole" or "tip-in" hesitation driveability deficiencies commonly requiring compensation in developmental EFC systems at lean mixtures.

In contrast, Figure 9 illustrates the use of a digital EAC algorithm, resident in the same controller and under the same transient conditions. In this case, acceleration and deceleration produce an exhaust A/F signature opposite to that for EFC. This occurs because the step input is fuel flow, and the response and transport lags of the fuel circuit occur simultaneously with, and in this case are more than compensated by, lags of the airflow sensor, controller and electro-hydraulic air servo [18].

Rather than producing the "hole" or "tip-in" hesitation observed for EFC, consideration of the EAC torque response - A/F characteristics of Figure 7 shows that the rich A/F excursion at acceleration produces an instantaneous increase in engine torque. Similarly, at deceleration the lean A/F excursion results from the fact that the engine and EAC input parameter is fuel flow, which exhibits lags shorter than those of the air circuit. Significantly, the fact that both air and fuel lags occur simultaneously indicates that compensation may be applied to the faster of the air and fuel circuits, thereby theoretically allowing optimum synchronization of the air and fuel circuits to be effected (if such performance is desired).

To illustrate these delay and lag relationships, a simplified representation of the salient transfer functions for both EFC and EAC systems are provided in Appendix A in Figures A-1 and A-2, respectively. From the form of these assumed functions (in Laplace notation), it can be seen that the response of open loop F/A ratio in the frequency domain is:

$$\text{EFC:} \quad \frac{F(s)}{A(s)} = \frac{Ke^{-T_f s}}{(\tau_c s + 1)(\tau_f s + 1)}$$

$$\text{EAC:} \quad \frac{F(s)}{A(s)} = \frac{Ke^{-T_f s}(\tau_c s + 1)}{e^{-T_c s}(\tau_f s + 1)}$$

It can be seen that EFC F/A response is dominated by fuel transport time (T_f) and the fuel controller (τ_c) and manifold fuel (τ_f) first-order lags. This frequency dependence of F/A cannot be completely compensated in the fuel control as a negative dead time $(-T_f)$ is not realizable. However, the EAC F/A response illustrates that, when the air controller contains dead time compensation $(T_c \equiv T_f)$ and a first-order fuel lag compensation $(\tau_c \equiv \tau_f)$ the F/A (s) = K, and the transient F/A response can be compensated. A subsequent section of this paper illustrates this significant capability in providing adaptive compensation $(\tau_c$ programmed as a function of engine coolant temperature) during the cold-start federal test procedure (FTP) driving schedule.

ENGINE AIR CONTROL TORQUE CHARACTERIZATION

The foregoing engine torque (T) interactions can be quantified in terms of airflow (W_a), fuel flow (W_f), engine speed (N), indicated specific fuel consumption (ISFC) and indicated specific air consumption (ISAC). From the definition of ISAC:

$$\text{ISAC} = K \frac{W_a}{T \cdot N}$$

Therefore:

$$\frac{W_a}{N} = K_1 (\text{ISAC})(T) \quad \left\{ \frac{g\ air}{revolution} \right\} \quad (2)$$

The parameter W_a/N, as developed for GMR experimental fuel-air management systems [20,21], has been dubbed the engine "Charging Function." It is a measurable quantity closely related to volumetric efficiency. For a constant ISAC, the Charging Function is an exact measure of indicated engine torque. Hence it is a good measure of engine torque and manifold pressure in the rich region, where the slope of the torque-A/F interaction curves, $[\partial T/\partial (A/F)]_A$, are essentially zero - implying nearly constant ISAC with A/F.

However, in the lean operating region of interest, $[\partial T/\partial (A/F)]_A$ (and hence, ISAC) is highly variable. Thus, an analogous parameter, the "Fueling Function" (W_f/N) has been utilized in this region [20,21] where $[\partial T/\partial (A/F)]_F$

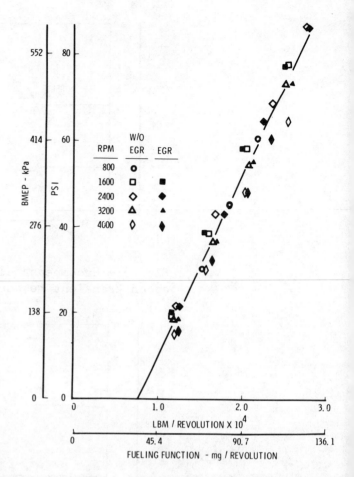

Fig. 10 - BMEP - fueling function correlation at MBT over broad load and speed ranges

approaches zero, and therefore ISFC is nearly constant. By a development similar to that for the Charging Function, the Fueling Function can be represented as

$$\frac{W_f}{N} = K_2 (\text{ISFC})(T) \quad \left\{ \frac{g\ fuel}{revolution} \right\} \quad (3)$$

Since engine ISFC at MBT spark timing and in the A/F region of best efficiency (and best torque with EAC) is also not a strong function of engine speed and load, it can be seen from the above expression that the Fueling Function is a rather direct measure of engine indicated torque. By way of example, using data obtained from a conventional 5.75 ℓ (350 CID) engine over an extensive load and speed range, those taken at MBT both with and without EGR at lean mixture ratios have been plotted in Figure 10 on a BMEP versus Fueling Function basis. Since $W_a/N = A/F (W_f/N)$, a plot of BMEP versus Charging Function (at constant A/F) would also exhibit the good correlation of Figure 10. This behavior is characteristic of spark ignition engines and has been recognized by Soltau and Senior [22]. Hence both the Charging Function and Fueling Function parameters are useful in engine systems for control of engine

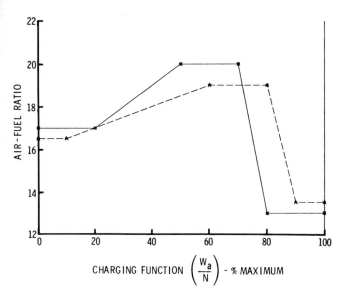

Fig. 11 - Early use of charging function for A/F control - two of the programmable functions

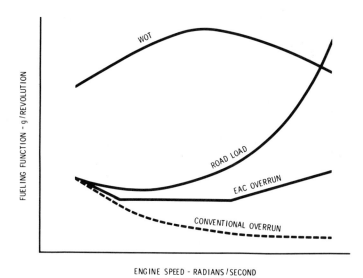

Fig. 12 - Fueling function maximum (WOT) limit, road load, and overrun signatures

load-related functions such as A/F, spark advance, EGR and transmission modulation. The analogous parameter in common use today is engine manifold vacuum or pressure.

Use of the Charging Function and Fueling Function as an A/F control parameter has been demonstrated in GMR experimental fuel-air metering control systems. Figure 11 illustrates the programmable characteristics of the A/F function generator as employed in the hydraulic air servo control systems for the IVT engine demonstration vehicle prepared for the GM Progress of Power Show [20,21]. In this application the measured Charging Function (absissa) resulted in an output A/F signal according to a previously established program. The program was adjustable as shown.

As shown in Figure 5, the operator input to the EAC system is fuel flow (nominally time-rate of fuel flow). Hence, EAC input is desired indicated engine power (at constant ISFC), and the EAC processor converts this input to represent indicated engine torque through division by engine speed. Thus the fuel flow corresponding to the desired engine power is directly admitted to the engine and the required airflow corresponding to the desired A/F (a function of the Fueling Function similar to that shown in Figure 11) is metered to the engine by the EAC processor via a feed-forward loop.

At "wide open throttle" (WOT), when the operator is requiring full engine output, the maximum Fueling Function admissible to the engine is limited by the engine WOT air-breathing characteristics. This is shown as the WOT line of Figure 12. Due to compromises imposed on the ideal valve actuation dynamics required for the broad speed and load ranges typical of automotive applications, engine WOT volumetric efficiency generally peaks in the mid-speed range. Hence the WOT Charging Function, W_a/N, also displays this characteristic. Engine WOT best power A/F is approximately 13:1. Since $W_f/N = (W_a/N)/(A/F)$, then the WOT Fueling Function also has this characteristic, as depicted in the Figure. Since, in the EAC application, the fuel rate is established by accelerator pedal position (Figure 5), a controller algorithm has been employed to (1) divide this input by engine speed, (N); (2) measure the dynamically changing WOT Charging Function; and (3) appropriately limit the maximum Fueling Function. By this means, overfueling of the engine is prevented.

The overrun, or "closed throttle" deceleration, Fueling Function is likewise modified as a function of speed in the EAC logic. In this way the minimum fueling rate to the engine can be maintained at a level sufficiently high (at high overrun speeds) to avoid the high manifold vacuum, incomplete combustion conditions which typically produce high emissions of unburned hydrocarbon with more conventional EFC systems. Figure 12 illustrates that the minimum, or overrun, Fueling Function can be scheduled to be sufficiently below the indicated road load requirement to provide adequate engine braking, yet can be maintained considerably above the typical curve (shown dashed) for conventionally carburetted systems. The very low Fueling Function values shown for the conventional systems result from throttle closure to idle position, thereby flowing essentially idle air and fuel flow rates, resulting in extremely low W_a/N and W_f/N values at high speeds. In EFC systems measures are often taken to increase the high-speed overrun Fueling Function. These have included decel-

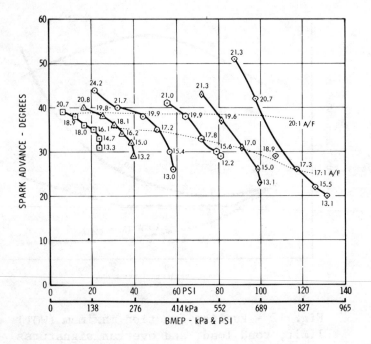

Fig. 13 — Conventional engine MBT spark advance, 1200 rpm

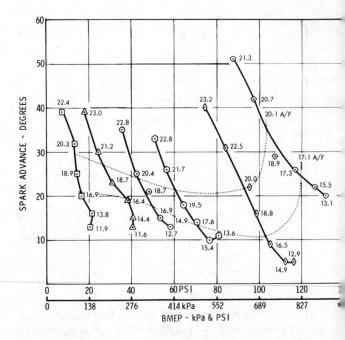

Fig. 14 — Intake valve throttled engine MBT spark advance, 1200 rpm

eration throttle "crackers," throttle return check valves and dashpots, incorporation of throttle bypass air whenever manifold vacuum exceeds a predetermined value, and/or fuel cut-off during high vacuum (low W_a/N) overruns.

TORQUE, EGR, AND SPARK ADVANCE INTERACTIONS

The total response of engine torque (T) to variations in fuel rate (F), air rate (A), exhaust gas recirculation rate (EGR) and spark advance (θ) may be simply described by the total differential:

$$dT = \frac{\partial T}{\partial F} dF + \frac{\partial T}{\partial A} dA + \frac{\partial T}{\partial (EGR)} d(EGR) + \frac{\partial T}{\partial \theta} d\theta \quad (4)$$

The partial derivatives are considered to be the influence coefficients with respect to each variable. From a control and driveability standpoint, it is highly desirable that the perturbation of a separate variable (other than fuel flow), such as by a control subsystem for EGR, has a minimum effect on either the torque or the other torque influencing parameters. The torque interactions of air and fuel flow have been addressed in a previous section. However, when EGR is employed for nitric oxide formation control, the manner of introduction determines in large measure the consequences of the "complex interactions of the many variables" noted by Draper and Li [9].

That the engine spark advance requirement is a strong function of mixture strength is shown by Figure 13. These data are the MBT spark advance requirements associated with the 6.56 ℓ (400 CID) engine fishhooks of Figure 1. It can be observed, as discussed earlier, that the torque (or BMEP) is only slightly sensitive to fuel rate at rich mixtures, but is significantly affected at lean mixtures. The MBT spark advance requirements for 17:1 and 20:1 A/F are depicted by the broken lines. The 17:1 requirement shows the typical increase in spark advance requirement as load is reduced, or the typical "vacuum" advance requirement. By comparison, Figure 14 shows the sensitivity of spark advance requirement to A/F and burn rate. These data are those obtained using the IVT throttling process [16], which produces a "fast burn" at weak mixtures. They were obtained simultaneously with the fishhook data of Figure 2. Again the 17:1 and 20:1 A/F spark advance requirements are indicated by the broken lines. These data indicate the relatively faster burn rate of the IVT process by the degree of retard exhibited at 17 and 20 A/F in the part load region, compared to the conventional throttling data of Figure 13.

Also notable is the greater effect of A/F on spark advance requirement with a fast burn process. This results from "compression of time" effects. That is, the ignition, propagation and completion of combustion all occur in a smaller crank angle (time) interval, thereby providing a greater average expansion ratio for the products. Since all these processes are compressed into a region closer to piston top-dead-center (TDC), the optimum spark advance timing is more critical with respect to influences such as A/F and load [23].

The influences of both air-fuel ratio and EGR on engine combustion speed, as evidenced by the MBT spark timing requirement, are shown in Figure 15. An eight-cylinder, 6.37 ℓ (389 CID) 10:1 compression ratio engine was operated at

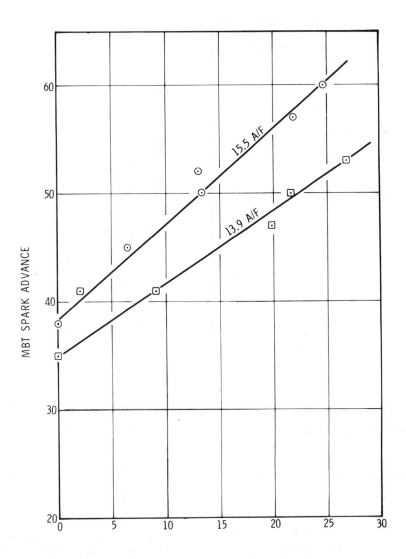

Fig. 15 - MBT spark advance as influenced by exhaust gas recirculation and A/F

constant load of 414 kPa (60 psi) IMEP at 1440 RPM. Since dilution of the combustible mixture by either excess air or exhaust gas produced both a lower reaction temperature and rate as the exhaust recirculation fraction was increased, this conventional engine required a significant increase in MBT spark advance for either of two air-fuel ratios. The increased spark advance requirement for the air-fuel ratio with the greater amount of excess air (A/F of 15.5) is apparent. Hence there is a similarity in effect on combustion rates between the diluents - air and EGR. This increased static spark advance requirement with weak mixture calibrations has generally been addressed in production vehicles.

The mixture of air and EGR admitted to the engine can be collectively considered to be the "oxidant." The data of Figure 15 can be replotted on an oxidant-fuel basis as shown in Figure 16. It is now apparent that at constant oxidant-fuel ratio the mixture of lowest air-fuel ratio (greatest exhaust gas dilution) requires somewhat greater spark advance, indicating lower reaction rates and temperatures occur in this case. At the weaker oxidant-fuel ratios, the differences in spark advance requirements are significantly reduced by normalizing the data in this fashion. It is subsequently shown that the method of EGR introduction to the engine can take advantage of these normalizing properties.

It is apparent that, from a spark advance control algorithm viewpoint, it would be desirable to decouple the spark advance algorithm from interaction with engine A/F and EGR. This would eliminate the deleterious positive feedback effect observed with conventional vacuum advance systems wherein, when mixtures lean or EGR increases, vacuum drops and spark timing retards - yet the engine requires more advanced timing, as shown above. Driving spark advance from the Fueling Function in lieu of engine vacuum provides this decoupling attribute. Experiments were run on the Driver/Vehicle Model (Appendix B) with the objective being to

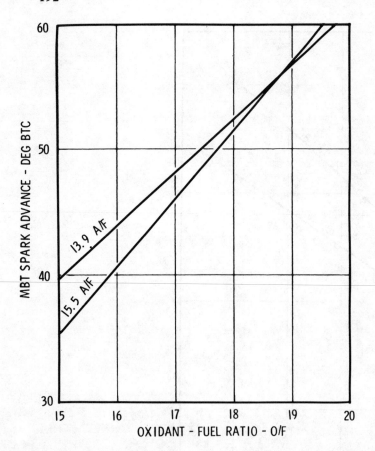

Fig. 16 - MBT spark advance as influenced by oxidant-fuel ratio

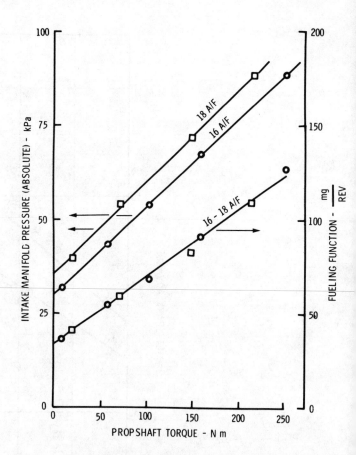

Fig. 17 - Lean A/F interaction on manifold pressure and decoupling effect of fueling function

synthesize engine systems hierarchical control laws which decouple these interactions. Using a conventional 5.7 ℓ engine, Figure 17 illustrates the increase in engine manifold pressure resulting from an increase in air-fuel ratio throughout a range of vehicle propshaft torques. Variations in EGR rate have similar effects on manifold pressure. As previously indicated, when the spark advance of conventional production systems is scheduled employing engine vacuum, weakening the mixture results in spark retard when in fact additional spark advance is required to maintain fuel economy and driveability. A hierarchical air-fuel ratio/spark advance controller has been implemented which divorces spark advance from engine vacuum [18]. Instead, the Fueling Function (fuel mass per engine revolution) has been employed as it effectively correlates with output torque over lean air-fuel ratios, as shown in the Figure.

Experiments were run comparing conventional and hierarchical modes of control. Results for the second cycle of the transient FTP are illustrated in Figure 18. The upper curve traces vehicle speed on that cycle. The second trace, showing the large spark retards induced in the production system by an A/F schedule change from 16 to 18, can be compared to the more consistent spark-advance schedule of the EAC-driven system below it. The relative change in fuel consumption obtained with the improved system (relative to the production system) is illustrated by the bottom trace. Thus, by replacing the manifold vacuum signal with an engine load parameter already used by the experimental Engine Air Control, a better spark advance schedule was obtained that improved fuel economy 5.0 percent on this cycle of the federal emission test.

THE NORMALIZED DILUTION PROCESS OF NITROGEN FIXATION CONTROL

The addition of EGR below the air metering element (or carburetor) but above the throttle can reduce both engine airflow and fuel flow (at essentially constant A/F). As a result, engine torque is reduced and the mixture is made weaker. These effects all contribute to an increased MBT spark advance requirement, which is usually not automatically compensated by the spark advance control. Hence this method of EGR introduction has a significant effect on both engine torque and efficiency.

Introduction of EGR below the throttle will, at light load, have a minimal effect on A/F and air and fuel flow rates, but the dilution of the mixture causes an increased MBT spark advance requirement. The common spark

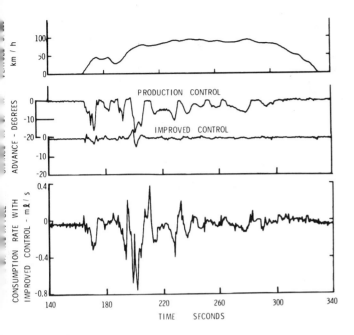

Fig. 18 - Improved fueling function - driven spark advance control decoupled 16:1 - 18:1 A/F schedule change interaction

advance control, "vacuum advance," will monitor a reduced manifold vacuum due to the EGR admission, and hence will retard the spark. Thus, under these "positive feedback" conditions, the spark control operates in the wrong direction, compounding the loss in torque, efficiency and vacuum.

These interactions, as well as the observed "similarity between the operation of an engine with relatively high amounts of exhaust recirculation and conventional air-fuel ratios to the operation of an engine with lean air-fuel ratios but without recirculation" [19] led to the adoption of an alternate method of EGR introduction called Normalized Dilution [19,23]. This method, shown in Figure 19, involves the introduction of EGR at a point upstream of the engine air metering element. It can be appreciated that, for a particular engine throttle setting, increases in the rate of EGR produced by the control valve implicitly result in an offsetting decrease in engine airflow. The air metering system actually meters the mixture of air and EGR flow in this case, and the total flow rate does not change appreciably with EGR modulation. The conventional method of EGR introduction behind the metering element is shown by the dotted line.

The Normalized Dilution process was shown to have a thermodynamic basis [19,23]. The associated mixture adiabatic flame temperature at 30 atmospheres pressure was calculated for a range of lean air-fuel ratios and EGR percentages. Figure 20 illustrates the combinations of A/F and EGR which yield the adiabatic flame temperatures calculated at both 17:1 and 20:1

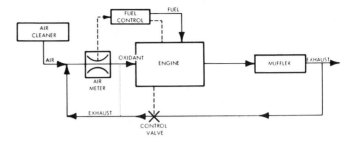

Fig. 19 - Normalized dilution method of exhaust gas recirculation meters oxidant (conventional EGR shown by dotted line)

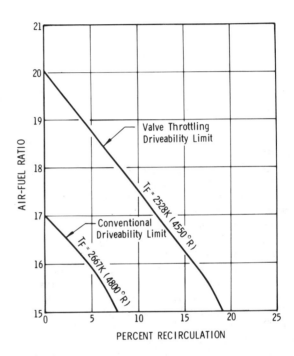

Fig. 20 - A/F - exhaust gas recirculation adiabatic flame temperature relationship generalized driveability interactions for two charge preparation geometries

A/F without EGR. It was observed [19,20,23] that these temperature isopleths generally defined combustion stability and driveability limit domains observed for conventional (circa 1967) and fast burn (IVT) spark ignition engines. The Figure shows that the fast burn engine domain is bounded by a characteristic adiabatic flame temperature of 2528 K (4550 R); considerably lower than the 2667 K (4800 R) temperature which characterized the conventional engine. As previously observed [19], these boundaries are quite generalized and are influenced by engine load, speed, geometry, etc. through their effects on charge dilution and combustion rates. Since the effects of both excess air and EGR on spark advance requirement have been shown to be similar, the combined effects are better characterized by

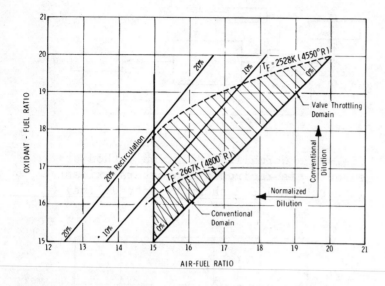

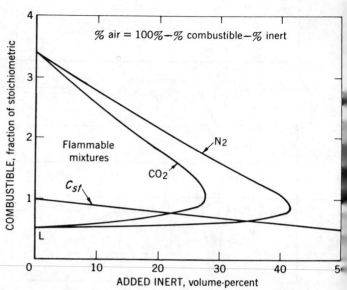

Fig. 21 - Generalized combustion stability limit interactions with A/F and O/F

Fig. 22 - Approximate limits of flammability of higher paraffin hydrocarbons (C_nH_{2n+2}, n≥ in carbon dioxide-air and nitrogen-air mixtures at 25°C and atmospheric pressure (from reference 25)

the oxidant-fuel ratio. Thus an increase in oxidant-fuel ratio through an increase in either excess air or exhaust gas recirculation results in a reduction in combustion temperature and reaction rates.

References 19 and 23 provided the plot of Figure 21: oxidant-fuel ratio versus air-fuel ratio, with percent EGR rate, or recirculation, as a parameter. Also shown are the constant adiabatic flame temperature boundaries characterizing the stable combustion domains established for engines with the combustion stability limits as calculated for Figure 20. The directions the EGR process paths take with conventional and Normalized Dilution are depicted by the arrows. As discussed above, the conventional introduction of EGR behind the metering element produces an increase in oxidant-fuel ratio at constant air-fuel ratio, and traverses a vertical path on the Figure. Such a vertical process line produces a rapid approach to the limiting reaction temperatures postulated to characterize the combustion stability limit.

Alternatively, a Normalized Dilution process path is traversed horizontally, at constant oxidant-fuel ratio. It can be seen that at an initial A/F of, say 19:1, about four times the percent recirculation may be introduced (before approaching the temperature limit) with Normalized Dilution than can be achieved with the conventional processes. Looked at another way, the Normalized Dilution process implicitly provides much of the A/F enrichment required for stable operation with EGR.

The adiabatic flame temperature combustion stability criterion disclosed in Reference 19 and described herein is not inconsistent with the literature. The White criterion states that the lower limit of combustion (obtained with downward propagation of flame) occurs at a constant flame temperature [24]. Zabetakis [25] developed an analogous constant flame temperature criterion for the flammability limit of the paraffin hydrocarbon series. More recently Quader [26,27] experimentally and analytically demonstrated in a single-cylinder engine that at the lean limit the adiabatic flame temperature and combustion duration reach limiting values. In this work [27] he also disclosed, in a plot analogous to Figure 21, that the total diluent-fuel ratio (oxidant-fuel ratio as defined herein) at the combustion limit increased with A/F in substantially the same manner as that defined by the 2528 K (4550 R) boundary of Figure 21.

Also indicative of the enrichment required with diluents is Figure 22, taken from Reference 25. Here the approximate limits of flammability of the higher paraffins (C_5 and greater) are shown for combustion in air with N_2 or CO_2 as diluents. This behavior is typical of gasoline fuel also [25]. The stoichiometric concentration of the combustible is shown to decrease linearly with dilution by the straight line, C_{st}. It can be seen that, with increasing dilution by N_2 (along the N_2 line), the flammable concentration of combustible, although nearly constant, actually exhibits an increase. At high values of dilution, this concentration crosses the stoichiometric value. This behavior is predicted by the constant adiabatic flame temperature criterion of Reference 19 and Figure 21. Quader [28], using ten percent by volume N_2 to replace engine intake air (a procedure duplicating the Normalized Dilution process), observed that the energy density of the engine charge was the same at the lean limit either with or without charge dilution. These results are consistent with the N_2 dilution flammability limit data of Figure 22, where at ten percent N_2 dilution, the limiting combustible concentration is essentially the same as that without dilution.

An early in-house study of the equilibrium and kinetic processes of engine combustion nitrogen fixation utilized the Bray Criterion [29] (a criterion which stipulates the freezing of kinetic processes during expansion based on a comparison of the instantaneous reaction rate with the instantaneous rate of shift of the equilibrium product species) to illustrate that, for rich mixtures the NO decomposition rate controlled the frozen NO exhaust concentration, whereas for lean mixtures the NO formation rate controlled the frozen NO concentration. For rich mixtures, it appeared that the quenched NO concentration was characterized by a fixed freezing temperature. This logic was induced from the analogous behavior of carbon monoxide/hydrogen quenching in engines, viz. the rich mixture carbon monoxide exhaust concentrations normalize with essentially a constant water-gas shift reaction quench temperature of bout 1806 K (3250°R), independent of the stoichiometry or associated adiabatic flame temperature. This fortunate characteristic is what permits the accurate determination of exhaust gas stoichiometry by classical, graphical techniques [30] or by computational methods [17].

The analogy here espoused between the freezing mechanisms for rich mixture CO and rich mixture NO pertains to the freezing of shifting equilibrium processes. However, for the lean mixtures of interest herein, a kinetic rate-limited, non-equilibrium NO formation process must be considered. Using simplified kinetics which have been validated earlier [31], these effects may be readily illustrated in Figure 23. Assuming the combustion process to be a step change in temperature, pressure and equilibrium composition (except NO) from the unreacted state to the burned state, the NO kinetic and equilibrium characteristics of an element of products are simply related. As an initial condition, the upper curve provides the kinetic NO formation rate for an assumed 16.4 A/F (0.9 equivalence ratio) and 2500 K (4500°F), and illustrates the equilibrium value to be about 7000 ppm.

Also on Figure 23 are NO kinetic data for lower temperature (2222 K (4000°R)) combustion processes achieved in two ways - one by leaning the mixture to 21.2 A/F (0.7 equivalence ratio) and the other by weakening the mixture via EGR. Note that the leaning process (not coincidentally) resulted in essentially the same equilibrium NO value. Leaning of the mixture produces two opposing kinetic forces -

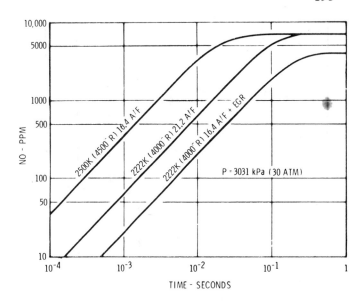

Fig. 23 - Nitric oxide kinetic and equilibrium values estimated for step-changes in reactants to the equilibrium states (except NO) indicated-simulation of stoichiometry and EGR effects

the predominant one of a monotonic reduction in products temperature and the opposite one of increasing the oxygen concentration. Although at equilibrium these effects are equal and opposite, the temperature reduction effect during the kinetic formation process (when the reverse, or dissociation, reactions are of little importance) shows that the reduced rates allow significantly less NO formation at any reaction time. At 10 ms, say, the resultant NO is reduced more than five fold.

With respect to the mixture weakening process (EGR), the same adiabatic temperature reduction has been achieved with the original stoichiometry. Thus the effect of temperature reduction alone manifests itself in reductions of both the equilibrium and instantaneous kinetic NO values. At the 10 ms time, the NO has been reduced over fifteen fold from the original high temperature condition, while the equilibrium value has been reduced by less than a factor of two. Comparing the two temperature reduction processes, the effect of leaning resulted in kinetically limited NO concentrations at 2222 K (4000°R) more than a factor of three higher than those predicted to result through the use of EGR. Hence, combustion temperature reductions via EGR (without increasing the oxygen concentration) analytically appear to be more effective in NO reduction than those via lean mixtures. This observation is supported by the data and conclusions of Toda, Nohiro, and Kobashi [32].

The NO kinetics are hereby seen to be strongly affected through the stoichiometry at constant temperature because NO is formed in

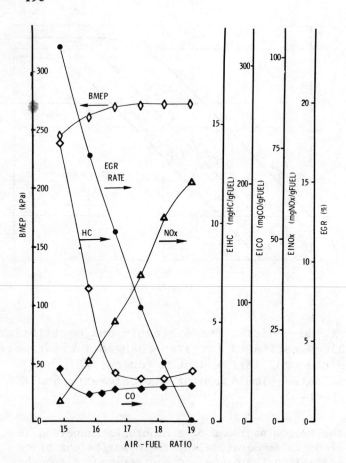

Fig. 24 - Variation of A/F, BMEP, and emission indices with the normalized dilution method of EGR introduction

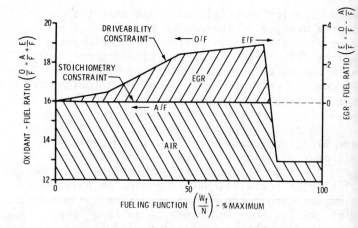

Fig. 25 - Driveability and stoichiometry constraints generally define allowable air and EGR flow via the O/F ratio and fueling function

the combustion products from available oxidizing species. Conversely, the combustion kinetics stability criteria discussed earlier are based on the much faster C-H-O kinetic reactions (compared to the NO formation mechanism being considered). Thus a "window" of reaction temperature is possible in which adequate exothermic combustion rates may be had, but in which the NO formation rate is minimized. This observation is supported by the classically observed low freezing temperature of the water-gas shift reaction and most recently by the kinetics studies of Way [33].

Since the combustion process for a premixed charge is highly exothermic, the necessary turbulent interchange of heat and mass are strongly dependent upon the turbulent scale and intensity induced in the reactants [16]. Thus the weak mixture tolerance (i.e., minimum adiabatic reaction temperature for adequate combustion stability) is highly dependent upon engine construction, giving the broad band of limiting mixture strengths (adiabatic flame temperatures) illustrated in Figures 20 and 21.

Assuming construction for weak mixture tolerance is had, the 10 ms, kinetically limited NO values of Figure 23 are the end points of a nonequilibrium formation process. However they do serve to show the trend line for the Normalized Dilution process. Thus, for the two cases, leaning and EGR addition to 2222 K (4000°R), calculated equilibrium and kinetically arrested NO values indicate that a monotonically decreasing NO concentration should result as EGR is used in conjunction with enrichment, maintaining a relatively constant adiabatic flame temperature and combustion stability throughout the Normalized Dilution process.

Another experiment, using Normalized Dilution on an 8:1 compression ratio version of the engine employed for Figure 2, likewise confirmed the potential advantages of this method [23]. Figure 24 (reproduced from Reference 23) shows the Normalized Dilution process at fixed speed, spark advance, and fuel rate. As the EGR rate was increased from zero to about 25 mol percent of airflow, the air-fuel ratio was automatically enriched since the engine air control was metering oxidant (air plus EGR) flow. (Figure 19 illustrates this mechanical arrangement.) It is again clear that the NO concentration (via the emission index - $EINO_x$) decreases monotonically throughout the air-fuel ratio region which usually provides the double-valued, bell-shaped characteristic.

Importantly, it can be seen that torque (BMEP) and brake efficiency (proportional to BMEP since fuel rate was constant) are insensitive to EGR rate up to about 15 percent EGR. These data substantiate the results of previous vehicle experiments. Also noteworthy is the fact that both CO and HC emissions exiting an oxidizing exhaust reactor are minimum throughout this range, and only increase with Normalized Dilution as the A/F enriches to values richer than about 16.5:1, where the reducing nature of the exhaust species become important. Hence, under the Normalized Dilution process, as A/F approaches the lower emission and efficiency constraint established by stoichiometry, the combustion stability (or driveability) constraint is likewise approached as characterized by the constant flame temperature criterion illustrated in Figure 21.

It can now be appreciated that this normalizing process also reverses the usual combustion stability/stoichiometry interaction wherein richer mixtures generally produce improved stability through increases in fuel concentration and thereby flame temperature. Figure 25 generally depicts how these constraints - stoichiometry and "driveability" - interact to define the allowable EGR control schedule, using "E/F" as an EGR-fuel ratio parameter. Thus maximizing the weak mixture tolerance of the engine (i.e., greater O/F ratio capability) allows greater E/F ratios (E/F = O/F - A/F) to be employed. Real-time characterization of "driveability" in this combustion stability context is demonstrated in a subsequent section, wherein the preceding interaction considerations are applied from an adaptive feedback control implementation viewpoint.

ENGINE AIR CONTROL IN A VEHICULAR SYSTEMS CONTROL HIERARCHY

In earlier sections of this paper the interactions among the primary engine output (torque) and the control parameters (fuel rate, air rate, EGR rate, and spark advance) were addressed. It was shown that, for lean, efficient engines, improved engine/vehicle response and stability can be had by employing fuel flow as the primary input (forcing function) and airflow as the controlled parameter. With respect to combustion stability limits and NO_x emission signature, the Normalized Dilution process was shown to provide a desirable coupling between airflow and EGR rate and was therefore also amenable to being scheduled as a function of fuel flow. Also the engine spark advance requirement was shown to be characterized by oxidant-fuel ratio and desirably scheduled as a function of fuel flow. This is the premise of this Hierarchical Engine System Control (HESC) philosophy - using fuel input as the forcing function with distributed control over airflow, EGR, spark advance, transmission (coupling, shift and clutch capacity), etc. The advantages of this approach will be apparent below.

With the control actions considered herein to be scheduled as primary functions of fuel flow, engine torque response (as represented by the total differential of Equation (4)) may be restated in terms both of the influence coefficients described earlier and the fuel-driven control actions in a linear fashion:

$$\delta T = \left(\frac{\partial T}{\partial F} + \frac{\partial T}{\partial A}\frac{\delta A}{\delta F} + \frac{\partial T}{\partial (EGR)}\frac{\delta (EGR)}{\delta F} + \frac{\partial T}{\partial \theta}\frac{\delta \theta}{\delta F}\right)\delta F \quad (5)$$

where the δA, $\delta(EGR)$ and $\delta\theta$ are the control actions scheduled as functions of the input fuel flow rate, δF. Naturally the control algorithm formulations include engine load functionality through the Fueling Function, as shown on Figure 25. Other influential parametric influences relating to engine speed and temperature, transmission gear ratio, barometric pressure, etc., are brought in via appropriate sensors, as is customary for all contemporary engine controllers.

Thus, the $\partial T/\partial F$ is the differential thermal efficiency, the $\delta A/\delta F$ characterizes the scheduled air-fuel ratio, the $\delta(EGR)/\delta F$ relates to the EGR schedule (which, with Normalized Dilution would be equal, but of opposite sign, to $\delta A/\delta F$), and the $\delta\theta/\delta F$ relates to the fuel-driven (via Fueling Function) spark advance schedule. Note that each of these fuel-driven control actions may now be adaptively phased (or synchronized) with the relatively slower input medium (fuel flow), as was discussed with respect to air control earlier. Separate experiments have identified the values of the adaptive time constants required to delay the response of spark advance and airflow to produce appropriate phasing with fuel flow at the engine intake valves.

An example of the identification process performed to establish the required adaptive gain schedule is best illustrated by the air-to-fuel phasing experiment. Immediately following a cold-start, the test engine/transmission was driven at a constant road load and speed on the dynamometer Driver/Vehicle Model while employing EAC resident in the test facility minicomputer (Appendix B). The exhaust emissions and engine parameters (Fueling Function, coolant temperature, etc.) were continuously monitored via the test cell data acquisition system. The input fuel rate was perturbed step-wise periodically to excite an air-fuel ratio step response signature in the engine exhaust during the transient warm-up process. The EAC-driven air valve (throttle) responded (via the EAC adaptive filter) to input fuel flow rate changes. For a fixed EAC air valve filter time constant, Figure 26 illustrates the time-wise exhaust A/F signature and associated engine coolant temperature recorded during the warm-up period. The maximum periodic A/F response amplitude envelopes are delineated by two curves - response to step increases and response to step decreases in fuel rate. Note that both these A/F excursions reverse polarity at a particular engine coolant temperature. Thus the lean, "tip-in" A/F spikes under cold engine conditions change to rich spikes as warmup ensues. The A/F envelope crossover point delineates the time, and hence coolant temperature, at which this time constant in the air servo response appropriately matches the fuel circuit response as seen at the engine intake valves and combustion chambers.

Replicate cold-start experiments employing a series of fixed air valve time constants established the effective fuel circuit first

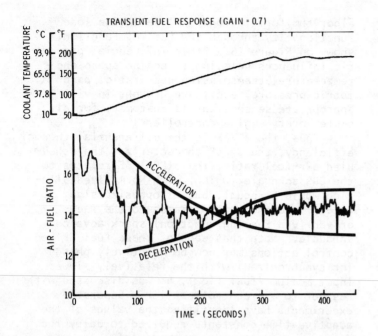

Fig. 26 - Exhaust A/F signature during engine warmup as it responded to periodic, step-wise changes in EAC fuel rate input

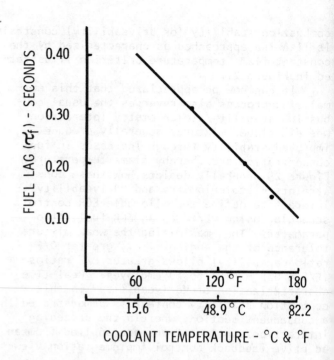

Fig. 27 - Engine warmup fuel lag defines adaptive EAC air-fuel compensation schedule

order lag/engine coolant temperature functionality shown in Figure 27. Hence these data also established the adaptive air valve compensation schedule required. These findings qualitatively agree with the results of Tanaka and Durbin [34] wherein they found no appreciable air flow lag, but a significant fuel flow lag with a time constant proportional to the ratio of inlet pipe length to engine speed. The fuel time constants they observed for their carburetted engine were longer than those shown here, which were obtained on a port-injected engine. Yet our results are highly significant in that the incorporation of the adaptive air compensation schedule of Figure 27 changed the engine from one which, under cold start, 17:1 A/F conditions, would backfire, sag and stall to one which provided relatively crisp response during the same cold start FTP driving schedule. The implications of this radical change in engine cold start behavior with adaptive air-fuel compensation are obvious with respect to potential reduction of engine choking requirements, HC and CO emissions, and fuel consumption.

In an analogous fashion, the fuel-driven spark advance hierarchy (Appendix B: IBM S/7 Controller - PPS-8 Microprocessor - modified HEI distributer) has had an appropriate filter lag term incorporated to reduce the rate of spark advance response to match that of the input fuel rate. This provision has modified what was a severe tip-in over-advance condition to a tractable rate of spark advance which also matches the rate of cylinder air and fuel charging. It can be appreciated that like com-

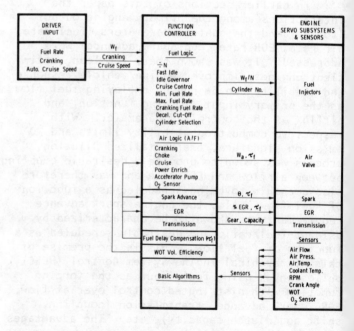

Fig. 28 - Engine air control system hierarchy

pensation can be employed for any air-fuel rate-related subsystem employed.

Figure 28 depicts a number of engine subsystems which have been and/or are amenable to being integrated within this Engine Air Control supervisory hierarchy. The fuel rate input is divided by engine speed, and this Fueling Function is passed without appreciable delay as a pulse width signal to the engine fuel injectors as the primary mode of control. Overriding fuel modification functions which have been implemented are (1) fast idle, (2) cruise control, and (3) idle speed governor (where the fuel rate is modified in an engine speed governor loop), (4) minimum Fueling Function (W_f/N) (where a "throttle return check" function serves to limit maximum overrun vacuum and exhaust HC release), (5) maximum Fueling Function (where a WOT air valve signal indicates fuel rate reduction to achieve programmed WOT A/F at the instantaneous engine volumetric efficiency), and (6) cranking fuel rate. Deceleration fuel cut-off and selection of the appropriate number of (and Fueling Function level for) a reduced number of active engine cylinders (see Appendix C) are readily incorporated fuel rate modification algorithms. Note that, for this control structure, these fuel rate functions eliminate the usually required actuators for these functions, most of which have heretofore been applied to the air circuit of EFC systems.

The air logic section of the controller depicted on Figure 28 also shows that the EAC air valve replaces separate actuators for the engine functions of cold enrichment (choke), cranking (vacuum break), open loop cruise A/F metering, power enrichment, accelerator pump (air-fuel phasing), and closed loop A/F authority based on oxygen sensor (O_2) feedback. The cranking, choke, air-fuel phasing, spark-fuel phasing, and open loop A/F scheduling implementation experiments are described in the next section. Additional EAC subsystems shown (spark advance, EGR, and the WOT volumetric efficiency fuel limiting logic) have likewise been implemented.

The engine subsystems and actuators itemized are those directly driven by the EAC controller (fuel injectors) and those hierarchical subsystems that contain subcontrollers which modify and perform logic on the EAC controller inputs (air valve servo-amplifier drives the servo; spark advance microprocessor separately inputs top dead center and RPM and outputs spark advance and dwell; EGR separately inputs manifold vacuum and exhaust backpressure and outputs a regulated EGR/exhaust flow ratio; transmission separately inputs gear ratio, WOT and vehicle speed and outputs shift initiation, hysteresis and clutch capacity). Hence this EAC engine system control structure is predicated on (1) centralization of sensor data and logic common to the subsystems and yet (2) maximization of the use of contemporary, distributed subsystem logic (pneumatic, hydraulic and/or electronic) where advantageous.

The enumeration of sensors in Figure 28, although not all inclusive, represents more than a minimum subset for the EAC validation experiments subsequently described. Fuel flow is implicitly sensed by the EAC controller from the well-defined electrical pulse-width/flow characteristic of contemporary solenoid fuel injectors. An inertialess air mass flow rate meter is desirable. This implementation utilizes a modified OEM, ultrasonic, vortex shedding, volume rate air meter, normalized to mass flow by the use of contemporary absolute pressure and temperature sensors (Appendix B). Alternatively, a speed-density algorithm, although less accurate, may be employed (using RPM, manifold pressure and temperature sensors) to eliminate the airflow sensor requirement and yet retain the precision of contemporary EFI systems.

A requisite speed (RPM) sensor may be either discrete or shared with the distributer subsystem. Coolant temperature data is used to schedule cold start spark advance, air-fuel ratio (choke function), EGR and adaptive air-fuel and spark-fuel phasing, as shown below. The "wide open throttle" (WOT) sensor (switch) is used in the adaptive limitation of maximum fueling rate as a function of instantaneous volumetric efficiency, as earlier described. The O_2-sensor is shown as an ancillary sensor for use if this control structure is employed with oxidizing-reducing (Phase II) catalyst systems. In this case, a speed-density EAC algorithm may better suffice in lieu of provision of an air meter.

Of the rather complete EAC system hierarchy simply depicted in Figure 28, the next section of this report provides experimental results obtained with implementation of the following functions:

Fuel Rate Input (EAC)

Fuel Logic:

    Cruise Control
    Idle Governor
    Min. W_f/N
    Max. W_f/N
    Cranking W_f/N
    Fast Idle

Air Logic:

    Cruise A/F
    Choke A/F
    Cranking A/F
    Power Enrichment A/F
    Adaptive Fuel-Air Phasing

Spark Advance:

    Cold Start Schedule
    Hot Schedule
    Adaptive Fuel-Spark Phasing

Driver Response Logic (Automatic Driver)

Vehicle Model (Real-Time Dynamometer Model)

IMPLEMENTATION OF ENGINE AIR CONTROL ON THE TRANSIENT DRIVER/VEHICLE MODEL

An experimental Engine Air Control System has been implemented on an engine dynamometer. The subset of functions shown on Figure 28 that

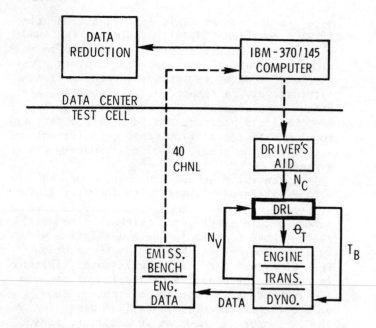

Fig. 29 - Driver response logic (DRL) and vehicle model (DYNO) provide a transient driver/vehicle model for engine-transmission control experiments

has been employed has been delineated above. The hardware is described in Appendix B. Since the premise of the EAC philosophy relates to transient response, stability and control interactions (as earlier developed), it was deemed requisite that validation of EAC be performed in the environment of transient engine/vehicle experimentation. Imposition of this constraint of course also required that the sensors, controllers and actuators be physically realized, optimized and implemented. Consequently numerous developmental and experimental programs ancillary to this validation experiment (and not described herein) were performed in order to achieve a real (albeit prototype) system.

A Driver Response Logic (DRL) circuit was developed which compares the vehicle model (dynamometer) speed with the desired (Driver's Aid) vehicle speed, thereby selecting the rates and amounts of application of "vehicle" throttle or brakes (see Figure 29). Only minor modifications were made to existing GMR dynamometer circuits for inclusion of "vehicle" brakes. With this vehicle model (grade, rolling, windage, inertial, and braking resistances), a large range of vehicle dynamics and aerodynamics can be modeled in real-time, allowing the research engine to be operated under (1) any arbitrary transient throttle schedule or (2) the FTP vehicle driving schedule. Hence, research engine/transmission configurations with unknown characteristics can be evaluated. In addition, the diagnostic capability of the GMR vehicle Fuel-Based Mass emissions [17] software is available to the research engine

test site. Thus full, transient FTP engine dynamometer evaluation of this experimental engine control configuration better relates to actual vehicle installation emission results than do existing steady-state test methods.

The Driver Response Logic and vehicle model were designed as adjunct analog controllers, autonomous from and not encumbering the EAC system experimental logic. The Driver/Vehicle Model adaptively and predictably adjusts engine throttle or "vehicle" brakes for any real-time variations in the powerplant performance (as would an actual driver) while applying the appropriate vehicular resistances at the propshaft in response to the full gamut of engine/transmission propshaft input torques (including coastdown following in-gear stall, etc.). This highly representative vehicle model, therefore, provides an early appraisal of (1) the capability of this experimental powerplant to follow adequately the FTP driving cycle, (2) the transient performance of the experimental EAC system, and (3) the actual, transient FTP exhaust emission and fuel economy potential of the experimental powerplant/control system.

Unencumbered by requirements for driver/vehicle simulation or data acquisition logic, the EAC system experimental hardware was procured and developed in a minimum configuration. The basic powerplant selected was a 5.74 ℓ (350 CID) production engine/automatic transmission featuring an electronic fuel injection system. A production catalyst was also fitted to the exhaust system. A vortex-shedding, ultrasonic, volume flow meter (Appendix B) was modified and adapted for instantaneous airflow sensing. The standard nickel wire resistance temperature sensors were retained. The associated electronic control unit (ECU) logic was for the most part disabled. Retained ECU functions were the sensor signal conditioning and excitation, injector drivers and the pulse width modulation (PWM) circuits.

The ECU injector PWM and sensor signal conditioning were interfaced with an IBM System/7 (S/7) digital minicomputer (12 K memory) which comprised the EAC supervisory controller (Figure 28). Resident within this controller were the control algorithms and calibrations which could be transferred to an on-board vehicular microprocessor controller, plus a minimal number of sub-programs for diagnosis and/or calibration purposes. This minicomputer was connected as a satellite to an IBM 370/145 host computer via a single coaxial cable and OEM Sensor Based Control Unit/Sensor Based Control Adapter (SBCU/SBCA). This extremely fast digital link allowed EAC control algorithm formulation in macro (MSP/7) and high level (Fortran) languages on the host computer.

The S/7 computer outputted to the engine air valve control system - an OEM servo-amplifier and dc servo which, in turn, positioned

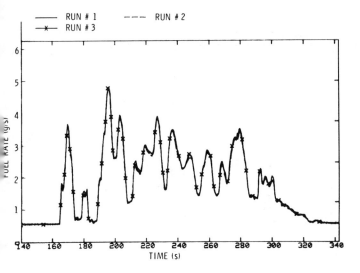

Fig. 30 - Repeatability of fuel rate (DRL output, EAC input) during three replications of the second FTP driving cycle

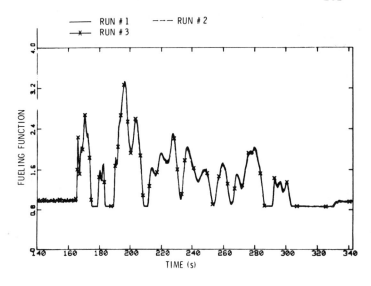

Fig. 31 - Repeatability of EAC fueling function

the butterfly air valve at the engine manifold. Also driven by the EAC algorithms resident in the S/7 were the fuel injectors (via the ECU PWM and driver circuits). For spark advance control, the S/7 controller interfaced with a Rockwell PPS-8 microprocessor which translated the spark advance command, established dwell, and subsequently drove a modified HEI distributer.

Initial calibrations for the EAC subsystems of airflow and spark advance were formulated from *a posteriori* knowledge of satisfactory functions gleaned during the initial setup. Satisfactory in this sense means that they were sufficient to allow continuous engine combustion throughout the FTP driving cycle. The Transient System Optimization procedure described below, and by Dohner [35] in more detail, iterated the calibration of the assumed form of the control laws to the optimal form, augmented the control laws with new functionality where indicated, and identified what minimum subset of new sensors, if any, would be required to achieve optimal control for minimum fuel consumption within arbitrarily established exhaust emission constraints over the FTP driving cycle.

Thus, the initial air valve control law was a simple 17:1 A/F over all loads (W_f/N) and speeds (RPM) of the driving schedule, ignoring cold enrichment and engine speed functionality (other than that implicit in the W_f/N function). The initial spark advance control law was mapped to a quadratic control law as a function of engine speed and Fueling Function (W_f/N) from the production centrifugal and vacuum advance calibration. The W_f/N vs. vacuum characteristic for a 16:1 A/F was utilized for this purpose. Hence, both the A/F and spark advance controls were divorced from engine vacuum in this EAC concept in order to minimize the deleterious, positive feedback control interactions described earlier.

Cold start FTP driving cycles were initiated using the simple initial EAC A/F and spark advance schedules indicated above. The Driver's Aid (an analog representation of the federal driving schedule intended to be matched by a real driver) was input to the Driver Response Logic (DRL). Based on instantaneous "vehicle" speed, the DRL output an engine accelerator position or vehicle brake signal to keep the vehicle speed matched to the input command. The accelerator pedal position signal was input to the A/D converters of the EAC digital controller Fuel Logic as shown on Figure 28. This accelerator signal (interpreted as a commanded fuel flow rate at the EAC input) is shown in Figure 30 as it responded in time during the second FTP driving cycle for three replicate experiments. This repeatability is typical of the Driver/Vehicle Model and greatly facilitated controller optimization using gradient techniques [35].

This accelerator signal was divided by engine speed, and the resultant Fueling Function was output through D/A conversion to the engine injector drivers through the ECU PWM signal conditioning. The replicate transient Fueling Function corresponding to Figure 30 is shown in Figure 31. It should be noted that, by dividing by engine speed, the transient excursions of this normalized Fueling Function parameter vary over about a 4:1 range, compared to the input fuel rate range of about 10:1.

The engine speed response for these same replicate tests is shown in Figure 32. It should be noted here that engine speed is not table-driven, but rather a state variable response within the central blocks of Figure 29 due to the following signal paths: (1) a Driver's Aid vehicle speed input (N_c), (2) a

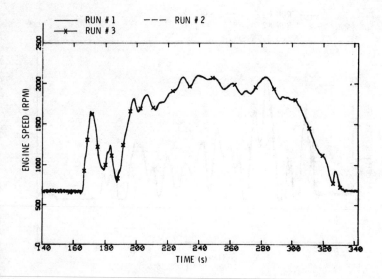

Fig. 32 - Repeatability of engine speed as a state variable response

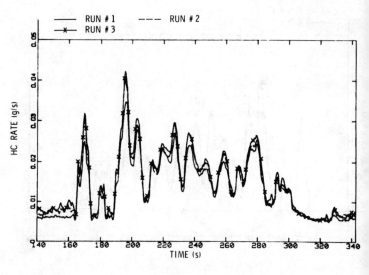

Fig. 33 - Repeatability of hydrocarbon emission rate

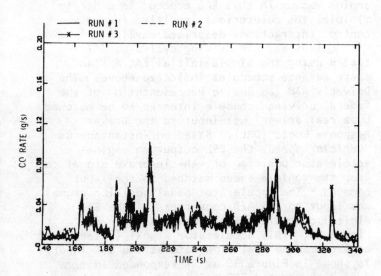

Fig. 34 - Repeatability of carbon monoxide emission rate

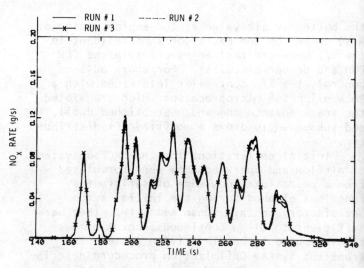

Fig. 35 - Repeatability of oxides of nitrogen emission rate

DRL controller accelerator output signal (Θ_T), (3) the EAC digital logic Fueling Function output (Figure 31), (4) the engine/transmission torque output, (5) the dynamometer vehicle model speed response (N_V), and (6) closed-loop vehicle model speed (N_V) fed back to the DRL Driver's Aid input comparator (N_C-N_V). Hence the repeatability in engine speed obtained in Figure 32 is a good measure of the repeatability and stability of the Driver/Vehicle Model which is necessary to allow discrimination these same attributes in the experimental HESC structure.

For these same replicate experiments, the engine-out exhaust HC, CO, and NO_x rates are provided for completeness in Figures 33, 34, and 35, respectively. Thus adequate repeatability of the exhaust emission analyzers,

their dynamic calibration, and the data acquisition systems were assured. These emission data were obtained continuously on a 2 Hz basis and computed using the Fuel-Based Mass emission procedure [17]. The magnitudes of these emissions are peculiar to this modified engine/vehicle hardware and not relevant to the validation of these control concepts.

With respect to the control actions (spark advance, air-fuel ratio) for this same set of FTP driving cycles, Figure 36 illustrates the spark advance signatures over this replicated second cycle. As discussed above, the spark advance control, divorced from engine vacuum to eliminate air-fuel ratio control interaction, was driven by the modified system input representing engine load (the Fueling Function) as well as by engine speed for the centrifugal

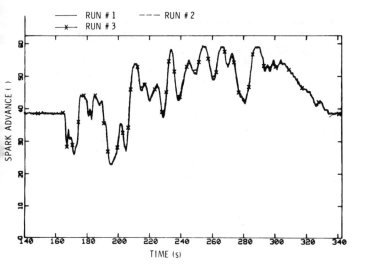

Fig. 36 - Repeatability of fuel-driven spark advance

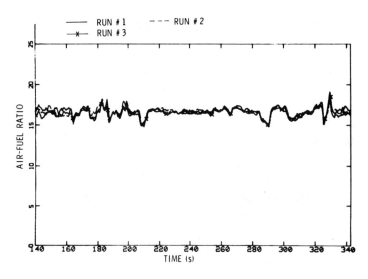

Fig. 37 - Repeatability of air-fuel ratio showing non-ideality of realizable control

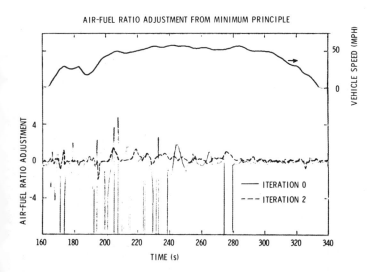

Fig. 38 - Initial and final prescribed A/F adjustments during transient system optimization

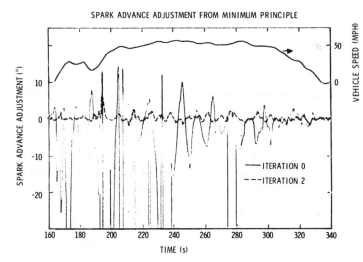

Fig. 39 - Initial and final prescribed spark advance adjustments during transient system optimization

advance function. Due to the small variance in these two spark-driving functions shown above, the resultant spark advance was likewise highly repeatable.

The corresponding air-fuel ratio signatures (as determined by exhaust gas analyses) are shown in Figure 37. Although this initial controller calibration was comprised of a constant A/F for all engine speeds and loads (Fueling Functions), it can be seen that, as in the physical realization of any controller, deviations from ideality occur. These highly repeatable deviations of a quasi-steady nature correlate with the Fueling Function signatures of Figure 31. Thus these deviations relate to combinations of imprecision in the signal conditioning and steady state calibrations of the experimental vortex-shedding air meter, engine speed sensor and fuel metering element

(dc voltage level corresponding to injector pulse time at the fuel injectors). Nevertheless, it is shown next that, via use of the Transient System Optimization procedure [35], over all loads and speeds optimal control laws have been synthesized which implicitly correct for these and other controller/engine/catalyst static and dynamic imperfections to yield maximum cold-start fuel economy within prescribed HC, CO, and NO_x emission constraints.

VALIDATION VIA TRANSIENT SYSTEM OPTIMIZATION

Having implemented both the Driver/Vehicle Model as a stable, repeatable, vehicle-like, transient environment and with the digital EAC/engine/transmission/catalyst experimental hardware initially calibrated as described above, cold-start FTP driving schedule emission

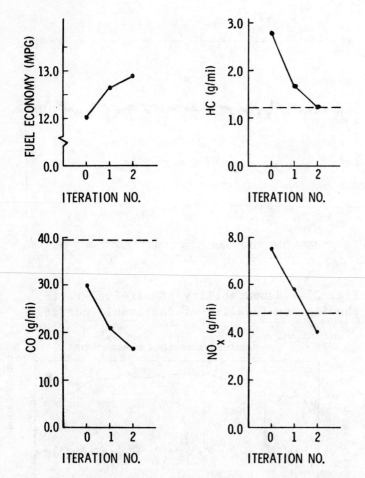

Fig. 40 - Maximization of fuel economy with convergence to emission constraints during transient system optimization

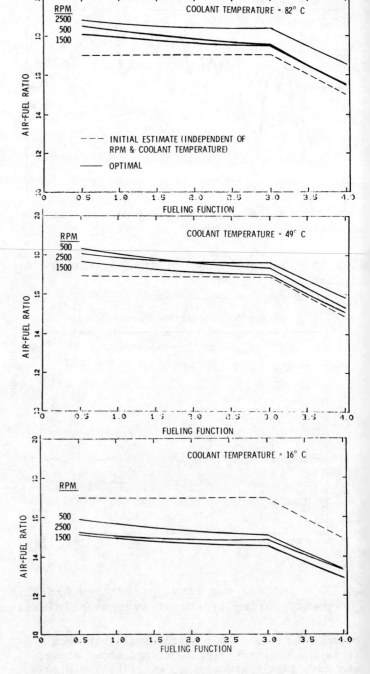

Fig. 41 - Initial (dashed) and optimal (A/F) feedback control function at three coolant temperatures

testing was initiated. Using the Transient System Optimization (TSO) procedure (an iterative, gradient technique described by Dohner [35]) sequential cold-start FTP driving cycles were performed to augment the control laws and optimize the controller calibration. For purposes of expediency, the first five cycles ("cold, transient" portion) of the FTP test were driven primarily. Approximately 80 percent of tailpipe HC and CO emissions occur during these cycles due to catalyst light-off delays and typical choke requirements. Hence the cold-start portion (5-cycle) emission constraints arbitrarily selected for this experiment were normalized to approximate overall FTP emission standards of 0.3 g/mi HC, 9 g/mi CO, and 3 g/mi NO_x.

The TSO procedure specifies continuously throughout time both direction and magnitude of simultaneous control adjustments (in this case A/F and spark advance) to be incorporated into the control laws in order to converge HC, CO, and NO_x emissions linearly to the constraints while maximizing fuel economy. Figures 38 and 39, for the second FTP driving cycle, illustrate the large prescribed adjustments both in A/F and spark advance required for the simple calibrations selected initially (iteration 0), and the small, residual adjustments prescribed for the optimized control laws implemented for the final iteration (iteration 2).

Simultaneously, the end-of-test, cold-start, HC, CO, and NO_x exhaust emissions rapidly converged to values not exceeding the constraints while maximizing fuel economy, as shown in Figure 40. Thus the residual adjustments of Figures 38 and 39 delineate the time-wise deviation of the final set of experimental control laws from ideality. Since the small

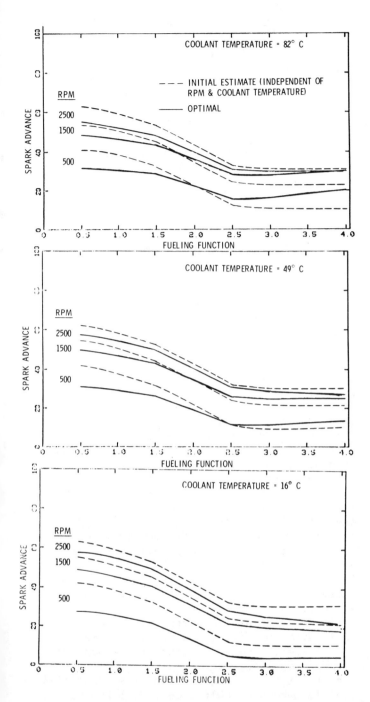

Fig. 42 - Initial (dashed) and optimal spark advance feedback control function at three coolant temperatures

magnitudes of these (which are representative of all cycles driven) lie within the tolerance bands of absolute precision expected of this experimental hardware, further iterations were deemed moot, and the final feedback control laws as implemented were thus judged to comprise the optimal set.

The original, simple A/F and spark advance control laws, shown to be deficient by the large required adjustments of Figures 38 and 39, were evolved to the optimal form via the TSO procedure by (1) identification and introduction of necessary additional functionality, (2) augmentation of the control structure, and (3) optimization of the controller calibrations, as described by Dohner [35]. Figures 41 and 42 show this evolution for A/F and spark advance, respectively. The additional functionality emerged via the regression analysis performed to transform the prescribed, time-wise adjusted control laws to time-independent, feedback control laws necessary to realize a practical control system. The feedback variables considered included those of the sensors originally incorporated into the structure as well as those additional laboratory sensors considered as possible requisites to achieve optimal control. Thus, the additional engine speed functionality for the A/F algorithm shown in Figure 41 was generated by this process. Likewise, the control structure was modfied to include an engine coolant sensor, and both A/F and spark advance control laws had this added function incorporated as a result of this identification process. Hence, both Figures 41 and 42 are shown for three coolant temperatures (16, 49, and 82°C), and these control laws are similarly modified for other engine coolant temperatures. By these means, the cold-start A/F enrichment (choke) schedule was implicitly evolved. The optimal form of the control laws thus existed in the EAC controller as second order Taylor series expansion representations of adjustments to the original control laws $((A/F)_o, (SA)_o)$, viz.:

$$A/F = (A/F)_o + 2.5 - 0.77 \frac{W_f}{N} - 1.44\, N + 0.89\, T_c$$
$$- 0.09\, (\frac{W_f}{N})^2 - 0.50\, (N)^2 + 0\,(T_c)^2$$
$$+ 0\,(\frac{W_f}{N})\,(T_c) + 0.56\,(N)\,(T_c)$$
$$+ 0.12\,(N)\,(\frac{W_f}{N})$$

$$SA = (SA)_o - 16.5 + 11.5 \frac{W_f}{N} + 10.86\, N + 0\, T_c$$
$$- 0.93\,(\frac{W_f}{N})^2 - 2.85\,(N)^2 - 3.52\,(T_c)^2$$
$$+ 3.5\,(\frac{W_f}{N})\,(T_c) - 4.1\,(N)\,(T_c)$$
$$- 2.0\,(N)\,(\frac{W_f}{N})$$

where
$\frac{W_f}{N}$ = normalized Fueling Function - g/rev

N = normalized RPM - rev/minute

T_c = normalized coolant temperature - °C

Normalization encompassed the scaling procedures typically required for digital representation and analog-to-digital conversion. It should be noted that by representing these control laws as quadratic algorithms, they may contain multi-dimensional functionality not readily amenable to the usual table look-up procedures. Due to the dimensionality required herein, this was found to be the case and the trade-off between computational time versus memory size requirement supported this approach. Also, quadratic representation is amenable to rapid calibration (gain modifications) of any of the matrix terms when contemporary, vehicular microprocessors are employed.

The TSO procedure elegantly provided a measure of the approach of this control structure/control law hierarchy to the ideal as earlier shown in Figures 38 and 39. For cold to hot transient driving, had not the necessary sensors been incorporated into the control structure and/or their functionality included in the control algorithms, the compromises in the control laws implicitly generated by the TSO procedure would have resulted in unacceptably large projected adjustments in the A/F and/or spark advance control laws during cold operation, and in similar but opposite polarity adjustments during hot operation. Such adjustments, the signature of a deficient control structure, could not have been reduced by further iterations, thereby identifying the deficiency. Since subsequent to the inclusion of the coolant temperature sensor and functionality Figures 38 and 39 do not evidence this characteristic, this experimental Engine Air Control - spark advance hierarchy has been thus validated to be sufficient in structural elements and control law formulation to provide the optimal control required for maximization of cold-start fuel economy within prescribed emission constraints. Consequently this experimental engine/transmission/catalyst/vehicle hardware has been enabled and shown to achieve its full cold-start, emission/economy trade-off potential through utilization of this hierarchical Engine Air Control - spark advance control structure.

Since this structure was optimized over the FTP driving schedule only considering the usual end-of-test emission constraints, Figure 41 illustrates that, as usual, optimal air-fuel ratio schedules for reduced NO_x at maximum economy are driven toward weak mixtures. Since practical, optimal control must also be constrained to the commercial driveability domain, a real-time "Torque Analyzer" was conceived, fabricated, and implemented to provide the "control constraint" data required by the TSO procedure in order to satisfy this essential requirement [35]. Using continuous pressure-volume signals from one (typical) engine cylinder, analog computations provided real-time imep data. These were converted to digital form, had statistical computations performed in a microprocessor, were converted to analog

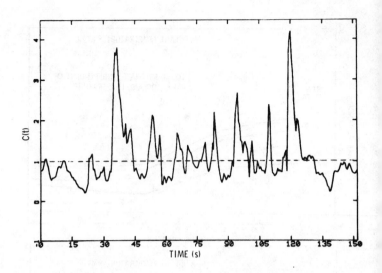

Fig. 43 - Driveability functions, $C(t)$, determined from fluctuations in engine cylinder pressure throughout the first FTP driving cycle

form, and were scanned in real-time by the data acquisition system.

These continuous imep average, variance, standard deviation and/or coefficient of variation values for engine imep therefore constituted the basis for imposing control constraints to A/F, spark advance, EGR, etc. throughout the transient driving schedule - somewhat analogous to the end-of-test emission constraint data provided continuously by the exhaust gas analyzers. Figure 43 illustrates the time-wise driveability function, normalized to eliminate necessary and desirable control actions, as obtained over the first, cold-start FTP driving cycle [35]. Since this function exceeds an arbitrarily prescribed, threshold value of unity over certain portions of the cycle, constraints to spark advance and/or air-fuel ratio control action are indicated.

Since the control structure was optimal through consideration of only the emission constraints, this engine/transmission/catalyst/vehicle hardware would provide a likewise unacceptable driveability signature no matter what control structure was employed to implement the optimal air-fuel ratio and spark advance control actions. Therefore this control constraint is considered to be necessary for specification of not only realizable, but practical, optimal control laws.

SUMMARY

Emerging capabilities of contemporary, digital microprocessor hardware gave impetus to synthesizing a coherent set of heretofore impractical engine systems control laws. Algorithms for control of A/F, EGR, and spark advance were especially formulated to decouple the usual transient interactions among these variables and engine torque, fuel consumption, exhaust emissions, and driveability. A significant reduction in the required number of engine function actuators typically required resulted thereby. A preliminary subset of these control laws was provided with initial *a posteriori* calibrations as preliminary settings for validation via transient experimentation.

A real-time Driver/Vehicle Model, incorporating analog electronics on an engine dynamometer, was developed to allow accurate transient calibration of an experimental engine/transmission and its digital controller. The Driver Response Logic (DRL) received both command (arbitrary or FTP) and actual (dynamometer) "vehicle" speeds as inputs, and applied either throttle (to engine controls) or "brakes" (to dynamometer) to follow the transient driving schedule. The dynamometer control circuitry responded in real-time to the arbitrary transmission torque inputs, modeling propeller shaft loading resulting from programmed vehicle parameters of grade, rolling, windage, and inertial resistances, as well as from the DRL vehicle braking commands.

Replicate, transient, cold-start federal test procedure vehicle driving cycles were "driven" between pertubations to the controller calibration parameters. Data were digitized at a two Hz rate throughout the transient experiments, reduced using a Fuel-Based Mass emission measurement technique, and compared (1) for system identification purposes and (2) to provide gradient data necessary to optimize the controller calibration. Using a Transient System Optimization procedure, Pontryagin's Minimum Principle was uniquely applied over all time of the transient experiments to establish the optimal form of the control structure, such that maximum fuel economy was realized while meeting arbitrary exhaust emission constraints. Since the domain of the optimization encompassed the transient implementation of the control laws, the procedure accounted for algorithm deficiencies, transport lags, sensor errors, controller imperfections and dynamics, and cold-start and thermal memory effects, as well as any uncompensated or unknown plant dynamics.

CONCLUSIONS

1. A new engine systems control structure based on an Engine Air Control (EAC) principle has been formulated to improve vehicle response and stability while minimizing actuator requirements as well as the undesirable control interactions heretofore associated with efficient, weak-mixture engines which employ lean air-fuel ratios and/or high EGR rates. It has been validated to be sufficient in structural elements and control law formulation to provide the optimal control that is requisite for maximization of cold-start fuel economy within specified emission constraints. Consequently the experimental engine/transmission/catalyst/vehicle hardware has been enabled and shown to achieve its full cold-start, emission/economy trade-off potential through utilization of this hierarchical EAC-spark advance structure.

2. The addition of a servo positioner at the throttle body of a contemporary, production engine/transmission/catalyst vehicle system enabled a digital Engine Air Control experiment. With this control structure the following functions have been enabled without requirement for any actuators:

Fuel Circuit

Fast Idle

Idle Speed Governor

Vehicle Cruise Control

Overrun Vacuum Limiter (Throttle Return Check or Dashpot)

Cranking Fuel Rate (Vacuum Break)

Deceleration Fuel Cut-Off

Cylinder Selection (Variable Displacement)

Air Circuit

Choke

Cranking A/F

Idle A/F

Part Throttle A/F

Power Enrichment

Adaptive Air-Fuel Phasing (Accelerator Pump)

Actuator requirements for these functions were obviated since they existed as electrical pulsewidth signals in the fuel circuit and as electrical air valve position signals in the air circuit. These findings indicate that a favorable cost/benefit trade-off between the requirement for an EAC air valve positioner and the resultant elimination of the above engine actuators may exist.

3. The engine spark advance control system, divorced from engine vacuum and fuel-driven by the EAC system, demonstrated significantly reduced interaction and coupling with engine air-fuel ratio. This hierarchical EAC-spark advance structure has been shown to be amenable to simultaneous controller/control law optimization for maximum vehicular fuel economy within specified emission constraints during the cold-start FTP driving schedule.

4. The EAC structure was shown to enable the adaptive phasing of engine airflow to fuel flow. Significantly large and variable fuel delays were identified throughout cold-start,

warmup and hot engine transients. The EAC air valve, adaptively delayed as a function of engine coolant temperature, transformed this engine (which at cold-start, 17:1 air-fuel ratio, would hesitate, sag and/or stall) to one which would perform relatively crisply under like FTP driving schedule conditions. These results indicate that the cold enrichment requirements (choke actuator function implicit in these control laws) for contemporary spark ignition engines can be mitigated and cold-start HC and CO emissions significantly impacted through implementation of the EAC adaptive air delay function.

5. A discrete airflow meter (ultrasonic vortex shedding) was employed as a sensor for these EAC experiments. Alternatively the engine speed and manifold pressure and temperature sensors of production engines are suitable to obviate this sensor requirement and yet make available to this EAC structure the same "speed-density" airflow measurement precision of the production systems.

6. The emerging use of microprocessors within vehicular environments brings significantly closer to realization practical application of the heretofore highly conceptual principles of this EAC-based hierarchical engine systems control structure.

ACKNOWLEDGMENTS

The contribution of A. R. Dohner was invaluable in the implementation of the digital controllers, analysis of data, and provision of the Transient System Optimization procedure. R. J. Drews and A. J. Stube provided the essential fabrication, assembly and operation of the many subsystems at the test site. M. J. Southern provided the secretarial services essential for the manuscript.

REFERENCES

1. W. Baxter, Jr., "Gasoline Automobiles," The Popular Science Monthly, LVII, 1900.
2. G. K. Barrett, "Gasoline Motors for Automobiles," Scientific American Supplement, n49, February 24, 1900.
3. A. L. Claugh, "Gasoline and Steam Vehicles Compared," The Horseless Age, 12, July 1903.
4. L. Bryant, "The Origin of the Automobile Engine," Scientific American, March 1967.
5. Nicholaus August Otto, Patent No. 2735, Germany, August 4, 1877.
6. C. Lyle Cummins, Jr., Internal Fire, Carnot Press/Graphic Arts Center, Oregon, 1976.
7. H. R. Ricardo, "Internal Combustion Engine," U. S. Patent No. 1,271,942; 1918.
8. C. S. Draper, Y. T. Li, and H. Laning, Jr., "Measurement and Control Systems for Engines," ASME Paper No. 49-SA-44, June 28, 1949.
9. C. S. Draper and Y. T. Li, "Principles of Optimalizing Control Systems and an Application to the Internal Combustion Engine," ASME Publication, New York, September 1951.
10. P. H. Schweitzer, C. Volz, and F. DeLuca, "Control System to Optimize Engine Power," SAE Paper No. 660022, January 1966.
11. P. H. Schweitzer, F. DeLuca, and C. Volz, "Adaptive Control for Prime Movers," ASME Paper No. 67-WA/DGP-2, November 1967.
12. P. H. Schweitzer, "Correct Mixtures for Otto Engines," SAE Paper No. 700884, November 1970.
13. P. H. Schweitzer, "Fishhooks and Carburetor Calibration," SAE Paper No. 700885, November 1970.
14. P. H. Schweitzer, "Control of Exhaust Pollution Through a Mixture Optimizer," SAE Paper No. 720254, January 1972.
15. R. L. Woods, "A Fluidic Fuel-Injection System Using Air Modulation," ASME Paper No. 73-WA/Flcs-3, November 1973.
16. D. L. Stivender, "Intake Valve Throttling (IVT) - A Sonic Throttling Intake Valve Engine," SAE Transactions, v77, 1968, Paper No. 680399.
17. D. L. Stivender, "Development of a Fuel-Based Mass Emission Measurement Procedure," SAE Paper No. 710604, June 1971.
18. D. L. Stivender, "Internal Combustion Engine Construction and Method for Operation with Lean Air-Fuel Mixtures," U. S. Patent No. 3,422,803, January 21, 1969.
19. D. L. Stivender, "Internal Combustion Engine Construction and Method for Improved Operation with Exhaust Gas Recirculation," U. S. Patent No. 3,470,857, October 7, 1969.
20. "GM Progress of Power - A General Motors Report on Vehicular Power Systems," GMR Technical information Department, May 7-8, 1969.
21. "GM Progress of Power Background Information," GMR Technical Information Department, May 1969.
22. J. P. Soltau and K. B. Senior, "Petrol Injection Control for Low Exhaust Emissions," Lucas Engineering Review, v6, n2, November 1973.
23. D. L. Stivender, Discussion of I. Mech. E. Paper: "Inlet Valve Throttling and the Effects of Mixture Preparation and Turbulence on the Exhaust Gas Emissions of a Spark Ignition Engine," by N. R. Beale and D. Hodgetts, Proceedings I. Mech. E., v190 1/76, January 8, 1976.
24. A. G. White, "Limits for the Propagation of Flame in Inflammable Gas-Air Mixtures. III. The Effect of Temperature on the Limits," J. Chem. Soc. v127, 1925.
25. M. G. Zabetakis, "Flammability Characteristics of Combustible Gases and Vapors," U. S. Bureau of Mines Bulletin, 627, 1965.

26. A. A. Quader, "What Limits Lean Operation in Spark Ignition Engines - Flame Initiation or Propagation?," SAE Transactions, v85, 1976, Paper No. 760760.

27. A. A. Quader, "Lean Combustion and the Misfire Limit in Spark Ignition Engines," SAE Transactions, v83, 1974, Paper No. 741055.

28. A. A. Quader, "Why Intake Charge Dilution Decreases Nitric Oxide Emissions from Spark Ignition Engines," SAE Transactions, v80, 1971, Paper No. 710009.

29. K. N. C. Bray, "Atomic Recombination in a Hypersonic Wind-Tunnel Nozzle," J. Fluid Mechanics, 6, 1-32, 1959.

30. B. A. D'Alleva and W. G. Lovell, "Relation of Exhaust Gas Composition to Air-Fuel Ratio," SAE Journal 38 90, 1936.

31. W. Cornelius and W. R. Wade, "The Formation and Control of Nitric Oxide in a Regenerative Gas Turbine Burner," SAE Paper No. 700708, September 1970.

32. T. Toda, H. Nohiro, and K. Kobashi, "Evaluation of Burned Gas Ratio (BGR) as a Predominant Factor to NO_x," SAE Paper No. 760765, October 18-22, 1976.

33. R. J. B. Way, "Methods for Determination of Composition and Thermodynamic Properties of Combustion Products for Internal Combustion Engine Calculations," Proceedings of I. Mech. E., Vol. 190 60/76, 1976.

34. M. Tanaka and E. J. Durbin, "Transient Response of a Carburetor Engine," SAE Paper No. 770046, February 28, 1977.

35. A. R. Dohner, "Transient System Optimization of an Experimental Engine Control System Over the Federal Emissions Driving Schedule," SAE Paper No. 780286, February 1978.

APPENDIX A - SIMPLIFIED TRANSFER FUNCTION REPRESENTATION

Figure A-1 illustrates a simplified transfer function representation of EFC (fuel injection, or carburetor) in terms of Laplace transforms. The intake manifold is comprised of a fast (negligible) first-order lag for the air circuit $(1/(\tau_a s+1))$. The fuel circuit is represented by a combination of a dead time and two first-order lags for the vapor and liquid fuel components of the manifold, $\exp(-T_f s)/[(\tau_a s+1)(\tau_f s+1)]$. The fuel controller is represented by a constant and lag, $K_{F/A}/(\tau_c s+1)$.

The air-fuel response at the engine $(A(s), F(s))$ to the input airflow disturbance $(R(s))$ is shown on the Figure as $A(s)/R(s)$ and $F(s)/R(s)$. Although the air response is represented by the fast first-order lag (negligible), the fuel circuit response, $K_{F/A} e^{-T_f s}/[(\tau_c s+1)(\tau_f s+1)(\tau_a s+1)]$, is considerably slower, being dominated by fuel dead time and fuel lag, $e^{-T_f s}/(\tau_f s+1)$. Thus we see in the A/F signature at the engine the typical EFC response, lean A/F at tip-in and rich A/F at overrun due to the relatively slow fuel response. Then to provide appropriate A/F phasing at the engine, the fuel circuit re-

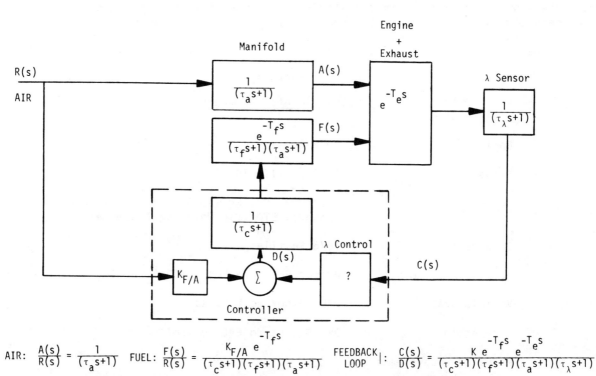

Fig. A-1 - Engine Fuel Control - simplified transfer function representation

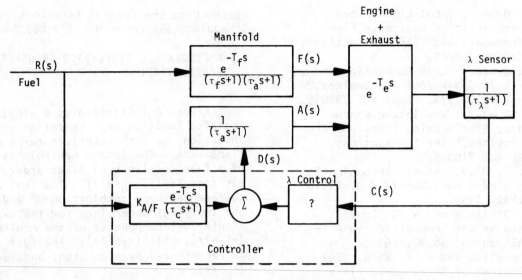

$$\text{FUEL:} \quad \frac{F(s)}{R(s)} = \frac{K\,e^{-T_f s}}{(\tau_f s+1)(\tau_a s+1)} \qquad \text{AIR:} \quad \frac{A(s)}{R(s)} = \frac{K_{A/F}\,e^{-T_c s}}{(\tau_c s+1)(\tau_a s+1)} \qquad \text{FEEDBACK LOOP:} \quad \frac{C(s)}{D(s)} = \frac{K\,e^{-T_e s}}{(\tau_a s+1)(\tau_\lambda s+1)}$$

Fig. A-2 - Engine Air Control - simplified transfer function representation

sponse must be accelerated (difficult due to the physics involved) or the airflow response compensated or slowed (not a feature of EFC but implicit in EAC systems).

Engine Air Control (EAC) is likewise represented in Figure A-2, wherein the same nomenclature is employed. Note that the input is fuel, the slowest parameter from the standpoint of dead time and lag. Hence, EAC controller action required to provide appropriate phasing of the air to fuel is readily accomplished by providing both dead time (T_c) and controller lag (τ_c) equivalent to that of the input fuel parameter. Hence tight step response of A/F is possible by slowing (compensating) the airflow circuit.

With respect to use of these EAC principles with a closed loop λ sensor (O_2) circuit, the response $C(s)/D(s)$ does not include fuel dead time or lag terms in the EAC configuration. Thus the closed loop A/F trim control may employ higher gains to achieve tighter control and improved response.

APPENDIX B

Experimental Apparatus

Engine:	1976 Seville
Transmission:	1976 Seville THM-350
Vehicle:	Adaptive 2041 kg (4500 lbm) Vehicle Model (General Electric Motoring Dynamometer)
Catalyst:	1976 Seville
Automatic Driver:	Driver Response Logic (DRL) (Analog Controller)
Vehicle Brakes:	DRL Output to Vehicle
Accelerator:	DRL Output to EAC Controller
Actuator Subsystems:	
Air:	Jordan Controls DC Servo Amplifier/Valve Positioner
Fuel:	1976 Seville ECU Injector Driver Circuits
Spark:	Rockwell PPS-8 Microprocessor/1976 Seville HEI Distributer

Sensors:

Coolant Temp.:	1976 Seville
Engine Speed:	F/V Conversion from HEI Distributer
Vehicle Speed:	General Electric Dynamometer dc Tachometer
Airflow:	J-Tec Ultrasonic, Vortex Shedding
Crankshaft Position:	Crankshaft Degree-Wheel/Magnetic Pickup
Air Sensor Temp.:	Microdot Resistance Thermometer
Air Sensor Press.:	Servonic Potentiometric
Fuel Flow:	Implicit in Seville Injector Pulse Width/Flow Calibration
Satellite Function Controller:	IBM System/7, 12 K Minicomputer
Host Computer:	IBM 370/145 Digital Computer

Emissions:

HC:	Hot Flame Ionization Detector
CO, CO_2:	Non-dispersive Infrared Analyzers
NO_x:	Chemilumenescent Analyzer
Fuel Flow:	Fluidyne Positive Displacement (0.01 mℓ resolution)
Algorithms:	GMR Fuel-Based Mass Emission Procedure

APPENDIX C - EAC ENGINE CYLINDER SELECTION (VARIABLE DISPLACEMENT) EXPERIMENT

An evaluation of the engine cylinder selection scheme mentioned above was made with the hardware of Appendix B. Reduction of engine manifold vacuum reduces pumping losses and improves vehicle fuel economy at all loads and speeds. The EAC structure allows the incorporation of an adaptive engine cylinder torque selection algorithm without recourse to any additional valve-train hardware and actuators as have been incorporated in many prior cylinder selection schemes. In order to quantify the fuel economy potential of this EAC-driven concept, an engine dynamometer experiment was run. Using the automatic driver (Driver Response Logic), Engine Air Control (EAC) and the 2041 kg (4500 lbm) Vehicle Model, a Seville engine/transmission was driven at 30 MPH in the cruise control mode.

Deactivation of one to six cylinders (via fuel cut-off) allowed the EAC control to transfer the brake power requirement to the remaining active cylinders. Figure C-1 illustrates that the EAC system thus automatically reduced engine vacuum from the normal road load value (8 x 8 mode) as cylinders were deactivated until very low vacuums were achieved (3 x 8 mode). The concomitant fuel economy and percentage increase are also shown in the Figure. The large fuel economy improvement between 4 x 8 and 3 x 8 operation points

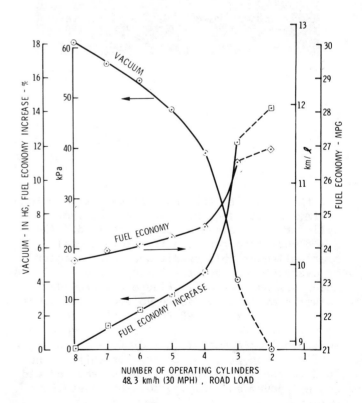

Fig. C-1 - EAC-driven cylinder selection (variable displacement) experiment - 48.3 km/h (30 mph) road load

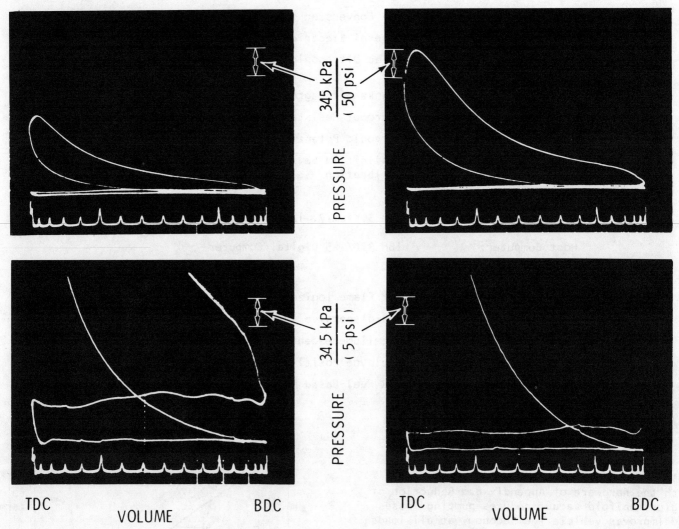

Fig. C-2 – EAC-driven cylinder selection (variable displacement) experiment – 8 x 8 and 4 x 8 modes at 1240 rpm, constant load

out the necessity for an adaptive system which minimizes, yet does not overfuel, active cylinders regardless of speed and load. The 2 x 8 mode data resulted from intentional overfueling at this condition. During coast or overrun maneuvers the 2 x 8, 1 x 8, or 0 x 8 modes are appropriate. Figure C-2 provides relative cylinder pressure-volume oscilliscope data acquired at a 1240 RPM, constant load condition. When the engine operating mode was switched from a normal (8 x 8) to a 4 x 8 configuration, the EAC controller shifted the load to the active four cylinders, and reduced engine vacuum and pumping loss by about 70 percent.

Automatic selection of the appropriate number of operating cylinders (NCYL) for minimum vacuum is basically a simple EAC algorithm:

$NCYL \geq (NCYL)_i \cdot (FUEL/REV)_i / (FUEL/REV)_{max}$.

Likewise, the basic algorithm to make automatic modulation of cylinder numbers and loads transparent to the operator is simply expressed: $(FUEL/REV)_i = (FUEL/REV)_8 \cdot 8/(NCYL)$. Consequently modulated engine operation from full load (8 x 8) to (1 x 8) or (0 x 8) operation is facilitated without the large throttle/torque interactions which would otherwise distract the driver. By these means, larger engines may be retained to provide both economy and performance.

At this juncture the concept has been demonstrated and its fuel economy potential assessed via experiment using the Driver/Vehicle Model. Obviously, complementary exhaust treatment research is required to make this concept viable in an emission control environment.

Energy Conservation with Increased Compression Ratio and Electronic Knock Control*

James H. Currie, David S. Grossman
and James J. Gumbleton
Engineering Staff, General Motors Corp.

INTRODUCTION

SPARK KNOCK CONTROL has been the focus of extensive research over the years by both oil companies and engine manufacturers. Efforts to improve the fuel economy and performance of spark ignited gasoline engines have been hindered by the occurence of knock which can be affected by such variables as engine tolerances, fuels, and vehicle operating conditions. In the past, spark knock control has been achieved by the use of fuel additives, fuel blends, or engine design modifications such as combustion chamber design, cooling system design, or retard of ignition timing. Currently, spark retard is utilized in many production engines to reduce the tendency for knock under adverse driving conditions such as heavy load, high ambient temperature, or low humidity. Accordingly, the engine may operate with reduced efficiency under less adverse conditions when retard is not required to control knock. With the advent of catalytic converters necessitating the use of unleaded fuels, the engine designer has limited the compression ratio so that the engine octane requirement is satisfied by the currently available unleaded fuels.

---

* Numbers in parenthesis designate references at end of paper.

*Paper 790173 presented at the Congress and Exposition, Detroit, 1979.

ABSTRACT

Previous investigations have shown that fuel economy gains are possible in vehicles with increased compression ratio engines that meet 1978 Federal emission standards using oxidizing converter-EGR emission control systems. There has been no incentive to raise compression ratios, however, since the vehicle gains are offset by energy losses in the refinery due to refining the higher octane unleaded fuel required by high compression ratio engines. This paper discusses the application of an electronic closed loop knock control system to a higher compression ratio engine to allow operation on 91 Research Octane Number fuel. Two cars with different compression ratios are compared with both oxidizing converter - EGR and 3-way oxidizing-reducing converter-EGR closed loop carburetor emission control systems.

Historically, raising the compression ratio of an engine has improved efficiency. However, there has been some concern, in addition to the concern for higher octane requirement that efficiency might not be improved when higher compression ratio engines were recalibrated to meet emission constraints. As a result, a study (1)* was recently completed at General Motors which indicated that with catalytic converter emission control systems, the traditional efficiency gains were possible at the current Federal Exhaust Emission Standards, (1.5 g/mi HC, 15 g/mi CO, 2.0 g/mi NOx) when raising compression ratio from 8.3:1 to 9.2:1.

This gain was accompanied by an increase in octane requirement, however, which could not be satisfied with 91 Research Octane Number (RON) unleaded fuel. Published information from the petroleum industry indicated that substantial energy losses occurred in the refinery when producing higher octane unleaded fuel. As a result, the energy losses in the refinery for making the higher octane fuels offset the efficiency gains in the vehicle. Consequently, the conclusion of this study was that there did not appear to be an incentive for increasing unleaded fuel octane levels to allow for the use of higher compression ratio engines with catalytic converter emission control systems.

This paper will describe an electronic closed loop knock control system that may satisfy the octane requirements of more efficient higher compression ratio engines, at current (1.5/15.0/2.0) and 1981 exhaust emissions levels (.41/3.4/1.0), with 91 octane fuel.

A knock control system (2) is currently being offered by the Buick Motor Division of General Motors on the Turbocharged V-6 engine. The scope of the paper will, therefore, be limited to the application of the knock control system to higher compression ratio engines.

CLOSED LOOP KNOCK CONTROL (CLKC) SYSTEM DESCRIPTION

The knock control system (Figure 1) was designed for use on engines with electronic ignition. It includes a detonation sensor or accelerometer, electronic controller and modified distributor.

The detonation sensor (Figure 2), is mounted on the intake manifold of the engine. The location of the sensor is carefully selected to insure that the knock vibration will be transmitted as equally as possible from all cylinders.

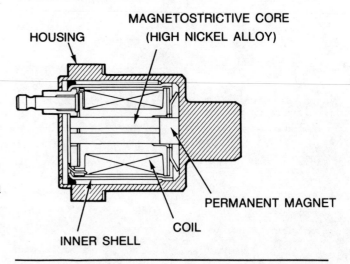

Fig. 2 - The detonation sensor is a magnetostrictive accelerometer

The knock sensor used in this study is a magnetostrictive transducer with an output voltage that is proportional to the vibration level at the "knock frequency". The frequency of the knock vibration in the engine has been found to be a characteristic of both the engine structure and sensor mounting for a particular engine family.

A typical frequency spectrum of the detonation sensor output, shown in Figure 3, illustrates the vibration characteristics of an

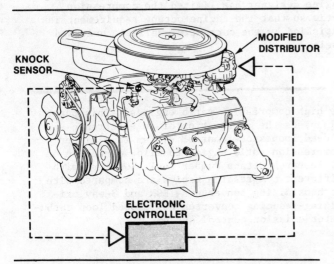

Fig. 1 - The closed loop knock control system consists of a sensor, electronic controller, and modified distributor

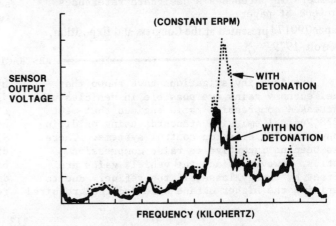

Fig. 3 - The frequency spectrum of output from the detonation sensor reveals the characteristic knock frequency for a particular engine family

engine when knock occurs. The solid line shows the output without knock and the dashed line shows the increased output at the characteristic frequency with knock. The electronic controller processes this signal from the detonation sensor so that knock can be differentiated from the normal engine vibrations or background noise that occurs without knock.

The logic used in the controller is illustrated by the block diagram in Figure 4.

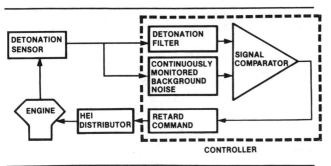

Fig. 4 - The electronic controller compares the instantaneous output of the detonation sensor to the average value to determine the amount of retard required when knock occurs

The output of the sensor is conditioned by the detonation filter circuit and also continuously averaged to establish the background noise level. The detonation signal and the background noise level are then compared. If no knock is present, the normal ignition timing, determined by the basic timing, centrifugal advance, and vacuum advance is passed directly to distributor. When knock occurs, the detonation signal is larger than the background noise level and a retard command is generated that is proportional to the intensity of knock encountered.

The retard command is used to produce a delayed or retarded ignition pulse to the distributor. The distributor ignition module has been modified to accept the retarded ignition pulse. When knock is reduced to desired levels, the controller restores the spark advance at a predetermined rate until the normal spark values are re-established. It is important to note that the system only retards and does not advance timing to seek out knock.

A typical recording of the varying spark retard generated by the knock control system is shown in Figure 5 for a high compression ratio

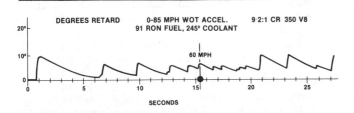

Fig. 5 - The knock control system retards the spark advance quickly to limit knock and readvances the spark at a much slower rate

engine operating under severe high ambient temperature and low humidity desert conditions during a full throttle acceleration with 91 octane fuel.

It can be noted that the spark is retarded rapidly to control knock and then readvanced slowly. This is a feature used to maintain smooth vehicle operation during knock control. More rapid changes in spark timing can cause uneven engine power resulting in surge which is observed by the driver as a jerky forward motion in the vehicle. Throughout the acceleration, the spark advance is retarded and readvanced while the system seeks to maintain a commercial knock level. In this example, the maximum retard for this condition was $11°$ during the test. The operation of the knock control system is dependent on the occurrence of knock. That is, a knocking cycle must occur to trigger the control system and generate retard. These "knock control" cycles, which occur just prior to control, may be heard as random trace to light knock in the engine during the control mode. It should, therefore, be emphasized that the system does not eliminate detonation, but controls the knock to commercial levels. More sensitive control of detonation below these levels can produce a "false retard" condition when retard occurs without knock in the engine. False retard should be avoided as it can result in a loss of vehicle performance and fuel economy.

The maximum spark retard available using the knock control system is determined by the physical characteristics of the distributor and vehicle driveability. For example, the maximum retard is physically limited by arcing or crossfire in the distributor cap which can result in a spark occurring in the wrong cylinder.

Likewise, large changes in spark retard could also contribute to the surge problem described earlier. The final vehicle spark calibration with a knock control system must, therefore, be evaluated under the most severe knocking conditions expected, to insure that the system has sufficient retard capacity to control knock while maintaining driveability.

APPLICATION OF THE KNOCK CONTROL TO HIGHER COMPRESSION RATIO ENGINE

A test program was performed to establish the potential for improving fuel economy with the knock control system and higher compression ratio at both current emission levels (1.5/15/2.0) and the 1981 emission levels (.41/3.4/1.0). Two vehicles were built and calibrated to meet targets for each emission standard. In each case, a baseline vehicle, at current production compression ratio, was compared to a similar vehicle with the compression ratio increased to 9.2:1. The vehicles at the current emission levels included an open loop carburetor and oxidizing catalytic converter, whereas the vehicles at 1981, emission levels utilized a closed loop stoichiometric carburetor and a 3-way catalytic converter.

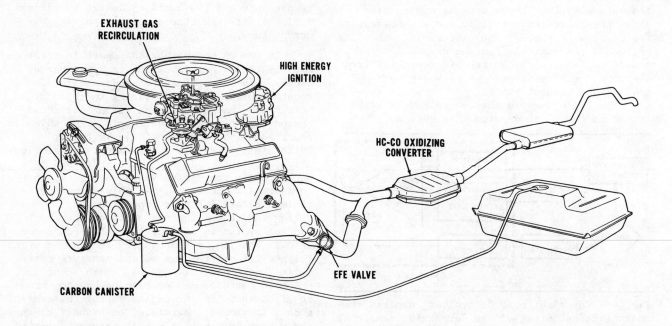

Fig. 6 - The catalytic converter - EGR emission control system represents the current production hardware

FUEL ECONOMY POTENTIAL AT 1.5/15/2.0 EMISSION STANDARDS - Two vehicles (4500 lb. inertia weight) were built with 8.3:1 and 9.2:1 compression ratio 350 CID engines. These were compared at the Federal Emission Standards of 1.5/15.0/2.0 on the 1975 Federal Test Procedure (3). Both cars were equipped with oxidizing catalytic converter-EGR emission control systems, as shown in Figure 6, and calibrated to meet the target levels shown in Figure 7. These target levels of .75/9.0/1.4-1.6 were established from experience with certification vehicles to allow for deterioration of the emission control system with mileage accumulation as well as for vehicle variability, and test variability. The knock control system was installed only on the 9.2:1 CR vehicle.

350 CID V-8 1975 FEDERAL TEST PROC.		4500 # I.W., 14.0 HP OXIDIZING CONVERTER–EGR		
AVERAGE OF 4 TESTS				
		TAILPIPE (GMS/MI)		
C.R.	AXLE	HC	CO	NOx
8.3:1	2.73	.62	3.1	1.38
9.2:1	2.41	.70	2.8	1.51
TARGET		.75	9.0	1.4-1.6
STANDARD		1.5	15	2.0

Fig. 7 - Exhaust emission targets for the 1978 Federal standards were met with both test vehicles

The carburetor, ignition and EGR systems were calibrated to meet the emission targets with each vehicle while maintaining commercial driveability. 91 RON fuel was used in both emission and fuel economy tests to include any affects of spark retard due to the knock control system. The centrifugal advance for the 9.2:1 compression ratio vehicle was adjusted from the 8.3:1 baseline to compensate for the effect of burn rate. The vacuum spark advance calibrations were adjusted so that both vehicles were within a 3% loss from optimum economy spark advance. Spark retard was not required on either vehicle to meet the hydrocarbon emission target although an increase in EGR rate was necessary for the 9.2:1 vehicle to maintain equivalent NOx emission levels.

At equivalent emission and performance levels, there was a 6.3% gain in EPA composite fuel economy (4) with the higher 9.2:1 compression ratio vehicle (Figure 8). During these emission and fuel economy tests, there was no significant amount of knock or retard from the knock control system with 91 RON fuel. Unleaded fuels of higher octane quality would, therefore, not be expected to improve fuel economy under these conditions. Performance was equalized by changing the axle ratio of these vehicles as shown. The axle ratio change allowed the improvement in thermal efficiency due to increased compression ratio to be realized as an increase in fuel economy only, as opposed to splitting the gain between performance and fuel economy.

Knock Control Under Severe Conditions - Under most ambient conditions, the 9.2:1 CR engine tested could operate on 91 RON fuel without spark retard. With more severe conditions of high ambient temperature, high coolant temperature and low humidity, the octane requirement of the higher compression ratio engine

| 350 CID V-8 | 4500 # I.W., 14.0 HP |
| 1975 FEDERAL TEST PROC. | OXIDIZING CONVERTER—EGR |

AVERAGE OF 4 TESTS

COMPRESSION RATIO	8.3:1	9.2:1	
REAR AXLE RATIO	2.73	2.41	
EPA FUEL ECONOMY			% GAIN
CITY	14.0	14.8	5.7
HIGHWAY	18.7	20.2	8.0
COMPOSITE 55/45	15.8	16.8	6.3

Fig. 8 - A 6.3% gain in fuel economy was realized for the composite 55/45 EPA fuel economy test with the 9.2 CR vehicle

would normally exceed 91 RON; however, the knock control system retards the spark as required to maintain commercial knock levels (Figure 9). Three operating conditions were chosen to represent the range of operation that might be encountered by the customer. During normal operating conditions as represented by the city driving schedule of the EPA emission test, the octane requirement of the 9.2 CR engine was found to be 84 for trace knock.

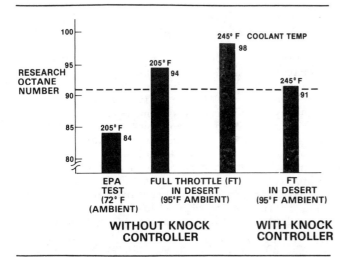

Fig. 9 - The octane requirement increased from 84 on the EPA test to 98 on a full throttle acceleration under desert conditions with the 9.2 CR vehicle

This requirement increased to 94 octane during full throttle operation at higher ambient temperatures with a normal coolant temperature of 205°F. At a higher coolant temperature of 245°F and high ambient temperature, the requirement of this engine reached 98 octane during full throttle operation. This last condition was obtained by artificially blocking the air flow through the radiator to elevate the coolant temperature to 245°F and represented the most severe operating conditions that might be encountered by the customer. This condition allowed for design margin for summer desert conditions.

Under these extreme conditions, without a closed loop knock control system, high octane unleaded fuel, in fact, 98 octane, would be required to assure satisfactory operation of the engine and emission control systems. However, with the knock control system, the engine was satisfied at commercial trace to light knock levels with 91 RON fuel under the most severe operating conditions. The amount of retard generated by the knock control system for the same range of operating conditions is illustrated in Figure 10. During operation on the EPA city schedule emission test there was no knock, and therefore no retard on 91 octane fuel.

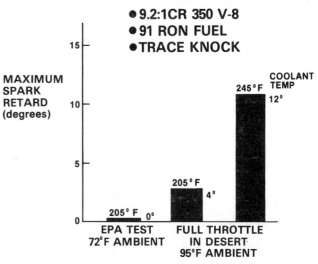

Fig. 10 - The maximum spark retard required to control knock increases under more severe ambient conditions

As conditions increased in severity, the amount of spark retard also increased to a maximum of 12° for control at commercial knock levels. This amount of retard was within the maximum retard capability of the control system and confirmed that the 9.2:1 CR engine used in these tests could be controlled to commercial knock levels using 91 RON fuel.

System Response To Fuel Octane - Another consideration in the application of the knock control system is its response to fuels that are available to the customer. In addition to engine design factors and operating conditions, the octane quality of the fuel represents a major effect on knock control system response. This was evaluated in a series of tests shown in Figure 11 with the 9.2 compression ratio engine under severe, high ambient, desert conditions. Performance and driveability were measured in these tests with fuels from 82 to 100 octane.

These results showed that as the octane of the fuel was reduced, the maximum amount of spark retard generated by the knock control increased until 19° retard was required to control knock with 82 octane fuel. The driveability was determined by measuring the variation in engine speed or surge level of the vehicle, expressed as millivolts, and was correlated with subjective surge ratings in the vehicle. As shown by the dashed line, noticeable surge and performance loss were observed with fuels below 88 octane when 12° of spark retard were required for knock control. These tests indicated that the 9.2 CR 350 V-8 was capable of operating on 91 RON fuel under all expected operating conditions and that knock could be controlled to commercial or trace levels with the closed loop system. They also indicated that the use of fuels with octane higher than 91 RON would not significantly improve the performance or driveability of this vehicle. These curves also illustrate that for the engine-vehicle system tested, the knock control capability was equivalent to about 10 Research Octane Numbers at about 12° retard without significant performance and driveability deterioration.

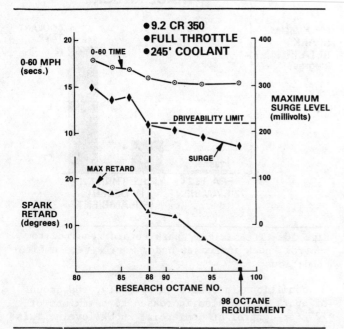

Fig. 11 - The octane quality of the fuel did not significantly effect driveability or performance above 88 RON for the 9.2 CR vehicle

FUEL ECONOMY POTENTIAL AT .41/3.4/1.0 EMISSION STANDARDS - Two vehicles (4000 lb. inertia weight) with 305 CID engines and 3-way catalyst emission control systems (Figure 12) were built to determine the fuel economy advantage of a higher compression ratio and a closed loop knock control at the 1981 Federal Emission standards of .41/3.4/1.0. The engines were built with 8.4:1 and 9.2:1 compression ratio, respectively, and calibrated within the same percent loss (approximately 1%) from best fuel economy spark timing. Both vehicles were equipped with closed loop knock control systems allowing the vehicles to be calibrated without a knock constraint. Fuel economy gains were therefore determined for the effect of compression ratio at equivalent emission levels. Rear axle ratios were selected to equalize the performance of the cars.

Emission Results - Emission targets of .25/2.5/.60 were established to allow for emission control system deterioration, vehicle variability, and test variability. These targets would not necessarily assure that these vehicles would meet the Federal Standards at 50,000 miles, however, since durability testing was not run as part of this program. Figure 13 illustrates that the exhaust emission goals for hydrocarbons, carbon monoxide, and oxides of nitrogen were met with each vehicle. In general, the higher compression ratio engine required more EGR for NOx control and more spark retard during cold start operation for HC control.

FUEL ECONOMY RESULTS - Figure 14 shows the comparison of fuel economy on the EPA city (5) and highway (6) schedules for the 8.4 CR and 9.2 CR vehicles. The 9.2 CR vehicle exhibited a 3.7% advantage in fuel economy on the EPA city test and a 5.3% gain on the EPA highway test. The overall fuel economy improvement for the composite 55/45 test was 4.3%. These results were obtained from the average of four tests on each vehicle and were all run with 91 RON fuel.

The fuel economy tests on the 9.2 CR vehicle were then run with clear Indolene test fuel (RON 98). This resulted in a small gain in fuel economy on both the city and highway tests (Figure 14) and the overall gain for the change in compression ratio was 5.9%. The reason for this difference appears to be due to the reduction in spark retard required to control knock with the higher octane fuel on the 9.2 CR vehicle.

Figure 15 illustrates the spark retard generated by the 9.2:1 compression ratio engine on 91 and 98 RON fuels during one of the more severe portions of the EPA city test. The frequency of occurrence of knock and amount of retard required to control knock was greater with 91 RON fuel. The increased retard had a minimal effect on the EPA fuel economy results as the spark was not retarded over the majority of operating conditions. The 8.4 CR vehicle did not have significant spark retard during the EPA test; therefore, no change in fuel economy would be expected due to increasing fuel octane over 91 RON.

The 305 CID, 9.2:1 CR vehicle with closed loop carburetion used in this study at .41/3.4/ 1.0 emission levels exhibited part throttle knock on the EPA emissions test, whereas no knock was evident under these conditions for the previously described 350 CID, 9.2:1 CR vehicle at the 1.5/15/2.0 emission levels. This difference may be accounted for by a combination of factors including the increased tendency for knock to occur at stoichiometric air/fuel mixtures, changes in the carburetor power air/fuel mixture cut-in point to accommodate the more stringent CO standard, and changes in the vehicle power/weight ratio.

Effect of Fuel Octane on Fuel Economy - To further evaluate the effects of higher compression ratio and fuel octane on vehicles equipped with closed loop knock controls, road fuel economy tests were run on both vehicles using

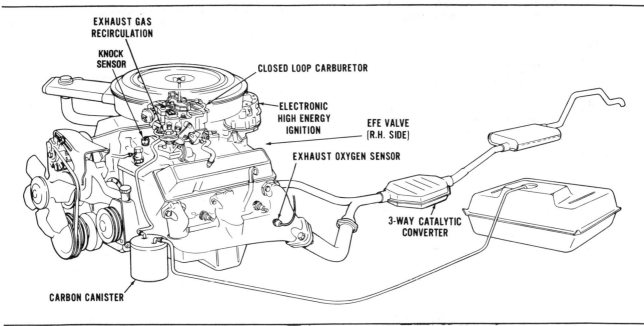

Fig. 12 - The emission control system used for the 1981 Federal standards included closed loop carburetor, 3-way catalyst and EGR

305-4	4000# I.W., 10.8 HP
CLOSED LOOP CARBURETOR	3-WAY CONVERTER - EGR
1975 FEDERAL TEST PROC.	91 RON FUEL
AVG. 4 TESTS	

		TAILPIPE (GMS/MI)		
C.R.	AXLE	HC	CO	NOx
8.4:1	2.41:1	.25	2.5	.52
9.2:1	2.28:1	.22	2.4	.63
TARGET		.25	2.5	.60
STANDARD		.41	3.4	1.0

Fig. 13 - The exhaust emissions of the 8.4 CR and 9.2 CR vehicles met the target levels for the 1981 Federal standards

305-4		4000# I.W., 10.8 H.P.		
CLOSED LOOP CARBURETOR		3 WAY CONVERTER-EGR		
AVERAGING OF 4 TESTS				
	COMPRESSION RATIO	8.4:1	9.2:1	
	REAR AXLE RATIO	2.4:1	2.28:1	
FUEL	EPA FUEL ECONOMY			% GAIN
91 RON	CITY	16.1	16.7	3.7
(CHEVRON 91)	HIGHWAY	22.7	23.9	5.3
	COMPOSITE 55/45	18.5	19.3	4.3
98 RON	CITY	—	16.9	5.0
(INDOLENE	HIGHWAY	—	24.4	7.5
CLEAR)	COMPOSITE	—	19.6	5.9

Fig. 14 - The 9.2 CR vehicle exhibited a 4.3% gain in fuel economy on the EPA composite 55/45 test

SAE Urban and I-55 fuel economy schedules (7) along with the GM City schedule. The amount of fuel economy improvement due to higher compression ratio varied depending on the driving schedule chosen. Also, there was a slight increase of about 1.5% in fuel economy with increasing fuel octane between 91 and 100 RON for the 9.2:1 CR vehicle. Octane quality did not appear to have an effect on fuel economy on the 8.4:1 CR vehicle.

Effect Of Fuel Octane On Performance - The full throttle performance, as measured by a 0-60 MPH acceleration time and the retard required to control knock is shown in Figure 17 with fuel octane varying from 82 to 100 RON. The performance of both vehicles was not affected greatly by a decrease from 100 to 90 RON fuel. Below 90 RON the acceleration times for both vehicles began to increase markedly. With the difference in axle ratios, the acceleration times of both vehicles were evenly matched irrespective of fuel octane.

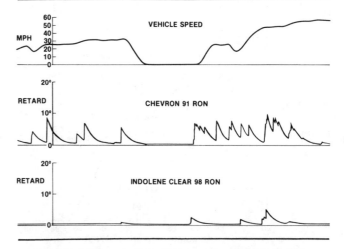

Fig. 15 - The amount of retard encountered by the 9.2 CR vehicle on the EPA City test was reduced when fuel octane was increased from 91 RON to 98 RON

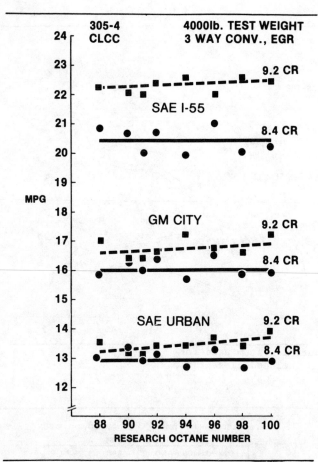

Fig. 16 - Increasing fuel octane resulted in a slight gain in fuel economy during road testing for the 9.2 CR vehicle

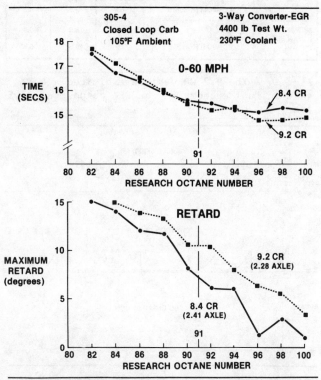

Fig. 17 - Full throttle performance was not effected greatly by a decrease in fuel octane from 100 to 90 RON

The maximum number of degrees of spark retard necessary to control knock to trace levels during the full throttle accelerations is also shown. The maximum amount of spark retard available was restricted to 15 degrees for driveability considerations and physical limitations in the distributor to prevent crossfire. The 9.2 CR vehicle did not reach the 15 degree limit until the fuel octane was reduced to 84 RON. This testing was performed under severe conditions at the General Motors Desert Proving Grounds with ambient temperatures in excess of 100°F.

Octane Requirement - The octane requirements at full and part throttle for the 9.2 CR and 8.4 CR vehicles were determined both with and without the knock control at the General Motors Desert Proving Grounds (Figure 18) with coolant temperatures of 225°F and 240°F. Without the electronic knock control, the full throttle octane requirement of the 9.2 CR vehicle with closed loop carburetor was similar to that of the 9.2 CR open loop carburetor vehicle used in the previously described testing at 1.5/15/2.0 emission levels. At wide open throttle (WOT) both vehicles were operating open loop at power air/fuel mixtures. The octane requirement of the 8.4 CR vehicle was greater than 91 because the spark calibration had been set closer to MBT spark advance than a vehicle without an electronic knock control. The part throttle requirements for these vehicles were high. The 9.2 CR vehicle part throttle requirement approached, or was equal to, the full throttle requirement. This was consistent with the observations made on the chassis dynamometer emission tests previously discussed. It is evident that the closed loop knock control with 15° maximum retard capacity would be capable of controlling knock at trace to light levels on these test vehicles even with the coolant temperature elevated to 240°F.

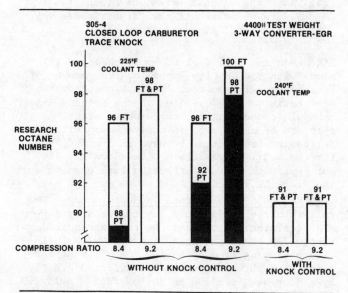

Fig. 18 - The octane requirement with 240°F coolant is within the control capability of the knock control system when 91 RON fuel is used

It is important to note that the application of the knock control to production vehicles requires customer acceptance of trace knock as a normal condition to achieve optimum fuel economy under certain modes of engine operation. Improvements to the system may reduce the knock to generally inaudible levels, however it is inherent in the control scheme to have a knock event initiate the retard mechanism.

CONCLUSIONS

Based on our studies of a 350 CID, 4500 lb. I.W. vehicle equipped with a closed loop knock control and oxidizing converter-EGR emission control system, increasing compression ratio from 8.3:1 to 9.2:1 at 1978 Federal emission levels (1.5/15/2.0) results in a gain of approximately 6% in fuel economy with 91 RON fuel.

A similar study of a 305 CID, 4000 lb. I.W. vehicle equipped with closed loop knock control, closed loop carburetor, and a three-way emission control system at the 1981 Federal emission levels (.41/3.4/1.0) showed a gain of 4.3% in fuel economy for an increase in compression ratio from 8.4:1 to 9.2:1 on 91 RON fuel.

The combined results indicate that the capability of the closed loop knock control system as applied to these engine-vehicle combinations appear to be:

1. 8 to 10 RON control - Based on both full throttle and part throttle octane requirements, the system can reduce the vehicle octane requirement for trace knock levels by 8 to 10 RON.

2. 15° maximum spark retard - To maintain acceptable driveability and distributor performance, the system was limited to a 15° maximum spark retard.

3. 3 to 6% fuel economy improvement - With 9.2:1 compression ratio and 91 RON fuel, the closed loop knock control can provide a 3 to 6% improvement in fuel economy depending upon emission control standards, driving schedule, ambient conditions, and engine design. Some engines, however, may be emission limited at higher compression ratios and may not achieve the same gains in fuel economy.

These tests illustrate that the knock control system only retards when required due to operating conditions and, therefore, permits fuel economy gains under the majority of driving conditions. It is expected, however, that other engines may require more or less retard to control knock at higher compression ratios depending on such design parameters as combustion chamber, fuel control system, emission control system or cooling system. Application of the knock control system would, therefore, require thorough testing to determine the fuel economy advantage with each engine family and the maximum compression ratio which can be used within the design constraints of the knock control system.

ACKNOWLEDGEMENTS

The authors gratefully acknowledge the significant contributions of John Auman, Edward Durham, Edward Riley, David Singer, and Walter Taylor to this study.

REFERENCES

1. J. J. Gumbleton, G. W. Niepoth, and J. H. Currie, "Effect of Energy and Emission Constraints on Compression Ratio", Paper 760826 presented at SAE, October 1976.

2. T. F. Wallace, "Buick's Turbocharged V-6 Powertrain for 1978", Paper 780413 presented at SAE, March 1978.

3. Federal Register, Part 86, "Certification and Test Procedures", Volume 40, Beginning Page 27590, June 30, 1975.

4. U.S. Department of Transportation and Environmental Protection Agency, "Potential for Motor Vehicle Fuel Economy Improvement, Report to Congress", October 24, 1974, pg. 17.

5. R. E. Kruse and T. A. Huls, "Development of the Federal Urban Driving Schedule", Paper 730553 presented at SAE, May 1973.

6. T. C. Austin, K. H. Hellman and C. D. Paulsell, "Passenger Car Fuel Economy During Non-Urban Driving", Paper 740592 presented at SAE, August 1974.

7. "Fuel Economy Measurement - Road Test Procedures", SAE J1082, April 1974.

A Fuel Economy Development Vehicle with Electronic Programmed Engine Controls (EPEC)*

Bruce D. Lockhart
Engineering Staff, General Motors Corp.

INTRODUCTION

WITH THE GROWING IMPETUS for energy conservation in the use of automotive transportation systems the incentive for technology gains is clear. A 30 percent reduction in fuel consumption through technology gains might allow the vehicle size and weight to remain at the 1979 level and still satisfy the 1985 fuel economy standard. Or with no technology gain the weight of the average vehicle might have to be reduced to less than 1100 kg (2400 lbs.)

The list of technology to reduce fuel consumption is continually being expanded through improvements in engine, drivetrain, and chassis efficiency and the use of electronic control systems. Many of these devices have had limited evaluations and are being further developed. What has been needed is to combine these devices to evaluate their total effect.

As the number of items requiring electronic controls for maximum benefit increases, the development time required to obtain acceptable driveability, emission levels, and fuel economy improvements also increases dramatically.

*Paper 790231 presented at the Congress and Exposition, Detroit, 1979.

---ABSTRACT---

A vehicle has been built to incorporate a number of fuel economy improvement components with an Electronically Programmed Engine Control (EPEC) system. The vehicle system was optimized for fuel economy within driveability and emission constraints and provided a 12.5 percent improvement in fuel economy on the combined FTP tests. An evaluation of the fuel economy benefit of the combined system and individual components is presented.

DISCUSSION

A vehicle was built to incorporate a number of fuel economy improvement items with an Electronically Programmed Engine Control System (EPEC). This vehicle was developed to extend the technique of applying electronic controls to inter-related, multi-function systems with common input parameters. The vehicle was optimized for fuel economy and driveability at the 1981 emission levels of .41 gm/mi HC, 3.4 gm/mi CO, and 1.0 gm/mi NOx. This vehicle was under development for nearly a year prior to the test runs presented here and represents a fuel economy potential, not a state-of-the-art for production.

The EPEC vehicle and the comparison vehicle were both 1978 full size vehicles with a 4.3ℓ V-8 engine, 2.56:1 axle ratio, and tested at 4000# inertia weight. This combination was chosen based on information from an optimum engine size study [1] indicating economy improves as engine size is reduced until a minimum performance level of 20-25 seconds 0-60 MPH is reached. These vehicles have a 0-60 MPH performance time of 17-18 seconds.

The fuel economy improvement features combined together in this one vehicle are categorized in four areas as shown in Table 1. Items that reduce road load horsepower requirements are: 1) Retracted pad 'drag free' front disc brakes, 2) Steel belted radial tires one size oversize from standard production and inflated to 30 psi pressure. 3) Lower viscosity (75 w) axle lube with friction modifiers.

Items that reduce engine and accessory parasitic losses include an electric cooling fan controlled by the micorprocessor and Mobil 1 low viscosity engine oil.

The following items improve engine operating efficiency and provide the required emission controls. These are: 1) Electronic sequential port fuel injection with air metering and individual cylinder fuel bias capability for improved control of mixture distribution. 2) Closed loop stoichiometric A/F control with an externally heated O_2 sensor, 3) A 260 IN^3 3-way bead catalytic converter. 4) Electronic programmed spark timing (EST). 5) Programmed variable orifice EGR (EEGR). 6) Programmed idle speed control (ISC) and cold fast idle requirements. 7) Best economy highway mode logic that allows a complete calibration change (A/F, EGR, spark advance) to operate the engine at best economy during extended cruise conditions. The best economy calibration for this engine is near the lean limit A/F (20:1) with 0-4% EGR and MBT spark for those conditions.

Automatic transmission losses were reduced by use of a program controlled torque converter clutch to combine a high percentage of useage with acceptable driveability.

Table 1 - Electronic Programmed Engine Control (EPEC)

I. VEHICLE

    FULL SIZE SEDAN – 4000 LB. I.W.
    260 CID V-8 ENGINE
    2.56 REAR AXLE RATIO

II. FUEL ECONOMY IMPROVEMENT FEATURES

1. ROAD LOAD HP REQUIREMENT REDUCTION
   - DRAG FREE BRAKES
   - GR78-15 TIRES @ 30 PSI
   - LOW VISCOSITY AXLE LUBE

2. ENGINE PARASITIC LOSS REDUCTION
   - ELECTRIC COOLING FAN
   - LOW VISCOSITY LUBRICANTS (MOBIL 1 OIL)

3. ENGINE EFFICIENCY – EMISSION CONTROL SYSTEM
   - EFI (AIR FLOW 8 PT. SEQUENTIAL)
   - A/F–CLOSED LOOP (HEATED O_2 SENSOR)
   - 3-WAY CATALYTIC CONVERTER (260 IN^3)
   - ELECTRONIC SPARK TIMING
   - VARIABLE ORIFICE EGR
   - IDLE SPEED CONTROL AND COLD FAST IDLE
   - BEST ECONOMY HIGHWAY MODE
     20:1 A/F 0% EGR ON HIGHWAY TEST

4. AUTOMATIC TRANSMISSION LOSS REDUCTION
   - PROGRAMMED TORQUE CONVERTER CLUTCH

ENGINE DYNAMOMETER DATA

To obtain calibration data for use in programming the microprocessor, engine dynamometer tests were conducted to determine the best economy spark timing, A/F ratio and EGR rate for the selected engine. The results of the engine dynamometer tests, as previously reported [2], indicated that lower fuel consumption is obtained when the engine is operated at A/F ratios near the lean misfire limit with a minimum of EGR. The best economy highway mode control allows the engine to operate at the best economy A/F ratio (20:1), with 0 to 4% EGR and MBT spark timing. Extending the lean misfire limit to 20:1 A/F ratio was achieved by electronically biasing the individual cylinder injectors to a minimum offset from the selected A/F ratio.

Except for the use of retarded spark timing during the first cycle of the FTP driving schedule, best economy spark timing was not compromised during the calibration and testing of the vehicle. After the first cycle of the FTP test the engine was operated at stoichiometric A/F (closed loop) with the best economy EGR rate that varied from 4 to 16% depending on engine speed and load conditions. Spark timing was maintained at minimum for best torque (MBT) for these conditions.

Although the engine dynamometer data indicated that lower fuel consumption could be obtained at A/F ratios leaner than stoichiometry, the vehicle system was calibrated at stoichiometry for the FTP schedule to meet the

1.0 gm/mi NOx level with a three-way catalyst system.

VEHICLE FUEL ECONOMY AND EMISSION TEST RESULTS

To provide an accurate representation of both the EPEC and the comparison vehicles for the vehicle dynamometer tests, coastdown tests were run to determine the required vehicle dynamometer load settings.(Figure 1) Coast times from 55-45 MPH and 35-25 MPH were measured to give two points on the road load Hp curve. Using an electric vehicle dynamometer it was possible to duplicate the coast times at both speeds. A different loading for the 30 MPH point is required than would be present with a hydraulic dynamometer using only the 50 MPH set point (3). In this case the comparison car was loaded higher than the hydraulic dynamometer curve at 30 MPH and the EPEC vehicle required a lower Hp setting for 30 MPH.

The shape of the road load Hp curve is a combination of a linear friction Hp and the Hp required to overcome the wind resistance which varies as the square of vehicle speed. At low vehicle speeds the rolling resistance is a major portion of the total road load horsepower requirement. As vehicle speed increases the speed squared wind resistance term becomes the dominant factor.

The changes included in the EPEC vehicle to reduce road load horsepower requirements reduce the friction horsepower with no change in the aerodynamic drag. These changes alter the shape of the road load horsepower curve as well as reduce the total horsepower requirement at 50 MPH.

The total EPEC system provided a combined 55/45 fuel economy gain of 12.5% (Table 2) over the base vehicle while also meeting the lower '81 emission requirements at low mileage.

Table 2 - EPEC (Electronic Programmed Engine Control)

SYSTEM	EMISSIONS–FUEL ECONOMY								% GAIN OVER '78 BASE
	CITY SCHEDULE				HIGHWAY SCHEDULE				
	HC	CO	NOx	F.E.	HC	CO	NOx	F.E. 55/45	
EPEC W/MODE DISC.	.25	1.6	.50	18.6	.06	.04	1.89	26.6	21.5	12.5%
CLOSED LOOP WITH EGR ON HIGHWAY					.07	.30	.38	24.9	21.0	10.0%

When the best economy mode was not used and the closed loop A/F calibration was run on the highway test, the 55/45 fuel economy gain dropped to 10% with a reduction in NOx emission levels on the highway test from 1.9 gm/mi to .4 gm/mi. The best economy mode provided a 6% reduction in fuel consumption on the highway test.

Fuel consumption tests were also run on the road for comparison with the vehicle dynamometer data (Figure 2). For a direct comparison of road to vehicle dynamometer fuel consumption data, the '75 FTP and highway tests were run on the road at a 75°F ambient temperature with the driver starting the cold engine and following on-board speed trace. Fuel consumption was measured and calculated by weighting the cold and hot portions of the '75 FTP. Both the EPEC vehicle and the comparison vehicle were tested. Although the fuel consumption was 1 to 6 percent higher on the test track, there was less than 1% difference in the total system gains.

Other road tests including the SAE Urban, the GM City, and the SAE Interstate 55 MPH schedules and a 3500 km trip indicate the system gains varied from 7% to 19% depending on the driving schedule. The SAE Urban and GM City schedule results bracketed the FTP City schedule in absolute fuel consumption values and in percent gain. The SAE Interstate 55 MPH scheduled compared favorably with the FTP highway schedule. Results of a 3500 km cross country trip, including all types of driving, yielded intermediate fuel consumption values and gains.

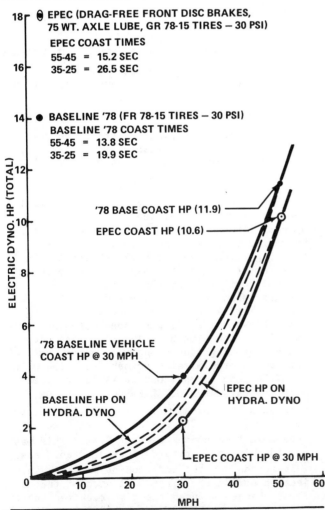

Fig. 1 - Dyno Hp Requirements to Duplicate Coast Down Times at 50 and 30 MPH 4000 lb. Full Size Vehicle W/O AC

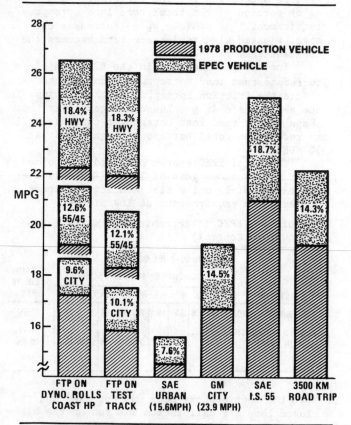

Fig. 2 - Fuel Economy Comparisons Road Schedules EPEC vs. 1978 Production

INDIVIDUAL COMPONENT EVALUATION

An evaluation of individual components used on the EPEC vehicle was made. To break down the total gain of 12.5% it was necessary to remove the items and evaluate the fuel consumption with and without the item functional. This was done for items such as the torque converter clutch control, best economy highway mode, and road load Hp improvements. The vehicle was also calibrated for 1978 emission levels and for 1981 emission levels to determine the effect of the lower emission requirements. The remaining gains were smaller than the error expected in test repeatability (3) and had to be estimated.

A comparison was made of the fuel economy improvement contributed by each item to estimates (4) published by DOT and by GM. Table 3 is a comparison of these estimates. The percent gains are shown over the production 1978 vehicle hardware. The accuracy of the vehicle test data is approximately ±1%.

The category of advanced control technology for this vehicle consisted of closed loop sequential port fuel injection, EST, programmed EGR, and 450 RPM idle speed control. When calibrated to 1978 emission levels these controls provided an estimated 1% increase in fuel economy. The addition of the best economy highway mode provided another 2.5% gain and is included in the total 3.5% gain for the advanced control technology. Estimates by DOT and GM were 3% and 5% respectively.

The addition of low viscosity lubes in the engine and rear axle account for approximately 1% gain. Because of the mileage accumulation required before full benefit can be realized, tests with and without the lubes were not made. The estimate of .5% for the programmed electric fan agrees with previous estimates of .5 to 2%.

Table 3 - Component Fuel Economy Improvements

COMPONENT	ESTIMATES DOT* % GAIN	GM** % GAIN	ACTUAL EPEC % GAIN
BASELINE ('78 PRODUCTION)	–	–	–
ADV. CONTROL TECHNOLOGY (EPEC)	3	5	3.5
• BEST ECONOMY MODE		2.5	
• OTHER (EFI, EST, EEGR, ISC)		(1.0)	
LOW VISC. LUBES	2	1	(1.0)
ELECTRIC FANS	2	0.5	(0.5)
REDUCED ROLLING RESISTANCE	5	5	4.0
LOCKUP CONV. CLUTCH	6	6	6.0
TOTAL AT 1978 STDS. (1.5, 15, 2.0)	18	17.5	15.0
LOSS DUE TO LOWER EMISSION STDS. (.41, 3.4, 1.0)	0	-8	-2.5
TOTAL AT 1981 EMISSION LEVELS	18	9.5	12.5

( ) INDICATES THE GAIN IS LESS THAN TEST REPEATABILITY ACCURACY AND WAS ESTIMATED
* THIRD ANNUAL REPORT TO THE CONGRESS JAN. 1979
** GMC ESTIMATED EFFECTS OF EXHAUST EMISSION STDS AND POTENTIAL HARDWARE DEC. 1978

The reduced rolling resistance on the EPEC vehicle contributed a gain of 4 percent in fuel economy. The gain of 4% shown on this vehicle is slightly below the estimated gains of 5% even though the coast horsepower curve was matched at two speed points to reflect the full benefit of reduced rolling resistance at low speeds. The gain would be less when using a hydraulic vehicle dynamometer.

The torque converter clutch provided a 6% improvement in fuel economy. The electronic control logic allowed the clutch to operate in any gear when the vehicle speed was above 24 MPH. Driveability was improved by using manifold vacuum and engine speed sensing logic to release the clutch during high load, low speed operating conditions that occur just before the transmission downshifts. The gain of 6% for the torque converter clutch agrees with the estimates of 6% by DOT and by GM.

When the emission levels were lowered to 1981 levels the EPEC system indicated an increase in fuel consumption of 2.5%. This was due to cold spark retard, increased cold fast idle speeds and the addition of port air injection during the first 2 minutes of the test. These changes reduced the time required for the catalytic converter to reach 260°C from 220 sec. to 140 sec. The greater flexibility of the control system minimized the loss in fuel economy and did not reflect the 8% loss estimated by GM. However, 50K durability tests have not been run on this configuration.

Combining technology into a single vehicle and then quantifying the results is a difficult and time consuming task. Some indication of the difficulty is shown with the following test summary. Approximately 75 EPA tests were required to achieve the final calibration that maximized fuel economy with acceptable driveability. An additional 125 tests were required to define the individual component contributions and there were 25 EPA tests required on the baseline vehicle for comparison purposes. The basic engine data required approximately 600 hours of dynamometer testing over a 5 month period.

CONCLUSIONS

The fuel economy gains of many individual items have been combined on one vehicle that represents a fuel economy potential, not a state-of-the-art vehicle for production. The control of these items has been integrated into one multi-function electronic control system with common input parameters.

The vehicle coast down tests show the reduction in road load friction Hp requirements on the EPEC vehicle affect the shape of the road load horsepower curve as well as the 50 MPH set point. An electric vehicle dynamometer was required to match these curves at more than one speed.

The total EPEC system provided a fuel economy improvement of 12.5% over the baseline comparison vehicle on the combined FTP city and highway tests while also meeting the lower 1981 emission levels at low mileage.

The total system gains for the FTP city and highway tests run on the track correlate very well to the vehicle dynamometer tests with a difference of less than 1%. The fuel consumption values were 1 to 6 percent higher for both vehicles when tested on the track. Additional track driving schedules show fuel economy gains for the EPEC vehicle range from 7% to 19% depending on the driving schedule.

As additional items are developed it is necessary to continue combining these devices to understand their total effect. The need to work with the total system is urgent, because the development period before acceptable driveability is achieved could be long and difficult.

REFERENCES

1. Roy C. Nicholson and George W. Niepoth, "Effect of Emission Constraints on Optimum Engine Size and Fuel Economy." SAE Congress and Exposition, February 1976. SAE Paper 760046.
2. Roy C. Nicholson, "Emissions and Fuel Economy Interactions." SAE Paper 780616 presented at SAE Passenger Car Meeting, Troy, MI., June 1978.
3. Ward W. Wiers, George W. Niepoth, and T. D. Hostetter, "Emission and Fuel Economy Measurement Improvements." SAE Congress and Exposition, February 1979. SAE Paper 790223.
4. DOT HS-803 777, "Third Annual Report to the Congress." Page 47. January 1979.

The Effects of Varying Combustion Rate in Spark Ignited Engines

R. H. Thring
Ricardo Consulting Engineers Ltd.

IT HAS BEEN SHOWN by calculation (1,2)* that, for given engine operating conditions, there is an optimum rate of combustion for minimum NOx emissions from spark ignited engines. In essence, this is because if the combustion rate is too slow, there is plenty of time for NOx formation, while if it is too fast, charge temperature becomes very high and more NOx is formed. It was considered that rates of combustion higher than those attained in current production spark ignition engines might be desirable, firstly because current rates of combustion were believed to be slower than the optimum for minimum NOx emissions and secondly because previous research had indicated that fast burn improved the lean mixture and EGR tolerances (3). Also, there were indications that fast burn improved economy.

Mayo (3) reported an experimental programme aimed at examining the effects of squish, spark gap position, swirl and charge velocity on combustion rate, emissions and fuel economy, using two V8 engines. A number of limitations were encountered with the work, and this programme was initiated as a continuation project. A single cylinder engine was used, and the effects of swirl and number of ignition points on combustion rate, emissions and fuel economy were examined. An improved data acquisition system was used enabling 300 pressure diagrams to be collected over a period of about five minutes.

OBJECTIVES

To study the effects of varying combustion rate by swirl and number of ignition points over the largest possible range of engine speed, load, equivalence ratio, EGR, octane number, compression ratio and spark timing, with a view to the reduction of engine baseline emissions, especially NOx, and fuel consumption.

DETAILS OF EQUIPMENT AND TECHNIQUES

THE ENGINE - The project was a fundamental one, and one which would require sophisticated instrumentation. Also it was going to study changes that might be small, and therefore would require a high degree of reproduci-

---

*Numbers in parentheses designate References at end of paper.
*Paper 790387 presented at the Congress and Exposition, Detroit, 1979.

ABSTRACT

It has been shown by calculation that, for given engine operating conditions, there should be an optimum rate of combustion for minimum NOx emissions from spark ignited engines. This paper gives experimental results from a single cylinder engine which confirm the theory, and show that, for a particular engine, the normal combustion rate needed reducing at zero EGR and increasing at high EGR rates, in opposition to its natural tendency to decrease. The effect on economy was a small loss at zero EGR, but an appreciable improvement at high EGR.

Cyclic variation and octane requirement studies are also included.

bility. It was therefore decided that a single cylinder research engine should be used. Compression ratio was an important variable, and it was required to use multiple spark plugs in the combustion chamber, and also a shrouded intake valve. For these reasons, a Ricardo E6 single cylinder variable compression ratio research engine was chosen. It was found to be possible by means of using a brass plate sandwiched between the cylinder head and cylinder block to arrange for four 10 mm access holes to the combustion chamber in addition to the two 14 mm access holes already existing. Two alternative combustion chambers were used; a bathtub type of chamber designed to have no squish, and an open disc chamber (Figs. 1 and 2). The four 10 mm access holes in each chamber were used for the installation of four 10 mm spark plugs, each of which could be switched in or out independently, to vary the rate of burning of the charge. There was no phase shift facility between the different spark plugs; those that were switched on fired simultaneously. The engine was already designed to incorporate the option of a shrouded intake valve. The engine specification is given in Table 1.

The two access holes in the cylinder head were used for a Ricardo balanced disc indicator and a Kistler piezo pressure transducer. The engine was equipped with electronic inlet port fuel injection, and an inlet mixing and damping chamber which had a volume equal to eight times the swept volume of the engine. An exhaust thermal reactor was used, with secondary air injection close to the exhaust valve.

A slotted wheel was mounted on the crankshaft and a chopper disc on the camshaft. The signals from the pick-ups were used to trigger the data acquisition system at the crank angles required for the recording of the cylinder pressures.

Data Acquisition and Processing - The instrumentation block diagram is given in Fig. 3. The hub of the system was an Intertechnique Histomat S. The analogue to digital converter, core store and display of the machine were used. Since the core store of the machine was too small to store more than six pressure diagrams, it was programmed to keep a running total of pressure diagrams until 300 were acquired. Variance was also calculated. The total diagram was punched on

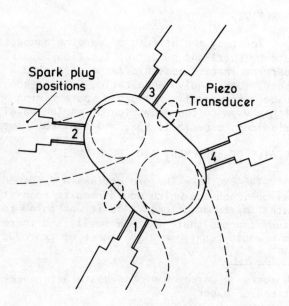

Fig. 1 - No-squish bathtub type combustion chamber

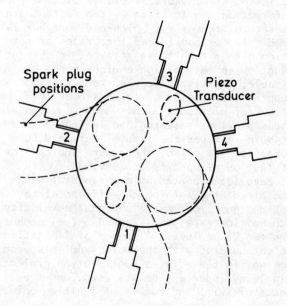

Fig. 2 - Open disc type combustion chamber

Table 1 - Ricardo E6 Brief Engine Specification

Bore	3 in. (76 mm)
Stroke	4⅜ in. (111 mm)
Swept Volume	31 in^3 (.506 litre)
Ratio con rod/crank radius	4.343
Inlet Valve Timing	9/38 crank degrees
Exhaust Valve Timing	13/38 crank degrees
Compression Ratio	up to 11 to 1, in this form

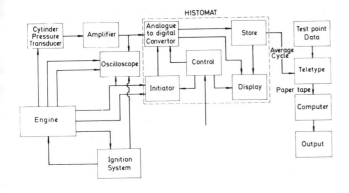

Fig. 3 - Instrumentation block diagram

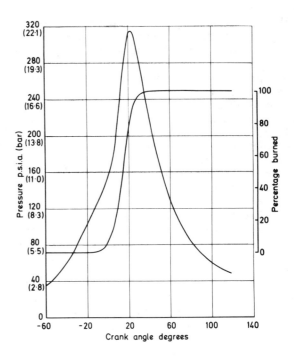

Fig. 4 - Typical computer output

to paper tape, and then fed into a Data General "Eclipse" computer. The reading from the balanced disc indicator was used to fix the absolute level of the pressure diagram, not obtainable from the piezo signal because of the finite time constant of the amplifier/transducer. The computer programme was used to calculate such results as maximum pressure rise rate, polytropic indices, etc., and also to calculate curves of per cent of charge burned versus crank angle; Fig. 4 is a typical example. The method used was that of Rassweiler and Withrow (4). The benefit of this method is that it uses no empirical correlation (such as the Eichelberg heat transfer correlation which is often used), the information required being entirely derived from the pressure data. Basically, the method is as follows: over one crank angle increment the pressure change measured is compared to the pressure change expected if there were no combustion. These differences are normalised to constant volume, and summed over the whole curve. The fraction burned at any point is the sum of the differences up to that point divided by the total. To calculate the expected pressure rise over an increment in the absence of combustion, a value is needed for the polytropic indices of the charge during combustion. Polytropic indices during compression and expansion were obtained by linear regression on the straight parts of the log P log V plot, immediately on each side of the burn region. The polytropic indices during the combustion period are different from these, so experiments were carried out to evaluate the effect of using the compression index, expansion index and a linear interpolation between the two, as values for the polytropic index during combustion. The resulting difference in burn time* was negligible so it was concluded that the lack of a genuine polytropic index during combustion was not a serious difficulty.

In order to check the absolute levels of the mean pressure diagrams, two balanced disc indicator readings were taken, one before and one after the pressure peak. The absolute levels of the mean diagrams were set up using the first reading, and the second reading compared with the value so obtained from the piezo transducer trace. It was found that the pressure value obtained from the balanced disc indicator was usually 1 to $3\frac{1}{2}$ lb/in^2 (.07-.24 bar) greater than the value obtained from the piezo signal, and this notwithstanding the fact that the balanced disc indicator signal was corrected for disc and gas column acceleration. Therefore some data were processed with an arbitrary 5 lb/in^2 (.35 bar) pressure error, and although the effect on the calculated values of the polytropic indices was large, the effect on the burn time was small.

RESULTS - ANALYSIS AND DISCUSSION

THE EFFECT OF NUMBER OF SPARK PLUGS - Table 2 shows some results obtained with five spark plugs in the engine, using the no-squish bathtub combustion chamber. The spark plug

---

*The expression "burn time" is used throughout this paper to mean the number of degrees crank angle required for combustion to progress from 10% to 90% of the mass of the charge burnt. This parameter was used as recommended by Mayo (3) because of ripples in the burn curve at the start and finish of the burn.

numbers correspond to those given in Fig. 1. The per cent change shown is based on plug number 5, which was in the standard position in the cylinder head. It can be seen that the spark plug positions near the exhaust valve were bad positions under these conditions, presumably because they were not well scoured by fresh charge during the intake stroke. This combustion chamber was found to have too low an HUCR (highest useful compression ratio) to carry out the required testing, and the remainder of the work was carried out with the open disc chamber (Fig. 2).

SPAN OF VARIABLES - In order to gain an appreciation of the effects of adjusting the input variables on the values of the output variables some experiments were carried out with the two extreme values of the input variables speed, load and number of ignition points (Table 3). The burn time was nearly halved by using four spark plugs while the BSNOx was increased by 13-113% depending on speed and load. Fuel consumption was reduced 7-28%.

Table 4 shows the results of varying equivalence ratio, manifold pressure, spark timing, speed and EGR rate over their extremes of range, using one spark plug and operating at spark timings fixed in terms of crank degrees rather than a fixed per cent load loss from best torque.

It can be seen that changing equivalence ratio had most effect on burn time in the range .9 to 1.0, while cyclic irregularity had a minimum at 1.1. Increasing load had very little effect on burn time or cyclic irregularity, and retarded ignition gave increased burn time and cyclic irregularity. The burn time nearly doubled and the cyclic irregularity nearly trebled over the speed range, but EGR had little effect.

EFFECT OF AIR/FUEL RATIO - Figure 5 shows some of the results of a survey over a range of air/fuel ratios. These results were all obtained at a constant BSNOx level of 12.8 g/bhp.h (17.2 g/kW.h), EGR being applied as required. The purpose of these experiments was to see if there was still an improvement in economy with fast burn even when the NOx emissions were brought back to their original levels by EGR. It can be seen that there was very little difference between the resulting fuel economies for the slow and fast burn conditions. There appeared to be a significant reduction in CO emissions at air/fuel ratios leaner than about 16/1, but HC emissions were worse except at air/fuel ratios leaner than about 17½/1.

The secondary purpose of these experiments was to see if there was an optimum exhaust oxygen concentration for minimum HC emissions across the air/fuel ratio range. The results indicated that the optimum oxygen level was not critical to within ± ½%, but it appeared that 4% oxygen was a reasonable overall level to use for minimum hydrocarbon emission.

STUDY OF COMBUSTION RATES -

Preliminary - Figures 6-9 show the results of a study of the effects of varying the rate of combustion in stages from fastest to slowest, at two different manifold pressures, 18 in and 26 in Hg (457 and 660 mm Hg). The rate of combustion was varied by the use of different combinations of the four spark plugs, the use of ordinary or extended electrode spark plugs, and the use of a shrouded intake valve. These tests were carried out at 1500 rev/min (25 rev/s), 7:1 compression ratio, zero EGR

Table 2 - Minimum Fuel Consumptions at 2000 rev/min (33.3 rev/s) 25 lb/in^2 (1.7 bar) BMEP

Spark Plug Number	BSFC lb/bhp.h (g/kW.h)	% Change in BSFC
1	.882 (537)	+ 2
2	.970 (590)	+ 12
3	.975 (593)	+ 12
4	.923 (561)	+ 6
5	.869 (529)	0
all five	.820 (499)	- 6

Table 3 - Span of Burn Times with Plain Inlet Valve

C.R. 7:1
Fuel 96 RON unleaded indolene
Zero EGR
Spark Timing retarded for 2% torque loss
MAP Manifold Air Pressure, in in.Hg (mm Hg)

10-90% BURN TIME, CRANK DEGREES

		Four Plugs		One Plug	
Speed rev/min (rev/s)		MAP 14 in. (356 mm)	MAP 26 in. (660 mm)	MAP 14 in. (356 mm)	MAP 26 in. (660 mm)
3000	(50)	18.6	18.3	34.1	34.1
1000	(16.7)	15.8	14.1	30.9	25.6

BSNOx g/bhp.h (g/kW.h)

		Four Plugs		One Plug	
Speed rev/min (rev/s)		MAP 14 in. (356 mm)	MAP 26 in. (660 mm)	MAP 14 in. (356 mm)	MAP 26 in. (660 mm)
3000	(50)	56.2 (75.4)	31.7 (42.5)	26.4 (35.4)	21.3 (28.6)
1000	(16.7)	20.2 (27.1)	27.4 (36.7)	17.9 (24.0)	21.7 (29.1)

BSFC lb/bhp.h (g/kW.h)

		Four Plugs		One Plug	
Speed rev/min (rev/s)		MAP 14 in. (356 mm)	MAP 26 in. (660 mm)	MAP 14 in. (356 mm)	MAP 26 in. (660 mm)
3000	(50)	1.658 (1009)	.619 (377)	2.282 (1388)	.680 (414)
1000	(16.7)	.749 (456)	.496 (302)	1.035 (630)	.534 (325)

BMEP lb/in^2 (bar)

		Four Plugs		One Plug	
Speed rev/min (rev/s)		MAP 14 in. (356 mm)	MAP 26 in. (660 mm)	MAP 14 in. (356 mm)	MAP 26 in. (660 mm)
3000	(50)	12.8 (0.883)	68.3 (4.71)	9.14 (0.630)	61.0 (4.21)
1000	(16.7)	29.6 (2.04)	89.6 (6.18)	21.6 (1.49)	85.2 (5.87)

Table 4 - Span of Variables

Compression Ratio 8:1
Plug Number 3
Fuel 100 RON
Standard Extended Electrode Spark Plug

Equiv Ratio	MAP psi (bar)	Spark Timing deg.C.A.	Speed rev/min (rev/s)	EGR %	ISFC lb/ihp.h (g/kW.h)	ISNO g/ihp.h (g/kW.h)	σ Pmax %	10-90% Burn Time Crank Degrees
0.8(lean)	8 (.55)	-35	1500 (25)	0	.351(214)	12.17(16.3)	10.96	26.6
0.9	" "	-30	" "	"	.364(221)	16.05(21.5)	10.33	25.1
1.0	" "	-30	" "	"	.362(220)	13.17(17.7)	7.05	20.9
1.1	" "	-30	" "	"	.408(248)	7.51(10.1)	6.85	19.7
1.2(rich)	" "	-30	" "	"	.448(273)	3.25(4.4)	6.99	19.5
1.0	6 (.41)	-30	1500 (25)	0	.370(225)	9.58(12.9)	8.70	25.3
"	11 (.76)	-20	" "	"	.387(235)	14.29(19.2)	9.82	22.9
"	14 (.97)	-15	" "	"	.412(251)	17.64(23.7)	10.09	27.0
1.0	8 (.55)	-45	1500 (25)	10	.388(236)	9.57(12.8)	9.26	22.4
"	" "	-15	" "	0	.403(245)	9.38(12.6)	10.48	28.3
"	" "	0	" "	0	.524(319)	4.30(5.8)	10.32)	-
1.0	8 (.55)	-30	750 (12.5)	0	.437(266)	12.18(16.3)	3.33	17.1
"	" "	-30	2250 (37.5)	"	.372(226)	11.77(15.8)	9.12	26.6
"	" "	-30	3000 (50)	"	.364(221)	10.61(14.2)	9.88	28.6
1.0	8 (.55)	-30	1500 (25)	10	.380(231)	6.63(6.2)	11.56	30.5
"	" "	-30	" "	20	.440(268)	0.69(0.9)	12.39	28.0
"	" "	-30	" "	25	.439(267)	0.51(0.7)	10.48	27.6

and a nominal air/fuel ratio of 16/1. The ignition timing was retarded from MBT to give 1½% IMEP loss.

ISNOx Results - It can be seen in Fig. 6 that reducing the burn time gave increased NOx emissions for all loads, fuels and combustion systems used in this part of the study. The reproducibility errors were at worst ± 5%. Theoretical predictions have shown that there should be a minimum in the curve of NOx versus combustion rate, so a comparison was made between theory and practice, and the results are shown in Fig. 7. The theoretical curve was taken from Reference 2. The experimental results shown in Fig. 7 are those given in Fig. 6 replotted to make them compatible. The Figures are not strictly comparable because some of the engine conditions were different, but it is clear that the reason no minimum was apparent in the experimental results was that the range of burn times explored was too narrow, and that the burn rates used were too rapid for these particular engine conditions.

It appeared from the results that the ISNOx emissions were not a unique function of burn rate, even at a given engine condition and correcting for any errors in setting the ignition timing. For example, it appeared that for a given burn time, spark plug number 2 gave significantly more NOx than the others. This was consistent with the expected result in that theoretical models show that a large proportion of the total NOx emitted comes from the first part of the charge to be burned, due to its continued heating by the combustion of the rest of the charge. Spark plug number 2 was the nearest one to the hot exhaust valve (Fig. 2).

Also, the ISNOx emissions were reduced at the light load condition by the addition of the shrouded intake valve. The reason for this was probably that the increased turbulence within the cylinder increased the gas to wall heat transfer coefficients enough to significantly reduce the sum of the time-temperature histories of many of the portions of the charge, where the charge density was relatively low, but that the reduction at the higher charge density was not significant. The Appendix gives a discussion of the results of static blowing tests on the cylinder head with both plain and shrouded intake valves.

The addition of the extended electrode spark plugs appeared in some cases to increase the NOx produced. The reason for this could be that bringing the points of ignition out into the chamber moved the position of the first part of the charge to be burnt (which is where a large proportion of the total NOx is produced) further away from the relatively cold walls of the combustion chamber, hence increasing the time-temperature history of this portion of the charge. It was partly because of these discoveries that it was decided to carry out a comparison between the use of multiple spark plugs and the use of a shrouded intake valve to attain a given burn rate over a range of engine speeds, loads and EGR rates.

ISFC Results - The ISFC results are given in Fig. 6. The reproducibility errors were at worst at about ± 5%. The ISFC at 26 in Hg (660 mm Hg) was 4 to 5% higher than at 18 in Hg (457 mm Hg) MAP possibly due to reduced combustion efficiency which was reflected in the CO and hydrocarbon emissions. An alternative explanation could lie in the friction values

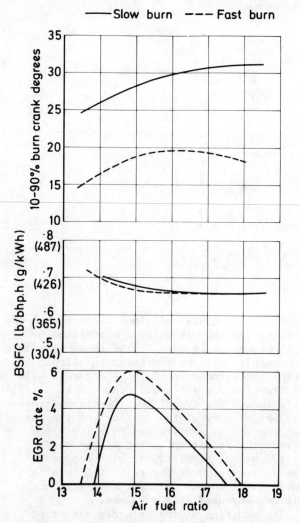

Fig. 5 - Burn time and BSFC versus air/fuel ratio

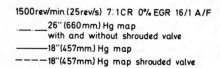
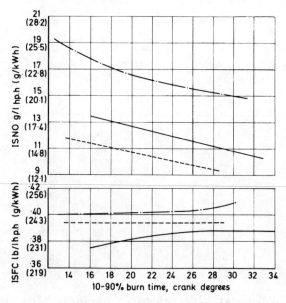

Fig. 6 - ISNOx and ISFC versus burn time

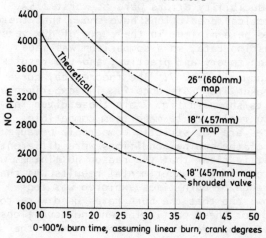

Fig. 7 - Experimental and theoretical NO

used, which were 8% different at the two manifold pressures. The discrepancy between dynamometer and directly measured IMEP's is discussed in Reference 6. The ISFC with the shrouded valve was increased by 2 to 3% at the light load and unchanged at the high load. The explanation could be that the heat transfer to the combustion chamber walls per unit mass of charge was significantly increased at the lower charge density while the change at the higher charge density was not enough to show up in the ISFC results. This would be because the heat transfer per unit mass of charge decreased with increasing charge density.

The ISFC figures showed very little change across the range of burn rates examined, although there appeared to be a slight improvement as the burn rate was increased. The change was small because the range of burn rates studied was limited, for example, the calculations in Reference 2 show an indicated thermal efficiency falling by only 5% across a range of 0-100% burn time of 60 degrees crank angle. The experimental results shown in Fig. 6 are in good agreement with this.

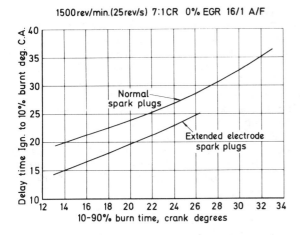

Fig. 8 - Delay time versus burn angle

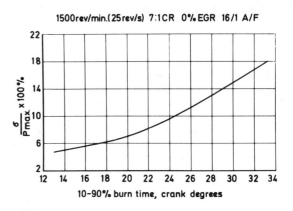

Fig. 9 - Normalised standard deviation at the peak pressure point

Delay Time and Cyclic Irregularity - Figures 8 and 9 show the results of delay time and cyclic irregularity. There was no distinct difference in delay time between any of the engine conditions tested during the study of combustion rates, except for the extended electrode spark plug results, where the delay time was reduced by 25-30% (Fig. 8). This reduction was reflected in the ignition timing. A possible reason for this reduction could be that the electrodes of the extended electrode spark plugs were running at a considerably higher temperature at the moment when the arc occurred, and therefore that the gas temperatures in the gap were higher during the combustion initiation period, hence increasing reaction rates and reducing delay time.

The cyclic irregularity results are shown in Fig. 9, which shows the standard deviation of the 300 cylinder pressures recorded at the crank angle where the pressure peak occurred on the mean diagram, normalised by dividing by the mean peak pressure and expressed as a percentage. The maximum standard deviation of cylinder pressure was found to occur at a crank angle slightly in advance of the peak pressure, the difference being virtually constant at about 6 degrees crank angle. Subject to scatter, the normalised standard deviations, both the maximum value and the value at the peak pressure point, were unique functions of burn time.

The reduction in cyclic irregularity with burn time could be because reduced burn time required later ignition timing, which meant that the combustion initiation period, which is the source of cyclic irregularity (7,8,9,10) occurred at a time when the charge pressure and temperature were higher than with a longer burn time. This theory is supported by the fact that a comparison between the standard deviation results and the ignition timing results revealed that over-advanced ignition timings yielded high values of standard deviation.

COMPARISON BETWEEN THE USE OF MULTIPLE SPARK PLUGS AND THE USE OF A SHROUDED INTAKE VALVE TO ATTAIN A GIVEN BURN TIME -

Object - The object of these experiments was to find out whether the use of two different methods of reducing burn time gave similar results. The implication was that if the results were similar, then burn time would be a good parameter to assist in the judgement of combustion chambers, while if the results were dissimilar, then combustion rate could not be considered without attention to the means of achieving it.

Method - Tests were carried out at a burn time in the centre of the range available, namely 23 degrees crank angle for 10-90% burn. Three combustion chamber configurations were chosen that gave burn times close to this figure:
(1) masked inlet valve and extended electrode spark plug number 1
(2) masked inlet valve and flush spark plug number 3
(3) plain valve and flush spark plug numbers 1 and 3

The tests were all carried out at 7:1 compression ratio and 16/1 air/fuel ratio. The ignition timing was retarded from MBT to give 1½% IMEP loss.

They were carried out in two parts:
Part 1 - The load was varied from 14 in to 25 in Hg (356 mm to 635 mm Hg) manifold pressure at 1500 rev/min (25 rev/s). The speed was then varied from 1000 rev/min to 3000 rev/min (16.7 rev/s to 50 rev/s) at a constant 18 in Hg (457 mm Hg) manifold pressure. This part was all carried out with zero EGR.

Part 2 - The results of Part 1 were extended by the addition of EGR in 5% stages up to a maximum level of the misfire tolerance minus 2%. This work was carried out at the same speed and load combinations as used in Part 1.

ISNOx Results - The differences in ISNOx between the three engine configurations tested were not consistent over the range of load, speed and EGR used. At high speeds there was little difference in ISNOx between the configurations, while at low speeds the plain valve configuration gave more NOx than the others. The explanation for this could be that at high speeds the turbulence levels were high enough both with and without the shrouded valve for the extra effect of the shrouded valve to make little difference, while at low speeds the shrouded valve made a considerable difference. It would follow from this that the NOx levels would also diverge as the load was reduced, and this was found to be the case.

The curves of ISNOx versus EGR rate were almost unchanged over the speed range and, although the absolute levels were lower, the curve shapes were very similar over the load range (e.g. 60% reduction in NOx for 10% EGR).

ISFC Results - There was no difference apparent in ISFC between the three configururations tested across the range of speed, load and EGR rate studied.

EXAMINATION OF THE EFFECT OF VARYING THE COMBUSTION RATE AT DIFFERENT EGR RATES -

Object - Theoretical work had shown that although ISNOx levels rose as the combustion rate was increased at zero EGR, the slope of the curve decreased as EGR was applied, and at an EGR rate of 25% a cross-over occurred with increased combustion rates causing a reduction in ISNOx levels at EGR rates greater than 25%. It was therefore considered desirable to carry out an experimental investigation to find out whether the predictions were borne out in practice.

Method - The results from this investigation can be divided into four parts:

Part 1 - This work was carried out at 1500 rev/min (25 rev/s), 71 lb/in^2 (4.9 bar) IMEP, 7:1 compression ratio, 16/1 air/fuel ratio, using a plain inlet valve and 100 RON unleaded gasoline. The ignition timing was retarded to give 1½% IMEP loss. The combustion rate was varied by changing the number of spark plugs in operation, the combinations used being chosen to give a reasonably uniform spacing between the fastest and slowest rates. The results of this work are shown in Fig. 10.

Part 2 - This work was carried out at 1500 rev/min (25 rev/s), 100 lb/in^2 (6.9 bar) IMEP, 7:1 compression ratio, 16/1 air/fuel ratio, using a plain inlet valve and 100 RON unleaded gasoline. The combustion rate was varied using the same spark plug combinations as were used in Part 1, and the EGR rates used were 20 and 25%. These values were chosen because the results from Part 1 had indicated that at high EGR rates it did not matter what the value of the combustion rate was, and therefore investigations were needed to find out whether there was an advantage to fast burn under different conditions. Some of the results of this work are given in Table 5.

Part 3 - This work was carried out at 10:1 compression ratio, the other conditions being the same as Part 1. Some of the results of this work are given in Table 6.

Part 4 - This work was carried out at stoichiometric air/fuel ratio and 18/1 air/fuel ratio, the other conditions being the same as Part 1. Some of the results of this work are given in Table 7.

Results - Figure 10 shows that NOx increased with increasing combustion rate over the range examined here at low EGR rates, while at high EGR rates the NOx decreased with increasing combustion rate. The cross-over point depended on load, and was about 16% EGR at 71 lb/in^2 (4.9 bar) IMEP and about 22% at 100 lb/in^2 (6.9 bar) IMEP at the engine conditions given in Fig. 10. The explanation for this is as follows: NOx formation depends on the time-temperature history of each element of the gas in the charge. At very short burn times the charge temperatures are high, giving high NOx levels. At very long burn times there is enough time for high NOx levels to form. The curve of NOx versus burn time therefore has a minimum value at some intermediate burn

Table 5 - ISNOx as a function of burn time at 100 lb/in^2 (6.9 bar) IMEP
1500 rev/min (25 rev/s)
16/1 A/F Plain Inlet Valve

EGR %	Burn Time	ISNOx g/ihp.h	(g/kW.h)
20	23.0	4.0	(5.4)
20	26.8	4.3	(5.8)
20	32.5	3.5	(4.7)
20	34.0	4.0	(5.4)
25	25.4	2.2	(3.0)
25	27.6	2.6	(3.5)
25	29.0	2.3	(3.1)
25	31.0	3.6	(4.8)

Table 6 - Effect of Compression Ratio on Burn Time
1500 rev/min (25 rev/s)
71 lb/in^2 (4.9 bar) IMEP
16/1 Air/Fuel Ratio

	Zero EGR		20% EGR	
C.R.	4 Ignition Points	1 Ignition Points	4 Ignition Points	1 Ignition Points
7:1	15.3	28.3	22.0	35.0
10:1	14.4	28.5	19.5	29.3
% reduction	6%	-1%	11%	16%

Table 7 - ISNOx as a Function of Burn Time at Different Air/Fuel Ratios
1500 rev/min (25 rev/s) 7:1 C.R. 70 lb/in² (4.8 bar) IMEP

Air/Fuel Ratio	EGR Rate %	Burn Time Crank Degrees	ISNOx g/ihp.h	(g/kW.h)
Stoichiometric	0	13.5	13.0	(17.4)
"	"	16.4	11.8	(15.8)
"	"	24.1	10.5	(14.1)
"	"	25.5	10.8	(14.5)
"	20	20.1	3.0	(4.0)
"	"	24.5	2.5	(3.4)
"	"	32.0	2.0	(2.7)
"	"	35.3	2.0	(2.7)
18:1	0	17.9	8.1	(10.9)
"	"	21.1	8.8	(11.8)
"	"	29.0	7.4	(9.9)
"	"	30.8	6.6	(8.9)
"	20	27.0	1.0	(1.3)
"	"	29.5	0.7	(.94)
"	"	32.3	1.0	(1.3)
"	"	33.3	1.1	(1.5)

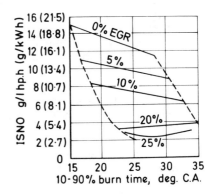

Fig. 10 - ISNOx versus burn time

time. The shape of the curve, its absolute levels, and the position of its minimum point depend upon engine conditions, including the EGR rate. As EGR is increased, the position of the minimum on the curves moves towards the shorter burn times. This is because the EGR reduces charge temperatures within the cylinder, and since the net rate of formation of NOx is exponentially dependent upon temperature, shorter burn times are required before the temperature is raised high enough for the increase in temperature with burn time to compensate for the reduction in reaction time available. Following from this argument, it would be expected that an increase in load would shift the minimum points for given EGR rates towards the longer burn times. This was confirmed by experiment; for example, at 20% EGR rate the minimum NOx point occurred at a 10-90% burn time of less than 23 degrees crank angle at 71 lb/in² (4.9 bar) IMEP, while at 100 lb/in² (6.9 bar) IMEP it occurred at a burn time of greater than 34 degrees crank angle. It was not possible to determine exactly where the minimum points did lie because the range of burn time available was not great enough.

When the compression ratio was increased to 10:1, the NOx values at zero EGR were very similar to those obtained at 7:1, but as the EGR rate was increased the NOx levels became progressively higher than the corresponding levels at 7:1 compression ratio at the same burn rate. The expected increase in NOx due to the higher compression ratio was evidently masked by the increased surface to volume ratio. This effect would be greater at zero EGR when charge temperatures were higher.

It was also observed that increased compression ratio gave decreased burn time, especially at high EGR rate (Table 6). This could be due to the associated increase in charge temperature and pressure, which would have more effect at high EGR where the baseline temperature would be lower.

The results at different air/fuel ratios (Table 7) indicate that the cross-over EGR value above which shorter burns reduce NOx was over 20% at 18/1 air/fuel ratio. This would be because of the higher charge temperatures at the richer air/fuel ratio giving the same effect as was observed when the load was increased.

OCTANE REQUIREMENT STUDY -
Object - It was considered that the rate of burning of the charge in an engine would have an effect upon the octane requirement of the engine. In view of the apparent incompatibility between the present trend towards low octane gasolines and the requirement for improvements in fuel economy, it was considered desirable to carry out experimental work to quantify this effect.

Method - The experiments carried out can be divided into two parts:

Part 1 - Experiments were carried out at 1500 rev/min (25 rev/s) and 16/1 air/fuel ratio using a plain inlet valve. Three fuels were used and two EGR rates.

Part 2 - Part 1 showed that the dependence of HUCR* on combustion rate was very much less than expected, particularly with 80 RON gasoline. Further experiments were therefore undertaken at the same speed, 1500 rev/min (25 rev/s), and air/fuel ratio 16/1, also with a plain inlet valve. Two fuels were used, and the engine was run at 26 in Hg (660 mm Hg) manifold pressure and zero EGR. The number of spark plug combinations was increased from the four used in Part 1 to every possible combination of the four spark plugs.

---

*HUCR (highest useful compression ratio) was determined by gradually raising the compression ratio of the engine while it was running at the desired test condition until knock was just audible.

Results - Figure 11 shows some of the results of HUCR versus burn time, using three different fuels. It can be seen that HUCR increased as the burn time was reduced, due to the reduced time available for pre-knock reactions to occur in the end-gas. However, this effect decreased as the RON of the fuel was decreased, and at 80 RON the effect had almost disappeared. The most likely explanation is as follows:

At high compression ratios, with the four peripheral spark plugs in a disc chamber the ratio of flame surface to cold wall surface area in the end-gas region was lower than it was at low compression ratios. Therefore the tendency to knock with four spark plugs operating was greater at the lower compression ratios associated with the lower octane fuels than it was at the higher compression ratios associated with the higher octane fuels.

The results at 26 in Hg (660 mm Hg) manifold pressure showed similar trends.

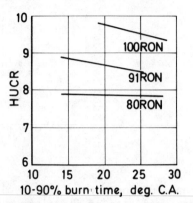

Fig. 11 - Highest useful compression ratio versus burn time

SUMMARY AND CONCLUSIONS

GENERAL - A Ricardo E6 single cylinder research spark ignition engine was converted to allow the use of four spark plugs in the combustion chamber. Cylinder pressure transducers were also fitted, and a data acquisition and processing system devised to enable such parameters as 10-90% burn time to be calculated. Experimental work was carried out to examine the effect of changing the combustion rate both by altering the number of spark plugs and varying the swirl. Parameters varied were speed, load, air/fuel ratio, EGR, octane number, compression ratio and spark timing.

It was found that at low EGR rates there was no benefit in fast burn. However at high EGR rates some reduction in NOx was gained with fast burn, especially at light load, associated with slight reductions in fuel consumption and CO.

COMBUSTION RATE - Multiple spark plugs increased the rate of combustion of the charge. Different spark plug positions gave different combustion rates. The spark plug position giving the slowest burn was not necessarily the one that was worst scoured during the intake stroke. The use of a shrouded intake valve increased the combustion rate. The use of extended electrode spark plugs sometimes increased the combustion rate, depending upon the spark plug position. Increased compression ratio decreased burn time, more significantly at high EGR rate conditions.

NOx EMISSIONS - ISNOx increased with combustion rate at zero EGR rate, and decreased with combustion rate at high EGR rates, over the range of combustion rates attainable with the combustion system used. This is in agreement with theoretical predictions. The changes found were in the region of 25% at zero EGR rate and 20% at 20% EGR rate. To obtain minimum NOx emissions at increasing EGR rates, it was desirable to increase the combustion rate in opposition to its natural tendency to decrease.

ISNOx emissions were not a unique function of combustion rate, some results being as much as 15% different from the mean value for a given combustion rate. In particular, combustion initiation near the exhaust valve gave high NOx values at high load, and the use of a shrouded inlet valve gave low values of NOx at light load.

The curves obtained of ISNOx versus EGR rate were almost unchanged over the speed range, and although the absolute levels were different, the curve shapes were very similar over the load range (e.g. 60% reduction in NOx for 10% EGR).

Ignition retard had a much greater effect on NOx at 18/1 air/fuel ratio than at stoichiometric.

ISNOx results as functions of burn time were found to follow similar trends at 10:1 compression ratio as at 7:1, but the absolute values were greater at high EGR rates.

FUEL CONSUMPTION - ISFC changed by 3-5% across the range of combustion rates studied, which is in good agreement with theory.

When operating at constant load, ISFC was found to decrease with increasing EGR rate, to give a minimum at about 20% EGR of up to 5% below the value at zero EGR.

HYDROCARBON EMISSIONS - ISHC was unchanged by increased combustion rate at low load, but increased by 20-25% at the high load, due to reduced exhaust temperature.

ISHC increased with increasing EGR rate when operating at constant load, probably due to the fall in exhaust temperature.

CO EMISSIONS - ISCO decreased with increasing EGR rate when operating at constant load, probably due to the longer burn time and the fact that the recirculated CO had a second chance to be oxidised.

IGNITION DELAY TIME RESULTS - Ignition delay time was found to decrease with decreasing burn time, there being no distinct differences between any of the engine conditions tested, except for the results obtained with extended electrode spark plugs, where the delay time was reduced by 15-30%.

Delay time was found to be insensitive to ignition timing. It would appear that the speed of the initiation reactions depended more upon the temperature of the surrounding surfaces (i.e. the spark plug electrodes) than upon the bulk temperature and pressure of the charge.

CYCLIC IRREGULARITY RESULTS - Cyclic irregularity, expressed as a normalised standard deviation of cylinder pressure at the peak pressure point, was a unique function of combustion rate. It decreased with increasing combustion rate. The maximum standard deviation occurred 4-8 degrees before the peak pressure.

There was a threshold level of standard deviation above which the engine became very unstable. At 1500 rev/min (25 rev/s), 70 lb/in^2 (4.8 bar) IMEP, the value of this threshold level was about 15%.

HIGHEST USEFUL COMPRESSION RATIO RESULTS - Increased combustion rate gave a higher HUCR although the effect was diminished with low RON fuels. This was thought to be the result of the lower surface to volume ratio of the combustion chamber at the lower compression ratios.

ACKNOWLEDGEMENTS

The author would like to thank the Directors of Ricardo Consulting Engineers for permission to publish this work, the Ford Motor Company for their sponsorship of the project, and also Mr. M.T. Overington and Dr. R.A. Haslett of Ricardo Consulting Engineers for many fruitful discussions throughout the duration of the project.

REFERENCES

1. P. Blumberg and J.T. Kummer, "Prediction of NOx Formation in Spark Ignited Engines - An Analysis of Methods of Control." Combustion Science and Technology, Vol. 4, 1971, p. 73.

2. P. Eyzat and J.C. Guibet, "A New Look at Nitrogen Oxides Formation in Internal Combustion Engines." Paper 680124 presented at SAE Automotive Engineering Congress, Detroit, January 1968.

3. J. Mayo, "The Effect of Engine Design Parameters on Combustion Rate in Spark Ignited Engines." Paper 750355 presented at Automotive Engineering Congress and Exposition, Detroit, February 1975.

4. G.M. Rassweiler and L. Withrow, "Motion Pictures of Engine Flames Correlated with Pressure Cards." SAE Transactions, Vol. 42, No. 5, 1936, p. 185.

5. R.A. Haslett and R.H. Thring, "Polynomial Expressions for the Ratio of Specific Heats of Burnt and Unburnt Mixtures of Air and Hydrocarbon Fuels." Ricardo DP 19417.

6. R.H. Thring, "The Discrepancy between Dynamometer and Directly Measured IMEP's." Ricardo DP 76/616.

7. S. Curry, "A Three-Dimensional Study of Flame Propagation in a Spark Ignition Engine." SAE Transactions, Vol. 71, 1963.

8. Sir Harry R. Ricardo, "The High Speed Internal Combustion Engine." Blackie, 1964.

9. E.S. Starkman, F.M. Strange and T.J. Dahm, "Flame Speeds and Pressure Rise Rates in Spark Ignition Engines." Pre-print for SAE Meeting, August 10-13, 1959.

10. R.E. Winsor and D.J. Patterson, "Mixture Turbulence, A Key to Cyclic Combustion Variation." SAE 730086.

APPENDIX

DISCUSSION OF STEADY STATE FLOW RIG TEST RESULTS

Air was blown through a clean and dirty inlet port, with and without the masked valve, and measurements of flow rate and swirl made for various valve lifts.

Ricardo swirl ratio is the predicted ratio of swirl to engine crankshaft speed. In D.I. diesel engines this value is typically 2.0, so in the E6 the masked valve gave a relatively high swirl (3.5) and the plain valve a relatively low swirl (0.65).

The values of the gulp factor obtained indicated that the restrictiveness of the cleaned port both with and without the mask was unlikely to reduce volumetric efficiency significantly at maximum load and speed. The carbon deposits in the port before cleaning affected port properties significantly.

Fuel and Lubricant Effects on Fuel Economy

Fuel Economy Improvements in EPA and Road Tests with Engine Oil and Rear Axle Lubricant Viscosity Reduction*

Malcolm C. Goodwin and Merrill L. Haviland
Fuels and Lubricants Dept.,
General Motors Research Labs

ONE OF GENERAL MOTORS MAJOR RESEARCH EFFORTS is to improve fuel economy in order to conserve petroleum resources and meet mandated fuel economy standards. To help achieve these goals, ways are being explored to reduce engine and driveline friction, part of which results from the effort required to shear and churn the lubricants involved. A number of investigators have found that fuel economy can be improved by using either multigrade or low-viscosity engine oils and rear axle lubricants; improvements ranged from a few percent to as much as 18 percent (1-11).* However, most previous studies

---

* Numbers in parentheses designate References at end of paper.
*Paper 780596 presented at the Passenger Car Meeting, Troy, 1978.

were conducted using either steady-state dynamometer or road tests. The results presented here show the effects on fuel economy of reducing engine oil and rear axle lubricant viscosities in EPA and various warmed-up and cold-start road tests, using a number of different power-to-mass ratio cars. In order to determine if our results, and those obtained by others, are reasonable, estimates based on friction reduction are made of maximum fuel economy improvements that might be expected by decreasing either engine oil or rear axle lubricant viscosity.

TEST LUBRICANTS, CARS, AND FUELS

TEST LUBRICANTS - Tests were conducted using the lubricants described in Table 1.

ABSTRACT

Effects of reducing engine oil and rear axle lubricant viscosities on fuel economy were determined in EPA combined City and Highway (EPA 55/45) tests and in road tests, using four different sized cars. In EPA 55/45 tests, fuel economy rating improvements averaged about 1.5 percent; warmed-up and cold-start road test fuel economy improvements averaged about 4 and 8 percent, respectively. For a specific engine oil viscosity reduction, warmed-up road test fuel economy increased with decreasing car mass and power-to-mass ratio. Warmed-up constant-speed fuel economy improvements obtained by lowering only the engine oil viscosity were about the same as those estimated from reductions in engine friction power. However, measured fuel economy improvements with low-viscosity rear axle lubricants were inexplicably higher than those estimated.

Although fuel economy results with low-viscosity lubricants were generally favorable, care must be taken to make sure the use of such lubricants will not reduce vehicle performance and durability under a variety of operating conditions. Accordingly, vehicle tests are underway.

Table 1 - Test Lubricants

Lubricant	Viscosity, cSt				SAE Viscosity Grade	Elemental Analyses, Mass %					Sulfated Ash Mass %
	0°C*	20°C*	37.8°C	98.9°C		Ca	Mg	P	S	Zn	
Engine Oils											
M	340	105	47.8	8.1	5W-20	0.21	<0.01	0.13	0.30	0.15	1.18
N	220	68	31.8	5.8	5W-20	0.37	<0.01	0.18	0.45	0.12	1.36
O	390	100	44.5	7.2	10W	0.24	<0.01	0.19	0.61	0.09	1.36
P	900	180	74.8	9.1	20W-20	0.37	<0.01	0.19	0.47	0.12	1.42
Q	520	160	71.4	11.6	10W-30	<0.01	0.10	0.14	0.36	0.15	Not Determined
R	820	220	96.7	14.7	10W-40	0.24	<0.01	0.21	0.60	0.12	1.31
Rear Axle Lubricants											
S	360	90	37.7	5.8	75W	<0.01	<0.01	0.45	1.6	0.29	Not Determined
T	3000	550	174	15.2	80W-90	<0.01	<0.01	0.31	1.4	0.24	Not Determined
U	8000	860	241	16.1	90	<0.01	<0.01	0.31	2.8	0.39	Not Determined
Automatic Transmission Fluid											
	370	100	46.6	7.4							

* Extrapolated from 37.8 and 98.9°C Viscosities

Engine oils were of SE quality, and gear lubricant GL-5. The high-viscosity lubricants (P, Q, R, T, U) were typical of those used for factory fill, and recommended for service by General Motors at moderate and high temperatures. The low-viscosity lubricants (M, N, O, S) are recommended for use in service at low and moderate temperatures.

Engine Oils - Engine oils N, O, P, and R are mineral oils containing the same additive treatment. Oil M is a synthetic engine oil. Elemental analyses are similar for all oils, except for oil Q which contains a magnesium compound. Oils N and R contain the same polymethacrylate VI improver, while mineral oil Q contains a polystyrene VI improver. Although SAE 5W-20 viscosity grade engine oils were used in this experimental program over a range of temperatures, they are not currently recommended by General Motors for ambient temperatures greater than 7°C.

Rear Axle Lubricants - Three different viscosity grade rear axle lubricants were used. The SAE 90 grade product (U) was used as factory fill from 1964 until 1976. Currently, the SAE 80W-90 lubricant (T) is being used, and the SAE 75W (S) may be used in the future. Lubricants S and T contain the same additive package.

Automatic Transmission Fluid - A current factory-fill automatic transmission fluid was used for all tests. Viscosities at 37.8 and 98.9°C were 46.6 and 7.4 cSt, respectively.

TEST CARS - The cars used are described in Table 2. Four-cylinder, six-cylinder, and medium and large eight-cylinder engines were selected. Some of the cars were instrumented with thermocouples to measure engine oil, rear axle lubricant, transmission fluid, engine coolant, and ambient air temperatures. Before starting a fuel economy test program with each car, the engines were tuned to meet recommended specifications. However, they were not tuned between lubricant changes. Neither the car air conditioner, heater, nor defroster were used during any of the fuel economy measurements.

TEST FUELS - A commercial unleaded gasoline was used in all road tests. The specific gravity of this fuel ranged between 0.723 and 0.740. Clear indolene (0.733 specific gravity) was used in all EPA tests.

PROCEDURES

LUBRICANT CHANGES - Prior to determining fuel economy with either high- or low-viscosity lubricants, the car engine sump and/or rear axle housing were drained and filled with test oil. After driving the car for 8 km to warm up and mix the test and residual lubricants, the engine and/or rear axle were again drained and filled with test oil. A new engine oil filter was installed after each engine oil drain.

FUEL ECONOMY MEASUREMENTS - In both EPA and road tests, fuel consumption and fuel temperatures were measured using a precision fuel meter, calibrated just prior to each test program. Observed fuel economy values were determined by dividing the nominal test schedule

Table 2 - Test Cars

Car	Engine Disp., L	Engine No. Cyl.	Engine Carb.	Rated Power, kW	Mass,* kg	Power-to-Mass Ratio, W/kg	Transmission	Axle Ratio	Tires (Steel-Belted Radial) Size	Pressure (Cold), kPa Front	Rear Road Tests	Rear EPA Tests
A	2.3	4	1 bbl.	58	1482	39	THM 250	2.92	BR78-13	165	178	193
B	3.8	6	2 bbl.	82	1633	50	THM 350	2.56	BR78-13	165	178	178
C	5.7	8	2 bbl.	108	2320	47	THM 350	3.08	HR78-15	165	178	193
D	6.6	8	4 bbl.	138	1895	73	THM 400	2.41	FR78-15	165	178	193

* Mass includes car, driver, 102 kg ballast (sandbags) and instrumentation for fuel consumption and temperature measurements.

distance by the amount of fuel consumed. The nominal test distances were: EPA City Cycle = 11.99 km (7.45 miles), EPA Highway Cycle = 16.51 km (10.26 miles), GM City-Suburban = 6.00 km (3.73 miles), and constant-speed = 7.18 km (4.46 miles). Actual EPA test distances varied about 0.2 percent due to driver variability and tire slippage on the rolls. GM City-Suburban and constant-speed test fuel flow measurements were made for one lap of the oval or circular test tracks, minimizing test variability due to wind speed and direction; test distances were repeated within an estimated 0.04 percent. These variations in distances did not significantly affect the variability in fuel economy measurements.

To help obtain good test repeatability, each fuel economy comparison between lubricant viscosities was made using one car and usually one driver. Moreover, each EPA test fuel economy comparison was made using the same chassis dynamometer and fuel meter. At the start of testing with each car, odometer readings ranged from 6 600 to 23 200 km. Each car had been broken in, either on the EPA Certification Schedule or in customer-type driving. However, break-in effect on fuel economy was not a major concern, since only about 300 km were accumulated on each car during EPA testing.

Except as noted, all observed fuel economy values were corrected to specific reference conditions: 15.6°C ambient temperature, 15.6°C fuel temperature, 90 kPa barometric pressure, and 0.737 specific gravity using procedures described in SAE J1082 Recommended Practice (12). EPA fuel economy values were determined from both fuel meter and carbon balance data (13). Only the fuel meter data are discussed because they were more repeatable. Comparisons of the fuel economy values derived from fuel meter and carbon balance data using car B are made in Appendix A. This comparison shows that the standard deviations of the measured EPA City Schedule values ranged from 7 to 56 percent less than the standard deviations derived from comparable carbon balance data. Standard deviations of the measured City and Highway Schedule fuel economies obtained using cars A, C, and D were nearly always less than standard deviations calculated from comparable carbon balance data.

DRIVING SCHEDULES - Fuel economy was determined in EPA 55/45 tests, in warmed-up and cold-start GM City-Suburban road tests, and in warmed-up 88 km/h constant-speed tests.

EPA 55/45 Tests - Only one EPA 55/45 test per car can be run each day, making lubricant comparisons very time consuming. As a result, only two to four City and Highway Schedules were run with each engine oil and rear axle lubricant combination (13,14,15). EPA City and Highway Schedule fuel economies were corrected for fuel temperature only.

Warmed-up Road Tests - GM City-Suburban and 88 km/h constant-speed road test procedures are detailed in Appendix B. Both tests are well-suited for fuel economy testing because they simulate driving conditions for a large segment of the driving public, they are repeatable, and a sufficient number of tests can be run in a short period of time to obtain statistically reliable results. Each of these tests could be run in less than 10 minutes. GM City-Suburban tests were run on an oval track, and the constant-speed tests were run on a circular track. Before measuring fuel economy, the car was driven for about 32 km at a constant speed of 88 km/h to stabilize engine, axle, and transmission sump temperatures. All warmed-up road tests were run at ambient air temperatures of 3 to 21°C. Ambient air temperatures generally differed by less than 8°C for each comparison of high- and low-viscosity lubricants.

Cold-Start Tests - The effect of lubricant viscosity on fuel economy was measured in cold-start GM City-Suburban driving tests using cars

A and C. Before each cold-start test, the car was soaked overnight for about 16 hours while the car was parked on an apron at the side of the oval track. Fuel consumption measurement was begun immediately after the car was driven onto the track pavement (about 10 seconds after start of the engine). Each test consisted of measuring the fuel consumed in traveling one lap of the track. Four to seven cold-start tests were run with each engine oil and rear axle lubricant tested. Ambient air temperatures ranged from -7°C to +10°C. Correction for ambient air temperature was not made because valid correction factors are not available for air temperatures less than -1°C.

RESULTS

EPA and road test fuel economies and statistical analysis of the data for the tests are given in Appendices A, C, and D. Some of the results detailed in these Appendices are extracted for inclusion in the following sections.

EPA TESTS

Engine Oil and Rear Axle Lubricant Viscosity Reduction - The combined effects of reducing engine oil and rear axle lubricant viscosities on EPA fuel economy are shown in Figure 1 for cars A, B, and D. EPA 55/45 improvements were 0.7, 1.9, and 2.3 percent for cars A, B, and D, respectively. With cars A and D, the change was greater in the City Schedule than in the Highway Schedule. Moreover, the difference was greatest in the first five cycles, when the lubricants were being warmed up. Using car B, fuel economy improvements with the low-viscosity lubricants were equal for the City and Highway Schedules, and improvement was least during the first five cycles of the City Schedule.

Engine Oil Viscosity Reduction - With car C (Appendix C), two EPA 55/45 tests each were conducted using either high- or low-viscosity engine oil, and the same rear axle lubricant, S. EPA 55/45 test fuel economy using synthetic engine oil M was 1.1 percent greater than that with SAE 10W-40 oil R. EPA City and Highway Schedule fuel economy improvements were 0.4 and 2.4 percent, respectively.

Rear Axle Lubricant Viscosity Reduction - As shown in Appendix C, with low-viscosity rear axle lubricant S in car B, EPA 55/45 test fuel economy was 0.3 percent greater than that with rear axle lubricant U. Engine oil Q was used in both tests with these axle lubricants. This improvement is based on the averages of three tests each with the high- and low-viscosity rear axle lubricants, and is not statistically significant at the 90 percent confidence level. Nevertheless, since the fuel economy improved 1.9 percent using both low-viscosity engine oil and rear axle lubricant, it is estimated that about 85 percent of the EPA 55/45 fuel economy improvement (1.6 percent) resulted from the engine oil viscosity reduction. This improvement is about the same as that obtained with engine oil M in car C.

Road Tests

WARMED-UP GM CITY-SUBURBAN TESTS - Figure 2 shows GM City-Suburban fuel economies obtained with high- and low-viscosity engine oils and rear axle lubricants, using cars A, C, and D. In these tests, all differences greater than about 0.8 percent are significant at the 95 percent confidence level. Decreasing only the engine oil viscosity (SAE 10W-40 to SAE 5W-20) improved fuel economy from 1.1 to 3.9 percent, for an average of 2.2 percent for the five possible comparisons. Reducing axle lubricant viscosity (SAE 90 to 75W) improved fuel economy 0.3 to 3.1 percent, for an average of 1.5 percent. Combining the effects of engine oil and rear axle lubricant viscosity reduction (cars A and D), GM City-Suburban fuel economy was improved an average of 3.8 percent. In these road tests, therefore, engine oil viscosity reduction accounted for about 60 percent of the total fuel economy improvement.

WARMED-UP 88 km/h CONSTANT-SPEED TESTS - The effects of engine oil and rear axle lubricant viscosity on constant-speed fuel economy are shown in Figure 3. Differences greater than about 0.8 percent are significant at the 95 percent confidence level. Reducing engine oil viscosity (SAE 10W-40 to 5W-20) improved fuel economy 2.4 to 3.9 percent for an average of 3.1 percent for the five comparisons. Rear axle lubricant viscosity reduction improved fuel economy 0.7 to 2.6 percent, for an average of 1.8 percent. Combining the lubricant effects (cars A and D) resulted in an average improvement of 4.8 percent. In these constant-speed tests, engine oil viscosity reduction accounted for about 60 percent of the total improvement, the same as that found in the GM City-Suburban tests.

COLD-START GM CITY-SUBURBAN TESTS - Engine oil and rear axle lubricant viscosity reduction effects on cold-start test fuel economy with cars A and C are shown in Figures 4, 5, and 6. Initial lubricant temperatures were nearly equal to the ambient temperature, since the cars had been soaked overnight at the test track just prior to running the tests (Appendix E).

Engine Oil Viscosity Reduction - Figure 4 shows that for either high- or low-viscosity engine oils, fuel economy increased linearly with increasing ambient temperature. The increases were about 0.08 km/L/°C for car A and about 0.04 km/L/°C for car C. With either car, fuel economy was highest for the lowest viscosity engine oil. At 0°C, the fuel economy improvements with the low-viscosity engine oils were 5.9 and 2.2 percent for cars A and C, respectively. At 0°C, the high-viscosity oils

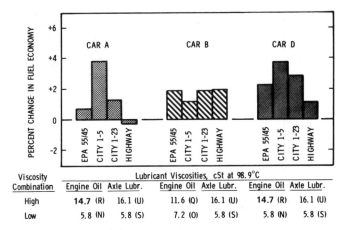

Fig. 1 - EPA 55/45 fuel economy - combined effect of engine oil and rear axle lubricant viscosity reduction

used in cars A and C were 3.7 and 2.3 times more viscous, respectively, than the low-viscosity oils.

Rear Axle Lubricant Viscosity Reduction - Figure 5 shows that fuel economy increased about 0.06 km/L/°C for car A and 0.04 km/L/°C for car C. Reducing rear axle lubricant viscosity was about as effective in improving cold-start fuel economy as was reducing engine oil viscosity. The fuel economy improvement with the low-viscosity rear axle lubricant increased slightly as the ambient temperature decreased. At 0°C, the fuel economy improvements using low-viscosity rear axle lubricant were 5.1 and 2.6 percent for cars A and C, respectively. The high-viscosity lubricant was about 22 times more viscous at 0°C than the low-viscosity lubricant.

Combined Engine Oil and Rear Axle Lubricant Viscosity Effect - Figure 6 shows the combined effect of engine oil and rear axle lubricant viscosity reduction on cold-start fuel economy. Fuel economy improvements with increasing ambient temperature were about 0.07 km/L/°C for car A and about 0.04 km/L/°C for car C. Use of low-viscosity engine oil and rear axle lubricant instead of high-viscosity lubricants improved 0°C fuel economy 11.0 and 4.8 percent for cars A and C, respectively, in the GM City-Suburban schedule.

ESTIMATES OF LUBRICANT VISCOSITY EFFECTS ON STEADY-STATE FUEL ECONOMY

Measuring the effects of engine oil and rear axle lubricant viscosity reduction on fuel economy is influenced by many variables, and since the effects are usually small, repeatable results are difficult to obtain. It is therefore of interest to estimate the potential fuel economy change that may be obtained with differ-

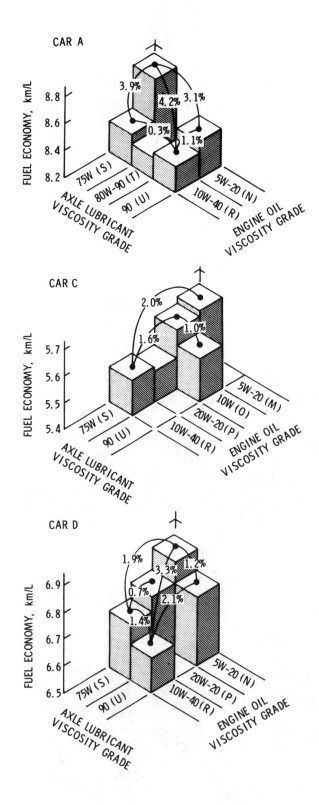

Fig. 2 - Engine oil and rear axle lubricant viscosity effects on fuel economy, GM City-Suburban driving schedule

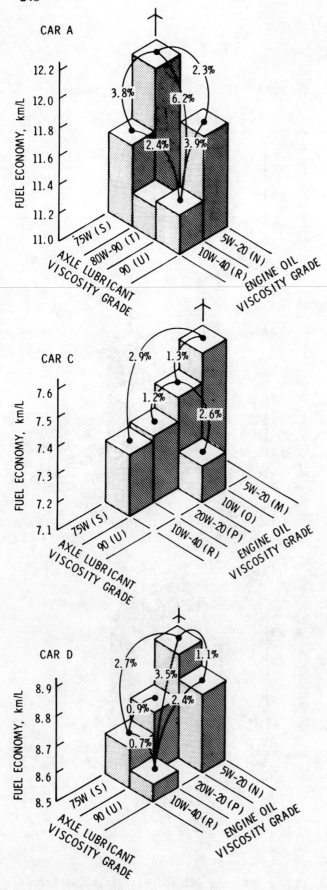

Fig. 3 - Engine oil and rear axle lubricant viscosity effects on fuel economy, 88 km/h constant-speed driving schedule

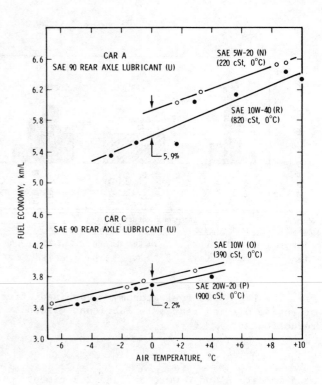

Fig. 4 - Cold-start fuel economy - effect of engine oil viscosity, GM City-Suburban schedule

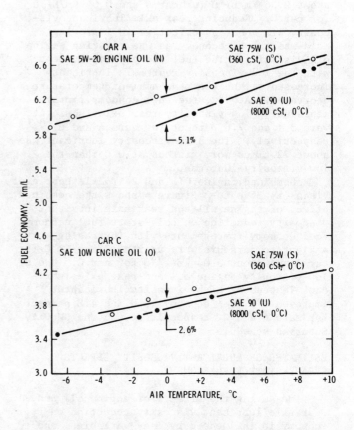

Fig. 5 - Cold-start fuel economy - effect of axle lubricant viscosity, GM City-Suburban schedule

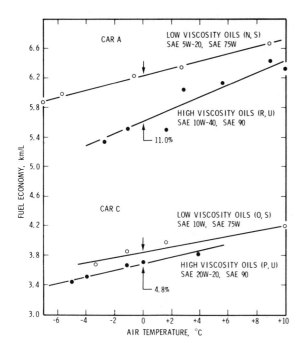

Fig. 6 - Cold-start fuel economy - combined effect of engine oil and rear axle lubricant viscosity, GM City-Suburban schedule

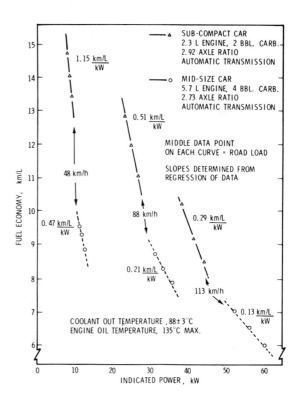

Fig. 7 - Vehicle fuel economy characteristics

ent viscosity lubricants, and compare these estimates with the actual measured values.

Engine fuel consumption decreases with decreasing indicated power, which is equal to the sum of brake power and friction power:

Indicated Power = Brake Power + Friction Power

In either an engine or axle, decreasing lubricant viscosities can decrease shear forces between surfaces separated by lubricant films, and decrease the power required to pump or churn the lubricant. Therefore, for a given road load operating condition, for a particular vehicle, reducing engine oil viscosity can reduce engine mechanical friction power, and reducing rear axle lubricant viscosity can reduce the engine brake power requirement.

Both effects result in lower indicated power and increased fuel economy as illustrated in Figure 7. These curves were plotted from part throttle engine-dynamometer test data (16). Fuel economy is shown as a function of indicated power, for three different speeds each, for cars equipped with either 5.7 or 2.3 L engines. Only small changes in indicated power, obtained by varying the throttle opening, were considered, since lubricant modifications are not expected to result in large reductions in engine or axle friction. Engine friction power, including pumping losses, was assumed to be equal to motoring power, measured about two minutes after the engine was run out of fuel while the engine was operating at stabilized road load for each speed. The throttle was opened wide and the ignition was shut off when friction measurements were made. Differences between part throttle and wide open throttle motoring friction power were neglected.

Fuel economy increased linearly with decreasing indicated power for both engines. For example, at 88 km/h, decreasing indicated power by 1 kW increased fuel economy 0.51 and 0.21 km/L for the 2.3 and 5.7 L engines, respectively. Using these relationships, the effects of engine oil and rear axle lubricant viscosity reduction on steady state fuel economy were estimated.

ENGINE OIL VISCOSITY REDUCTION - A large number of engine oils sold in the U.S. for passenger car use are SAE 10W-40 grade products whose viscosities are about 15 cSt at 100°C. The lowest engine oil viscosity grade currently recommended by General Motors for use in all gasoline-powered passenger cars in warm weather operation is SAE 20W-20, whose minimum viscosity at 100°C is 5.6 cSt, or about one-third that of the SAE 10W-40 oils. Therefore, estimated fuel economy gains that may be achieved by a two-thirds reduction in 100°C engine oil viscosity is about the most that can be expected for these lubricants. These estimates are shown in Table 3.

Road load speeds considered were 48, 88, and 113 km/h. Motoring friction power ranged from 4.8 to 17.5 kW, and from 3.0 to 12.0 kW for the 5.7 and 2.3 L engines, respectively.

Table 3 - Estimated Fuel Economy Improvement Achievable by Reducing Engine Oil Viscosity (Constant Speed, Road Load, 67 Percent Viscosity Reduction

Speed, km/h	Motoring Friction Power, kW (Includes Pumping Losses)	Mechanical Friction Power,* kW	% Reduction in Mechanical Friction by Using Low Viscosity Engine Oil**	Power Saved, kW	km/L Gained kW Saved (Fig. 7)	Fuel Economy Improvement, km/L	Road Load Fuel Economy, km/L (Fig. 7)	Fuel Economy Improvement, %
Mid-Size Car, 5.7 L Engine, 2 bbl. Carb., Automatic Transmission, 2.73 Axle Ratio								
48	4.8	3.2	30	1.0	0.47	0.47	9.3	5
88	11.5	7.7	30	2.3	0.21	0.48	8.3	6
113	17.5	11.7	30	3.5	0.13	0.46	6.6	7
Sub-Compact Car, 2.3 L Engine, Automatic Transmission, 2.92 Axle Ratio								
48	3.0	2.0	30	0.6	1.15	0.72	13.9	5
88	7.5	5.0	30	1.5	0.51	0.76	11.9	6
113	12.0	8.0	30	2.4	0.29	0.70	9.2	8

* Mechanical Friction Power Assumed to Equal 2/3 Motoring Friction Power, Reference 1

** From References 1 and 9

It was assumed that two-thirds of the motoring power resulted from mechanical friction and the remaining third from pumping losses (1).

As summarized in reference 1, a number of studies have been made showing the effect of engine oil viscosity reduction on engine friction. For the analysis here, a 30 percent reduction in engine mechanical friction was selected, since this value has been reported for a two-thirds reduction in engine oil viscosity for Newtonian oils, SAE 30 vs SAE 5W. Also, more recent data show that with Newtonian oils, a one-half reduction in 98.9°C engine oil viscosity produced a 20 percent decrease in mechanical friction power over a wide speed range (9).

A 30 percent reduction in mechanical friction at the three road load conditions produced power savings of about 1 to 3.5 kW for the 5.7 L engine, and 0.6 to 2.4 kW for the 2.3 L engine. Multiplying the power-saved values by the slope of the appropriate curve in Figure 7 resulted in average fuel economy improvements of 0.47 and 0.73 km/L for the cars equipped with 5.7 and 2.3 L engines, respectively. These improvements averaged about 6 percent for each of the cars.

Even though the mechanical friction in the 2.3 L engine was lower than that for the 5.7 L engine, the fuel economy gain per kW saved was higher. These results show that lowering engine oil viscosity for improved fuel economy can be effective in both small and large engines.

REAR AXLE LUBRICANT VISCOSITY REDUCTION - Estimated maximum road load fuel economy improvements achievable by decreasing axle lubricant viscosity are shown in Table 4. Currently, SAE 90 or 80W-90 rear axle lubricants are widely used in factory fill and service.

Rear axle lubricant operating temperatures partly control the minimum viscosity desired for best fuel economy. In these studies, warmed-up rear axle lubricant temperatures in constant-speed driving tests averaged about 60°C (Appendix D). Published axle efficiency data at about this temperature show that the lowest viscosity axle lubricant for improved fuel economy was an SAE 75W grade product (10). The 98.9°C viscosities of a large number of SAE 90 rear axle lubricants are about 3 times those of 75W lubricants. For the reduction from SAE 90 to SAE 75 viscosities, it has been shown that axle efficiency was increased by about 1 percent, reducing the brake power requirement by 1 percent, also (9,10). As shown in Table 4, the resulting power savings increased from about 0.1 to 0.4 kW as speed increased from 48 to 113 km/h for each car. These power savings produced average fuel economy improvements of about 0.6 and 0.8 percent for the mid-size and compact cars, respectively.

COMPARISON OF ESTIMATED AND MEASURED RESULTS - Estimated and measured road load fuel economy improvements obtained by decreasing engine oil and rear axle lubricant viscosities are compared in Table 5. The constant-speed

Table 4 - Estimated Fuel Economy Improvement Achievable by Reducing Rear Axle Lubricant Viscosity (Constant Speed, Road Load, 67 Percent Viscosity Reduction)

Speed, km/h	Brake Power Requirement, kW	Reduction in Brake Power Requirement By Using Low Viscosity Rear Axle Lubricant,* %	Power Saved, kW	km/L Gained / kW Saved (Fig. 7)	Fuel Economy Improvement, km/L	Road Load Fuel Economy, km/L	Fuel Economy Improvement, %
Mid-Size Car, 5.7 L Engine, 2 bbl. Carb., Automatic Transmission, 2.73 Axle Ratio							
48	7.0	1.0	0.07	0.47	0.03	9.3	0.3
88	22.2	1.0	0.22	0.21	0.05	8.3	0.6
113	38.7	1.0	0.39	0.13	0.05	6.6	0.8
Sub-Compact Car, 2.3 L Engine, Automatic Transmission, 2.92 Axle Ratio							
48	7.2	1.0	0.07	1.2	0.08	13.9	0.6
88	19.8	1.0	0.20	0.51	0.10	11.9	0.8
113	32.9	1.0	0.33	0.29	0.10	9.2	1.1

* From References 9 and 10.

Table 5 - Comparison of Road Load (88 km/h) Fuel Economy Improvements Using Low-Viscosity Engine Oil and Rear Axle Lubricant

Car Type	Engine Displacement, L	Oil Type	98.9°C Viscosity Reduction, %	Fuel Economy Improvement, %	
Engine Oil Effect					
Sub-Compact	2.3	Newtonian	67	6	Estimated
Mid-Size	5.7	Newtonian	67	6	Estimated
Car A	2.3	Non-Newtonian	60	3.8	Measured
Car B	5.7	Non-Newtonian	60	2.9	Measured
Rear Axle Lubricant Effect					
Sub-Compact	2.3	Newtonian	72	0.8	Estimated
Mid-Size	5.7	Newtonian	72	0.6	Estimated
Car A	2.3	Newtonian	64	2.3	Measured
Car B	5.7	Newtonian	64	2.6	Measured

road test data in Figure 3 for cars A, C, and D show that reducing engine oil viscosity by about 60 percent improved fuel economy about 3.8, 2.9, and 2.5 percent, respectively. If the engine oil viscosities had been reduced by 67 percent as for the fuel economy estimates, the improvements would probably have been slightly greater than those shown. Also, the non-Newtonian characteristics of the engine oils used in the road tests may have been responsible for the smaller fuel economy gains. Nevertheless, the agreement is good between the estimated and road test data.

Reducing rear axle lubricant viscosity by about 64 percent improved fuel economy in road load car tests more than that estimated for a 72 percent viscosity reduction. The reason for this large discrepancy between measured and estimated fuel economies is not known, but may be that at the road test axle lubricant temperatures investigated, axle efficiency was improved more than 1 percent.

The fuel economy improvements brought about by decreasing engine oil and rear axle lubricant viscosities can be added together. Therefore, the *maximum* fuel economy gain obtainable at road load is about 5 to 7 percent, based on estimated and measured results.

Fuel economy improvements obtained at road load by reducing engine oil and rear axle lubricant viscosity are expected to be greater than those in most other types of driving, because engine friction power is a relatively high percent of indicated power. Moreover, at low and medium speeds, road load, forces between engine and axle rubbing surfaces are generally low, allowing the formation of hydrodynamic lubricant films. During high temperature conditions when viscosities are low or during vehicle accelerations under moderate and heavy throttle conditions, loads between rubbing components can be high enough to either prevent the formation of lubricant films, or cause them to break down. Under such boundary lubrication conditions, lubricant friction modifiers or low-friction surface films must be relied upon to reduce friction and improve fuel economy. Using blends of friction modifiers in low-viscosity lubricants is believed to be the best way to obtain maximum fuel economy improvements for a wide range of operating conditions.

DISCUSSION

EPA TESTS - Of all tests used in this investigation, reducing engine oil and rear axle lubricant viscosity improved EPA 55/45 fuel economy the least. For three different cars, fuel economy improved an average of 1.6 percent. Fuel economy improvement was modest for several reasons. Low-viscosity lubricants may have *decreased* fuel economy during some of the 44 accelerations (3.7 accelerations per km) in the City Schedule, because lubricant films may not have been maintained for some of the low-viscosity lubricants, particularly during the latter part of the test when lubricant temperatures were high (Appendix C). Moreover, during acceleration, the percent of fuel consumed in overcoming friction is less than that under steady state or road load conditions. Therefore, even if film lubrication were to be maintained, the lower engine friction obtained with low viscosity oils would not improve fuel economy very much.

With cars A and D, all or most of the EPA City Schedule fuel economy improvement occurred in the first five cycles, when the lubricants were being warmed up and the viscosity differences between lubricants were greatest. Some of the rubbing engine and axle components were likely to have been separated by lubricant films. Decreasing lubricant viscosity under such conditions is an effective way to reduce engine and axle friction and increase fuel economy.

With car B, the low-viscosity lubricants were effective in improving fuel economy in practically all cycles of the City Schedule. With car C, the low-viscosity synthetic engine oil improved only the Highway Schedule fuel economy. The reasons for the differences among cars in responsiveness to lubricant viscosity reduction are not known, but are probably related to vehicle design parameters, such as clearances and unit loads between rubbing surfaces, surface finishes, and lubrication systems, and to EPA test procedure variability.

With car A, EPA Highway Schedule fuel economy with the low-viscosity lubricant combination was worse than that with the high viscosity combination. Although highway operation is generally considered to be mild, there are 24 accelerations (1.5 accelerations per km) in the EPA Highway Schedule. Moreover, the high engine and axle lubricant temperatures reduced the viscosity difference between the high- and low-viscosity lubricants (see Appendix C). Therefore, film lubrication between

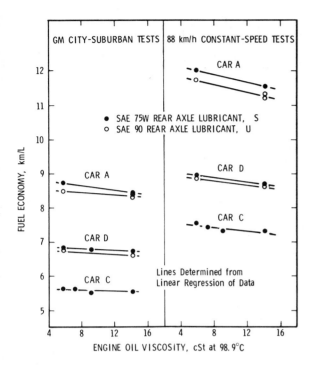

Fig. 8 - Effect of engine oil viscosity on fuel economy

rubbing components may not have been maintained with the low-viscosity lubricants, resulting in higher friction associated with boundary lubrication. This is one of the reasons we are concerned about using low-viscosity lubricants. The inability to maintain hydrodynamic lubrication is a major cause of excessive wear of critical engine parts (17). Other concerns include excessive oil consumption associated with volatile low-viscosity mineral oils, and excessive internal leakage of engine components such as the oil pump and hydraulic lifters. These problems may be economically solved by blending synthetic lubricants of low viscosity and low volatility with mineral oils of low volatility and high viscosity.

Possibly, engine oils meeting the SAE 5W-30 requirements or those of a new viscosity classification such as 7 1/2W-25 should be used for best over-all fuel and oil economy and wear performance. However, with such oils, EPA 55/45 fuel economy might be less than that with the SAE 5W-20 and 10W oils evaluated in this investigation.

WARMED-UP GM CITY-SUBURBAN AND CONSTANT-SPEED TESTS - Fuel economy improvements averaged 3.8 and 4.8 percent with low-viscosity lubricant combinations in warmed-up GM City-Suburban and 88 km/h constant-speed tests. Severity of operation for the GM City-Suburban test is believed to lie between that of the EPA City and Highway Schedules. There are 11 accelerations (1.8 accelerations per km), and the lubricant temperatures were generally slightly less than those in the EPA City Schedules (see Appendices C and D). If the ambient temperatures had been higher, lubricant temperatures would also have been higher, possibly resulting in a lower fuel economy improvement with the low-viscosity lubricants.

Effect of Viscosity - Data from Figures 2 and 3 are plotted in Figure 8 to show the effect of 98.9°C engine oil viscosity on fuel economy. Results are shown for two rear axle lubricants. The curves shown were obtained using a least squares linear regression of the data for each lubricant combination. Average GM City-Suburban fuel economy improvements for cars A, D, and C were, respectively, about 0.03, 0.02, and 0.01 km/L, for each centistoke engine oil viscosity was decreased. At a constant speed of 88 km/h, Figure 3, the average slopes were 0.06, 0.03, and 0.02 km/L per centistoke decrease in engine oil viscosity. Since the curves do not tend to level off at the lower viscosities, the minimum 98.9°C viscosity for best fuel economy may be less than 5 cSt. However, these tests were conducted at relatively low ambient temperatures (3 to 21°C). Additional data are required to determine the minimum 98.9°C engine oil viscosity for best fuel economy at high ambient temperatures.

Effect of Mass and Power-to-Mass Ratio - The slopes of each curve in Figure 8 are shown in Figure 9 as a function of car mass and power-to-mass ratio. In addition, results from a previous study using a large car (2 475 kg) with a high power-to-mass ratio (68 W/kg) are included (8). These data indicate that, for improved fuel economy, an engine oil viscosity reduction favors small cars with low power-to-mass ratios. However, this trend was not obtained in the EPA tests. Moreover, as discussed previously, percentage fuel economy improvement estimates made for engine oil and rear axle lubricant viscosity reductions were about the same for mid-size and subcompact cars equipped with 5.7 and 2.3 L engines, respectively. Additional work is needed to better understand the relationships between friction reduction and fuel economy improvement for different sized cars and different driving modes.

COLD-START GM CITY-SUBURBAN TESTS - Reducing engine oil and rear axle lubricant viscosities improved fuel economy the most in cold-start GM City-Suburban tests. A similar result was obtained in an earlier investigation using a high power-to-mass ratio (68 W/kg) car (8).

In the cold-start GM City-Suburban tests, fuel economy improved with increasing ambient temperature for several reasons: 1) fuel vaporization improved; 2) time-to-choke-opening decreased; and 3) general vehicle warm-up time decreased (tire and driveline friction decreased, and engine oil and axle lubricant

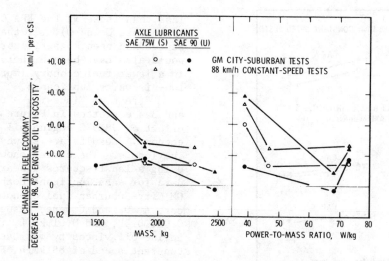

Fig. 9 – Effect of engine oil and rear axle lubricant viscosities on fuel economy for different cars

viscosities decreased). Of these, only engine oil and axle lubricant viscosities were test parameters, differing the most at low ambient test temperatures. At an ambient temperature of 0°C, cold-start GM City-Suburban test fuel economies were about two-thirds those obtained under fully-warmed-up conditions. The improvement in fuel economy with increasing ambient temperature is about the same as that obtained previously, and is believed to result primarily from lubricant viscosity reduction due to warm-up (1,8).

Engine Oil Viscosity Reduction - For cars A and C at 0°C, high-viscosity engine oils were 3.7 and 2.3 times, respectively, more viscous than the low-viscosity oils. Accordingly, the fuel economy improvement with car A (5.9 percent) was greater than that with car C (2.2 percent).

Rear Axle Lubricant Viscosity Reduction - At 0°C the high-viscosity rear axle lubricant was about 22 times more viscous than the low-viscosity lubricant. The large fuel economy improvements, 5.9 and 2.6 percent, obtained with cars A and C, respectively, reflected this large viscosity difference, as did the warm-up characteristics shown in Appendix E. In steady-state, low-load operation at 0°C, a 95 percent viscosity reduction improved axle efficiency from about 72 to 86 percent, for an engine brake power requirement reduction of about 0.8 kW (10). The effect of this power reduction on fuel economy at low temperatures is not known because low temperature engine performance characteristics are not known. However, under warmed-up 48 km/h road load conditions, a power reduction of 0.8 kW would improve car A fuel economy about 0.9 km/L or about 6 percent (from Figure 7).

Combining the effects on fuel economy of engine oil and rear axle lubricant viscosity reduction resulted in impressive gains (4.8 to 11.0 percent) in cold-start fuel economy. Therefore, the use of low-viscosity lubricants for low-temperature operation is especially desirable. However, before low-viscosity lubricants can be safely used for factory fill and recommended for service, high- and low-temperature durability testing with such lubricants must be satisfactory. Such tests are underway.

SUMMARY

The effect of reducing engine oil and rear axle lubricant viscosities on fuel economy has been determined in a variety of tests, using four cars whose power-to-mass ratios ranged from 39 to 73 W/kg.

Warmed-up engine oil and rear axle lubricant viscosities were reduced by 60 percent. In cold-start tests, engine oil and rear axle lubricant viscosities were reduced by about 60 and 95 percent, respectively. Reducing both engine oil and rear axle lubricant viscosities improved fuel economy as shown in summary Table 6.

About 60 to 85 percent of the improvements obtained in warmed-up tests resulted from the engine oil viscosity reduction. Engine oil and rear axle lubricant viscosity reductions were equally effective in the cold-start tests.

Generally, fuel economy improvement with low-viscosity lubricants increased with decreasing lubricant temperature, decreasing engine and axle loads, and, in road tests, with decreasing car size. Under warmed-up, steady-state, road load conditions, reducing engine

Table 6 - Summary of Results

Test	No. Cars	Average Percent Fuel Economy Improvement
EPA 55/45	3	1.6
City	3	2.0
Highway	3	1.0
Constant-Speed, 88 km/h	2	4.8
GM City-Suburban, (Warmed-up)	2	3.8
GM City-Suburban, (Cold-Start, 0°C)	2	7.9

oil viscosity improved fuel economy about 2.5 to 4 percent. These results are slightly less than estimated maximum values, based on engine power savings resulting from reduced engine friction. Estimated fuel economy improvements averaged about 0.7 percent for rear axle lubricant viscosity reduction, and were less than some of the measured values. The reasons for this discrepancy are not known.

CONCLUSION

Reducing either engine oil or rear axle lubricant viscosities can significantly improve vehicle fuel economy, both on the road in customer-type service and to a lesser extent in EPA tests. However, performance and durability tests should be conducted with low-viscosity lubricants to prove that they will be satisfactory in customer service before they are recommended for general use.

ACKNOWLEDGMENTS

By paying close attention to many details of the various test procedures, Jim Linden, Jack Collins, Dave Coleman, and Dave Hansen obtained data which showed significant fuel economy differences among lubricants of different viscosities. We appreciate their efforts and diligence.

REFERENCES

1. J. A. Whitehouse and J. A. Metcalf, "The Mechanical Efficiency of I.C. Engines - A Review of Published Information," Report No. 1958/5, Published by the Motor Industry Research Association (MIRA), Warwickshire, Great Britain, June 1958.

2. C. W. Georgi, "Some Effects of Motor Oils and Additives on Engine Fuel Consumption," 1954 SAE Transactions, Vol. 62, pp. 385-391.

3. J. B. Bidwell and R. K. Williams, "The New Look in Lubricating Oils," 1955 SAE Transactions, Vol. 63, pp. 349-361.

4. "The Efficient Use of Automotive Fuels: A Collection of Four Reports," Prepared by the British Technical Council of the Motor and Petroleum Industries, Chairman: H. J. C. Weighell, February 1976.

5. W. H. Richman and J. A. Keller, "An Engine Oil Formulated for Optimized Engine Performance," SAE Paper 750376, SAE Automotive Engineering Congress and Exposition, February 1975.

6. M. F. Smith, Jr., N. Tunkel, H. E. Bachman, and W. J. Fernandez, "A New Look at Multigraded Diesel Engine Oils," SAE Paper 760558, SAE Fuels and Lubricants Meeting, St. Louis, Missouri, June 1976.

7. J. C. Ingamells, "Fuel Economy and Cold-Start Driveability with Some Recent Model Cars," SAE Paper 740522, Combined Commercial Vehicle and Fuels and Lubricants Meetings, Chicago, Illinois, June 1974.

8. E. D. Davison and M. L. Haviland, "Lubricant Viscosity Effects on Passenger Car Fuel Economy," SAE Paper 750675, Fuels and Lubricants Meeting, Houston, Texas, June 1975.

9. W. B. Chamberlain and T. J. Sheahan, "Automotive Fuel Savings Through Lubricants," SAE Paper 750377, SAE Automotive Engineering Congress, Detroit, Michigan, February 1975.

10. P. A. Willermet and L. T. Dixon, "Fuel Economy - Contribution of the Rear Axle Lubricant," SAE Paper 770835, SAE Passenger Car Meeting, Detroit, Michigan, September 1977.

11. B. M. O'Connor, W. S. Romig, and L. F. Schiemann, "Energy Conservation Through the Use of Multigraded Gear Oils in Trucks," SAE Paper 770833, SAE Passenger Car Meeting, Detroit, Michigan, September 1977.

12. "Fuel Economy Measurement -- Road Test Procedure," J1082a, SAE Recommended Practice, SAE Handbook 1978, Part 2, pp. 24.122-24.129.

13. Federal Register, Vol. 38, Number 151, August 7, 1973.

14. Federal Register, Vol. 37, Number 221, November 15, 1972.

15. Federal Register, Vol. 41, Number 100, May 21, 1976.

16. GM Engine Test Code, Sixth Edition, Distributed by Engineering Standards Department, Engineering Staff, General Motors Corporation, Warren, Michigan, Copyright 1975.

17. J. L. Bell and M. A. Voisey, "Some Relationships Between the Viscometric Properties of Motors Oils and Performance in European Engines," Joint SAE-ASTM Symposium, "The Relationship Between Engine Oil Viscosity and Engine Performance," (SAE SP 416 and ASTM STP 621) held during the SAE International Automotive Engineering Congress and Exposition, Detroit, Michigan, February 1977.

APPENDIX A

Table A-1 - Comparison of Measured and Calculated Fuel Economies in EPA Tests

Car B Lubricants		Fuel Economies, mpg					
		Calculated from Carbon Balance			Measured with Fuel Meter		
Engine	Axle	City Schedule - Cycles			City Schedule - Cycles		
		1-5	19-23	1-23	1-5	19-23	1-23
O	S	16.93	19.89	17.62	15.90	19.39	17.52
O	S	16.65	19.54	17.34	16.26	19.73	17.82
O	S	16.44	19.79	17.29	16.05	19.56	17.66
O	S	16.87	19.48	17.31	16.30	19.17	17.55
Mean		16.72	19.68	17.39	16.13	19.46	17.64
Standard Deviation		0.22	0.20	0.15	0.19	0.24	0.14
Q	U	16.75	19.71	17.45	15.88	19.15	17.31
Q	U	16.29	19.15	16.96	15.70	19.01	17.20
Q	U	16.59	19.52	17.32	16.22	19.01	17.42
Mean		16.54	19.46	17.24	15.93	19.06	17.31
Standard Deviation		0.23	0.28	0.25	0.26	0.08	0.11
Q	S	16.70	19.53	17.32	16.07	19.25	17.41
Q	S	16.96	20.27	17.93	16.03	19.45	17.60
Q	S	16.41	19.52	17.14	15.82	19.15	17.17
Mean		16.69	19.77	17.46	15.97	19.28	17.39
Standard Deviation		0.28	0.43	0.41	0.13	0.15	0.22

Cars	Lubricants	Standard Deviation, mpg			
		City Schedule, Cycles 1-23		Highway Schedule	
		Carbon Balance	Fuel Meter	Carbon Balance	Fuel Meter
Car A	N,S	0.04	0.04	Not Measured	0.02
Car A	R,U	0.23	0.13	Not Measured	0.26
Car C	M,S	0.12	0.05	Not Measured	0.12
Car C	R,S	0.26	0.26	Not Measured	0.35
Car D	N,S	0.20	0.10	0.47	0.41
Car D	R,U	0.29	0.02	0.08	0.14

APPENDIX B

GM CITY-SUBURBAN FUEL ECONOMY DRIVING SCHEDULE

After warm-up, the GM City-Suburban driving schedule is run for one lap on the 6.00 km oval track at the GM Proving Grounds, Milford, Michigan as shown in Table B-1.

88 km/h CONSTANT-SPEED FUEL ECONOMY DRIVING SCHEDULE

After warm-up, the 88 km/h constant-speed driving schedule is run for one lap on the 7.18 km circular track at the GM Proving Grounds, Milford, Michigan as follows:

At 88 km/h driving speed, start fuel meter at a selected beginning point. Maintain constant speed during test run with throttle movement the minimum required to maintain desired speed. Stop fuel meter at the end of the lap and record fuel consumed, elapsed time, and fuel temperature. At least ten repeat tests should be run.

The average driving speed is 88 km/h and approximate running time is 293 seconds per lap.

Table B-1 - GM City-Suburban Fuel Economy Driving Schedule

Distance, km	Driving Maneuver*
0.00	Start fuel meter and timing device, accelerate to 32.2 km/h at an initial acceleration of 1.8 m/s^2.
0.16	Accelerate to 40.2 km/h at 0.9 m/s^2.
0.48	Stop at a constant deceleration of 1.8 m/s^2, accelerate to 40.2 km/h at 1.8 m/s^2.
0.80	Decelerate to 16.1 km/h at 1.2 m/s^2, accelerate to 48.3 km/h at 1.8 m/s^2.
1.01	Stop at 1.8 m/s^2, idle 25 seconds, accelerate to 48.3 km/h at 1.8 m/s^2.
2.41	Decelerate to 40.2 km/h at closed throttle, accelerate to 56.3 km/h at 0.9 m/s^2.
3.22	Stop at 1.8 m/s^2, accelerate to 64.4 km/h at maximum acceleration (wide open throttle).
4.51	Stop at 1.8 m/s^2, idle 25 seconds, accelerate to 40.2 km/h at 1.8 m/s^2.
4.83	Decelerate to 8.0 km/h at 1.2 m/s^2, accelerate to 40.2 km/h at 1.8 m/s^2.
5.15	Stop at 1.8 m/s^2, idle 25 seconds, accelerate to 56.3 km/h at 2.4 m/s^2.
5.47	Decelerate to 40.2 km/h at closed throttle, accelerate to 56.3 km/h at 0.9 m/s^2.
6.00	Begin braking at 1.8 m/s^2 to arrive at stop at 6.00 km distance. Stop fuel meter and record fuel consumed, elapsed time and fuel temperature.

At least six repeat tests should be run.

* Only initial acceleration values are indicated. Hold initial throttle position constant until specified speed is reached. Driving maneuvers, except for the final stop, are initiated at the points indicated.

The average driving speed is 39 km/h and approximate running time is 558 seconds per cycle.

APPENDIX C - EPA FUEL ECONOMY TEST RESULTS (FUEL METER DATA)

CAR A

	EPA 55/45	City Schedule - Cycles			Highway Schedule
		1-5	19-23	1-23	
LOW-VISCOSITY LUBRICANT COMBINATION: N,S					
Mean Fuel Economy, mpg	19.33	14.77	17.64	16.50	24.45
Number of Tests	2	2	2	2	2
Standard Deviation, mpg	0.05	0.15	0.12	0.04	0.02
95% Confidence Interval, mpg	19.26 to 19.40	14.56 to 14.98	17.40 to 17.88	16.46 to 16.54	24.27 to 24.63
Mean Engine Oil Temperature, °C End of Test	-	74	101	101	115
Mean Axle Lubricant Temp., °C End of Test	-	34	59	59	78
HIGH-VISCOSITY LUBRICANT COMBINATION: R,U					
Mean Fuel Economy, mpg	19.19	14.23	17.52	16.29	24.53
Number of Tests	4	4	4	4	4
Standard Deviation, mpg	0.09	0.13	0.16	0.13	0.26
95% Confidence Interval, mpg	19.09 to 19.29	14.09 to 14.37	17.22 to 17.82	16.16 to 16.42	24.27 to 24.79
Mean Engine Oil Temperature, °C End of Test	-	79	102	102	116
Mean Axle Lubricant Temp., °C End of Test	-	41	65	65	85
LOW-vs HIGH-VISCOSITY COMBINATIONS					
Difference between Means, mpg	0.14	0.54	0.12	0.21	-0.08
Fuel Economy Improvement, %	0.7	3.8*	0.7	1.3*	-0.3

*Significant at 90% Confidence Level

APPENDIX C - EPA FUEL ECONOMY TEST RESULTS (FUEL METER DATA)

CAR B

	EPA 55/45	City Schedule - Cycles			Highway Schedule
		1-5	19-23	1-23	
LOW-VISCOSITY LUBRICANT COMBINATION: O,S					
Mean Fuel Economy, mpg	20.49	16.13	19.46	17.64	25.55
Number of Tests	4	4	4	4	4
Standard Deviation, mpg	0.19	0.19	0.24	0.14	0.49
95% Confidence Interval, mpg	20.30 to 20.68	15.94 to 16.32	19.22 to 19.70	17.50 to 17.78	25.07 to 26.03
Mean Engine Oil Temperature, °C End of Test	-	Not Measured	Not Measured	Not Measured	Not Measured
Mean Axle Lubricant Temp., °C End of Test	-	"	"	"	"
HIGH-VISCOSITY LUBRICANT COMBINATION: Q,U					
Mean Fuel Economy, mpg	20.11	15.93	19.05	17.31	25.06
Number of Tests	3	3	3	3	3
Standard Deviation, mpg	0.10	0.26	0.08	0.11	0.18
95% Confidence Interval, mpg	19.99 to 20.23	15.63 to 16.23	18.96 to 19.14	17.19 to 17.43	24.86 to 25.26
Mean Engine Oil Temperature, °C End of Test	-	Not Measured	Not Measured	Not Measured	Not Measured
Mean Axle Lubricant Temp., °C End of Test	-	"	"	"	"
LOW-vs HIGH-VISCOSITY COMBINATIONS					
Difference between Means, mpg	0.38	0.19	0.41	0.33	0.49
Fuel Economy Improvement, %	1.9*	1.2	2.2*	1.9*	2.0

APPENDIX C - EPA FUEL ECONOMY TEST RESULTS (FUEL METER DATA)

CAR B, ENGINE OIL Q

	EPA 55/45	City Schedule - Cycles			Highway Schedule
		1-5	19-23	1-23	

LOW-VISCOSITY REAR AXLE LUBRICANT: S

	EPA 55/45	1-5	19-23	1-23	Highway Schedule
Mean Fuel Economy, mpg	20.17	15.97	19.27	17.39	25.03
Number of Tests	3	3	3	3	3
Standard Deviation, mpg	0.21	0.15	0.12	0.22	0.15
95% Confidence Interval, mpg	19.93 to 20.41	15.80 to 16.14	19.14 to 19.40	17.14 to 17.64	24.86 to 25.20
Mean Engine Oil Temperature, °C End of Test	—	Not Measured	Not Measured	Not Measured	Not Measured
Mean Axle Lubricant Temp., °C End of Test	—	"	"	"	"

HIGH-VISCOSITY REAR AXLE LUBRICANT: U

	EPA 55/45	1-5	19-23	1-23	Highway Schedule
Mean Fuel Economy, mpg	20.11	15.93	19.05	17.31	25.06
Number of Tests	3	3	3	3	3
Standard Deviation, mpg	0.10	0.26	0.08	0.11	0.18
95% Confidence Interval, mpg	19.99 to 20.23	15.63 to 16.23	18.96 to 19.14	17.19 to 17.43	24.83 to 25.29
Mean Engine Oil Temperature, °C End of Test	—	Not Measured	Not Measured	Not Measured	Not Measured
Mean Axle Lubricant Temp., °C End of Test	—	"	"	"	"

LOW-vs HIGH-VISCOSITY   Rear Axle Lubricants

	EPA 55/45	1-5	19-23	1-23	Highway Schedule
Difference between Means, mpg	0.06	0.04	0.22	0.08	-0.03
Fuel Economy Improvement, %	0.3	0.3	1.2*	0.5	-0.1

*Significant at 90% Confidence Level

APPENDIX C - EPA FUEL ECONOMY TEST RESULTS (FUEL METER DATA)

CAR C, REAR AXLE LUBRICANT S

	EPA 55/45	City Schedule - Cycles			Highway Schedule
		1-5	19-23	1-23	

LOW-VISCOSITY ENGINE OIL: M

	EPA 55/45	1-5	19-23	1-23	Highway Schedule
Mean Fuel Economy, mpg	12.89	10.46	12.44	11.05	16.21
Number of Tests	2	2	2	2	2
Standard Deviation, mpg	0.00	0.09	0.07	0.05	0.12
95% Confidence Interval, mpg	12.89	10.27 to 10.65	12.34 to 12.54	10.97 to 11.13	16.04 to 16.38
Mean Engine Oil Temperature, °C End of Test	—	82	108	108	117
Mean Axle Lubricant Temp., °C End of Test	—	33	51	51	68

HIGH-VISCOSITY ENGINE OIL: R

	EPA 55/45	1-5	19-23	1-23	Highway Schedule
Mean Fuel Economy, mpg	12.75	10.66	11.99	11.01	15.83
Number of Tests	2	2	2	2	2
Standard Deviation, mpg	0.30	0.13	0.30	0.26	0.35
95% Confidence Interval, mpg	12.33 to 13.17	10.53 to 10.79	11.57 to 12.41	10.64 to 11.36	15.34 to 16.32
Mean Engine Oil Temperature, °C End of Test	—	82	108	108	118
Mean Axle Lubricant Temp., °C End of Test	—	33	51	51	68

LOW-vs HIGH-VISCOSITY COMBINATIONS

	EPA 55/45	1-5	19-23	1-23	Highway Schedule
Difference between Means, mpg	0.14	-0.20	0.45	0.04	0.38
Fuel Economy Improvement, %	1.1	-1.9	3.8	0.4	2.4

*Significant at 90% Confidence Level

APPENDIX C – EPA FUEL ECONOMY TEST RESULTS (FUEL METER DATA)

CAR D	EPA 55/45	City Schedule – Cycles				Highway Schedule
		1-5	19-23	1-23		
LOW-VISCOSITY LUBRICANT COMBINATION: N,S						
Mean Fuel Economy, mpg	15.26	10.83	15.04	13.11		19.10
Number of Tests	2	3	3	3		2
Standard Deviation, mpg	0.08	0.24	0.27	0.10		0.41
95% Confidence Interval, mpg	15.15 to 15.37	10.56 to 11.10	14.73 to 15.35	13.00 to 13.22		18.53 to 19.67
Mean Engine Oil Temperature, °C End of Test	–	78	101	101		115
Mean Axle Lubricant Temp., °C End of Test	–	Not Measured	Not Measured	Not Measured		Not Measured
HIGH-VISCOSITY LUBRICANT COMBINATION: R,U						
Mean Fuel Economy, mpg	14.92	10.43	14.53	12.74		18.87
Number of Tests	2	3	3	3		2
Standard Deviation, mpg	0.06	0.13	0.16	0.02		0.14
95% Confidence Interval, mpg	14.84 to 15.00	10.28 to 10.58	14.35 to 14.71	12.72 to 12.76		18.67 to 19.07
Mean Engine Oil Temperature, °C End of Test	–	78	104	104		118
Mean Axle Lubricant Temp., °C End of Test	–	Not Measured	Not Measured	Not Measured		Not Measured
LOW- vs HIGH-VISCOSITY COMBINATIONS						
Difference between Means, mpg	0.34	0.40	0.51	0.37		0.23
Fuel Economy Improvement, %	2.3*	3.8*	3.5*	2.9*		1.2

*Significant at 90% Confidence Level

APPENDIX D - Fuel Economy Test Results

Lubricant Combination SAE Viscosity Grade			GM City-Suburban Tests				Avg. Temp. °C			88 km/h Constant-Speed Tests				Avg. Temp. °C		
Engine	Axle	Codes	No. Tests	Mean km/L	Standard Deviation	95% Conf. Interval	Air	Eng. Oil	Axle Lube	No. Tests	Mean km/L	Standard Deviation	95% Conf. Interval	Air	Eng. Oil	Axle Lube
Car A																
5W-20	75W	N,S	17	8.74	0.04	8.72 to 8.92	4	90	50	24	12.02	0.07	11.99 to 12.05	3	93	65
5W-20	90	N,U	10	8.48	0.08	8.43 to 8.53	8	90	55	26	11.75	0.11	11.71 to 11.79	6	93	67
10W-40	75W	R,S	12	8.41	0.05	8.38 to 8.44	4	90	50	24	11.58	0.13	11.53 to 11.63	3	100	65
10W-40	80W-90	R,T	8	8.36	0.04	8.33 to 8.39	6	90	55	10	11.22	0.05	11.19 to 11.25	5	100	67
10W-40	90	R,U	14	8.39	0.04	8.37 to 8.41	8	90	55	20	11.31	0.09	11.27 to 11.35	3	100	67
Car C																
5W-20	75W	M,S	9	5.64	0.06	5.60 to 5.68	10	100	55	17	7.53	0.05	7.51 to 7.55	15	105	58
10W	75W	O,S	18	5.62	0.05	5.60 to 5.64	4	95	55	26	7.42	0.04	7.40 to 7.44	4	100	58
20W-20	75W	P,S	12	5.51	0.04	5.49 to 5.53	14	95	55	17	7.34	0.05	7.32 to 7.36	10	100	58
10W-40	75W	R,S	6	5.53	0.03	5.51 to 5.55	9	95	55	22	7.32	0.08	7.31 to 7.33	7	105	58
10W	90	O,U	6	5.56	0.05	5.52 to 5.60	4	95	57	17	7.23	0.05	7.21 to 7.25	4	100	63
Car D																
5W-20	75W	N,S	11	6.85	0.08	6.80 to 6.90	11	103	Not Meas.	12	8.96	0.11	8.90 to 9.02	3	96	Not Meas.
5W-20	90	N,U	7	6.77	0.05	6.73 to 6.81	21	98	"	16	8.87	0.18	8.78 to 8.96	18	102	"
20W-20	75W	P,S	6	6.77	0.05	6.73 to 6.81	6	93	"	7	8.81	0.09	8.74 to 8.88	6	98	"
10W-40	75W	R,S	7	6.72	0.07	6.67 to 6.77	15	98	"	11	8.73	0.07	8.69 to 8.77	9	104	"
10W-40	90	R,U	7	6.63	0.05	6.59 to 6.67	18	100	"	12	8.66	0.09	8.61 to 8.71	15	105	"

Appendix E

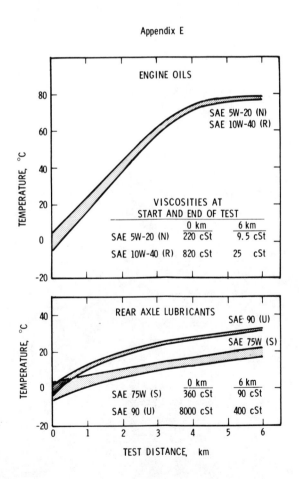

Fig. E-1 - Lubricant temperatures and viscosities, cold-start tests, cars A and C

Analysis of Fuel Economy

Fuel Economy of Alternative Automotive Engines – Learning Curves and Projections[*]

Roy Renner
Consultant, Sonora, CA
Harold M. Siegel
South Coast Technology, Inc.
Santa Barbara, CA

FORECASTERS STILL CAN'T AGREE on future automobile powerplants. Possibly the "engine of the future" will, as today, actually be a plurality of engines, each well suited for its own application. Engines appropriate for the larger passenger cars could be quite different from those found most suitable for smaller cars. Vehicles intended for stop-and-go service have far different requirements than those designed primarily for long distance travel and could use different engines.

Over recent years, a great many forecasts have been published on the properties of alternative automotive engines. Commonly, these forecasts have been based primarily on theoretical cycle calculations. Such estimates are often very optimistic; and while they may be appropriate as research goals, the probability of reaching such goals in practice and in the near term may not be high.

This paper describes another approach in making technological projections which is based on a study and review of learning curves. Human endeavors in general can be characterized as a series of learning experiences. The authors have constructed historical learning curves based on recorded fact for alternatives such as the Diesel, gas turbine, Stirling, and rotary engines. Since most developments proceed along evolutionary paths, projections of these learning curves were then extended to the 1990 era using a range within plausible slopes. The main focus of this paper is on fuel economy which is the bottom line, along with costs and investments, in evaluating alternate engine systems. Most of the work described in this paper is based on a study performed for the Office of Passenger Vehicle Research, Technology Assessment Division, National Highway Traffic Safety Administration, of the U. S. Department of Transportation. Information was developed mainly from the open literature, and conclusions were synthesized by a panel of engineers working under the direction of South Coast Technology, Inc. The views presented in this paper are those of the authors, and these views may differ from positions taken by the public agency sponsor.

This paper represents only a tentative beginning in defining automotive technology learning curves. It should be noted that the fuel economy curves may not be strictly comparable with each other since the vehicle designs, performance levels, and transmissions

[*]Paper 790022 presented at the Congress and Exposition, Detroit, 1979.

---ABSTRACT

This paper describes a distinct approach in making technological forecasts. From historical data, fuel economy learning curves have been constructed for alternatives such as the Diesel, gas turbine, Stirling, stratified charge, and rotary engines. Assuming that evolutionary development will take place, projections of these learning curves are extended to the 1990 era. The investigation found that no engine is likely to exceed the fuel economy of the Diesel in the next 10-15 years. However, serious questions are being raised regarding the feasibility of controlling nitrogen oxides and unregulated emissions in future Diesel engines.

differ. It is suggested that readers with special knowledge and insight in this field try their own hand at establishing learning curves and projections into the future. Since the authors do not claim to have the last word on this subject, they would like to hear other views or to obtain confirmation of the starting points and trends shown. This paper will have succeeded if it promotes a useful discussion of these issues, or if it proves useful to the readers in evaluating projects they may be working on.

WHAT IS A LEARNING CURVE?

Observation reveals that early learning experiences are marked by rapid progress. After a time, the rate of progress seems to slow down, and the slope of the learning curve decreases. Figure 1 shows the generalized shape of learning curves.(1)* We all know that learning experiences can often lead to discouraging plateaus. In many endeavors, it has been shown that a new learning curve can be superimposed on the base curve. This phenomenon was described several decades ago from observations of how telegraphers gained proficiency.(1) The beginner learns Morse code letter by letter. A renewed learning curve represents a capability of recognizing whole words or combinations of words as received in code.

Learning curves can be plotted from the historical development of heat engines. From the writings of Sir Dugald Clerk and others, it is possible to trace the development of thermal efficiency of internal combustion engines.(2) Such a curve is represented by Figure 2. Were this learning curve to be examined in more detail, doubtless it would be found to be a superposition of many overlapping curves of new learning, each representing individual breakthroughs in understanding or technology. Evolutionary improvements can still continue, with reductions in friction, leakage, and heat losses. As every engineer knows, ultimate achievements will be limited by the Second Law of Thermodynamics.

Figure 3, representing industrial and heavy vehicle gas turbine achievements and projections fits the classic concept of a learning curve.(3-5)

It has been correctly pointed out by George Thur of the U. S. Department of Energy that the true independent variable of an engine learning curve is the amount of money and effort expended in development, rather than calendar time alone. Time-based forecasts in this paper are predicated on R & D expenditures continuing at an adequate level. Such forecasts would be quite naive, were it not for the fact that tomorrow's test vehicles will incorporate a technology base which is becoming visible today.

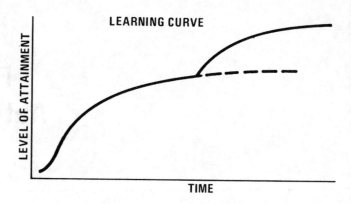

Fig. 1 - General shape of learning curves. Note that progress levels off, but breakthroughs can renew the curve

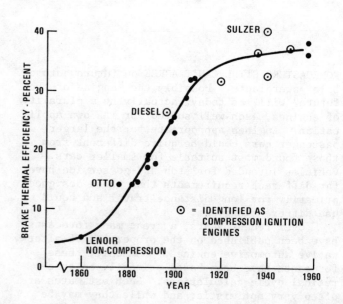

Fig. 2 - One hundred years of internal combustion engine improvement. Data mainly from large stationary engines

BASELINE ENGINE

As a point of departure, the conventional carbureted, spark ignition automobile engine was used. Table 1 lists the average fuel economy for 1978 model automobiles over a range of inertia weight classes. These data are for the combined city/highway fuel economy as published in the 1978 EPA-FEA Gas Mileage Guide.(7) Only gasoline fueled autos were included in Table 1. Values for ton-mpg* were computed by multiplying the inertia weight of the vehicle in U. S. tons by the combined fuel economy in miles per U. S. gallon.

---

* Numbers in parentheses designate references at end of paper.

Table 1

EXAMPLES OF 1978 AUTOMOBILE FUEL ECONOMY DATA
(Source: Ref. 7)

Inertia Wt., lbs	Average of Combined City/Highway Fuel Economy, mpg	Mass x Fuel Economy Product, Ton-mpg
2000	34.5	34.5
2500	28.1	35.1
3000	23.2	34.8
4000	17.8	35.6
5000	13.7	34.3

From Table 1, it will be noted that the product of mass x fuel economy is approximately equal to a constant, about 35 ton-mpg. This product may be thought of as a measure of vehicular thermal efficiency, which can be used to compare the fuel economy merit of vehicles over the weight range of interest to us. The figures in Table 1 are for vehicles meeting the 1978 emission standards (49-state) of 1.5 g/mi HC, 15 g/mi CO, and 2 g/mi NOx. In many instances, the same cars will meet California standards of 0.4/9/1.5.

Figure 4 represents our projected baseline for the uniform charge spark ignition engine. The improvement in fuel economy around 1975 was not the result of a basic engine improvement, but was caused by the introduction of new emission control technology and other vehicle changes. In extending this baseline to 1985, it was determined by the study panel that an engine-related fuel economy improvement of 10% over present practice can be achieved. A further assumption is that by 1990, an additional increment of 5% improvement can be realized. Two curves are shown on Figure 4. The upper curve will apply, we believe, to NOx emission levels down to 1.0 g/mi attainable by 1985. The lower curve shows a reduction in fuel economy which may represent NOx control down to the 0.4 g/mi level by 1985, assuming that level is both mandated and achievable. Essentially, the lower curve represents the growth potential of the 3-way catalyst system but with the efficiency constraints of stoichiometric operation. The baseline curves shown represent the average, rather than the best engines; and it is expected that fewer poor engines will be included in the average by 1985.

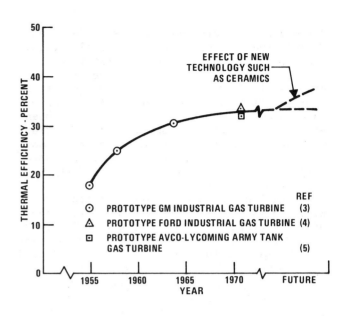

Fig. 3 - Trend in peak thermal efficiency for industrial and heavy vehicle gas turbine engines

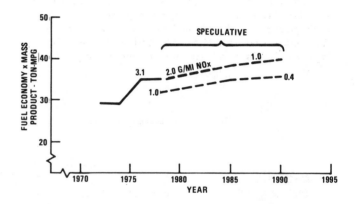

Fig. 4 - Baseline fuel economy for uniform charge spark ignition engine. Numbers near curves represent levels of NOx control

PRESENT STATUS OF ALTERNATIVES

Table 2 characterizes the present standing of a number of alternative automotive engines. Many of the engines represented are experimental and, hence, are very immature when compared with the conventional spark ignition engine.

---

* The authors have refrained from translating "ton-miles/gallon" into S.I. metric units, for fear that the mass-distance product would be misinterpreted as a literal unit of work or energy. Actually, the ton-mile has traditionally been viewed as a measure of transportation utility.

Moreover, there are considerable differences in emission control level, transmission type, fuel density, and acceleration performance.

Many of these same engines are also compared in Figure 5, but with some adjustments from the Table 2 values. Some attempt has been made to normalize the fuel economy data to an equivalent performance and an equivalent transmission basis. Figure 5 is, therefore, a more appropriate (but still not completely adjusted) comparison of the present standing of the various powerplants.

The first adjustment incorporated into Figure 5 is for the acceleration performance of the vehicle. All vehicles were normalized to a 0-60 times of 13 seconds. The adjustments for internal combustion engines were estimated from curves of fuel economy vs. acceleration capability developed by Nicholson and Niepoth of General Motors.(16) While the referenced correlations were not intended to represent the general case, at least they permit approximate adjustment in the right direction. External combustion engines (Stirling, steam) were left unadjusted, since information is lacking to confidently make a fuel economy correlation. In such engines, power is manifested more in the heat exchangers than in the expander (engine). Designs capable of higher peak power could, therefore, show less than the usual drop in thermal efficiency at the lighter loads used in fuel economy testing.

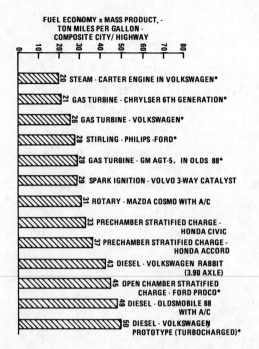

Fig. 5 - Comparison of alternatives: present fuel economy, adjusted for differences in performance and transmission. Considerable disparity exists in maturity and level of emission control. Present gasoline engines yield about 35 ton-miles per gallon. Examples marked (*) are not in production

Table 2 - Fuel Economy and Emissions of Automobiles Powered by Alternative Engines

Automobile	Engine	Test or Inertia Wt., lbs	Trans.*	Test: Air Conditioned?	Year Tested	Exhaust Emissions, g/mi HC	CO	NOx	Combined City/Hwy mpg	Ref.
1978 VW Rabbit	90 CID Diesel	2250	M-4	No	1977	0.3	1.0	1.1	44	7
1978 VW Rabbit	89 CID Gas	2250	M-4	No	1977	1.3	10.3	1.8	29	7
1978 Olds 88	350 CID Diesel	4500	A-3	Yes	1977	0.6	1.5	1.6	24	7
1978 Olds 88	350 CID Gas	4000	A-3	Yes	1977	0.5	6.1	1.8	19	7
1977 Volvo 244	130 CID Gas (3-way cat.)	3000	A-3	Yes	1977	0.31	4.0	0.25	19.5	8
1978 Mazda Cosmo	80 CID Rotary	3000	M-5	Yes	1977	0.9	7.7	1.7	22	7
Ford Test Vehicles	6.6 L PROCO Strat. Charge	5000	A-3	NA	1977?	0.23	0.2	0.76	17.9	9
1978 Honda Civic	91 CID CVCC	2000	M-4	NA	1977	NA	NA	NA	38	10
1973 Plymouth	Chrysler Gas Turbine	4500	A-3	NA	1974	0.7	3.8	2.9	8.9	11
Audi NSU Ro 80	VW Gas Turbine	3750	A-3	No	1977	0.4	3.4	2.0	14	12
1977 Olds 88	GM AGT-5 Gas Turbine	4000	NA	NA	1977	NA	NA	NA	14.7	13
VW Squareback	Carter Steam	2750	M-4	No	1974	0.4	1.1	0.33	15.9	14
Ford Torino	Philips Stirling	4500	NA	NA	1977	0.58	2.9	0.56	12.6	15
1977 Opel Rekord	United Stirling	3376	A-3	No	1978	0.12	0.43	0.37	19.4	71

* M-4 = Manual four-speed transmission, A-3 = Automatic three-speed, etc.
NA = Data not available at time of publication.

A second adjustment was made for transmission type. Both manual and automatic transmissions are represented among the cars listed in Table 2. All data were adjusted to automatic transmission equivalent for Figure 5. While there are exceptions, the fuel economy penalty for an automatic transmission usually varies from about 2% to around 20% (17), with the greater penalties commonly applying to the smaller cars. For our adjustment, we used a flat 10% as being representative.

Figure 5 remains unadjusted in terms of other important influences, such as the level of emission control. There may be some justification in ignoring differences in fuel density, since federal fuel economy standards are expressed in terms of miles/gal. rather than in miles per unit of fuel energy.

ALTERNATIVES: HISTORY AND PROJECTIONS

In this discussion, we will first examine the internal combustion engine (ICE) alternatives, then the external combustion engines (ECE); and, finally, we will make some mention of combined cycles.

STRATIFIED CHARGE ENGINES (ICE) - Two types of stratified charge engine are identified. The first type, typified by the three-valve Honda CVCC engine, is a prechamber engine. The second type, represented by the Ford PROCO and Texaco TCCS engines, is of the open chamber design.

Prechamber stratified charge engines have a divided combustion chamber. Ignition takes place in a prechamber containing a rich mixture, and the ignited prechamber contents then "torch off" a lean charge in the main chamber. In addition to production engines being built by Honda, similar engines have been built and tested by General Motors and Volkswagen. References 18-23 describe and characterize this engine.

Open chamber engines depend, on the other hand, upon swirl and other charge dynamics for the proper separation of the rich and lean mixtures within the cylinder. Open chamber engines are characterized in References 9 and 23-27.

Stratified charge engines can be built either as piston or rotary engine configurations, and the rotary engines will be discussed shortly.

Figure 6 shows the apparent learning curves for stratified charge engines. Like most curves in this paper, the data base is carried back only to the early 1970's, when EPA CVS-75 and Highway Driving Cycle data first became available. In those few cases where highway fuel economy numbers were not available, combined mpg figures were estimated by multiplying the urban cycle (CVS-75) number by a ratio which is characteristic of later tests of similar engines. Projected portions

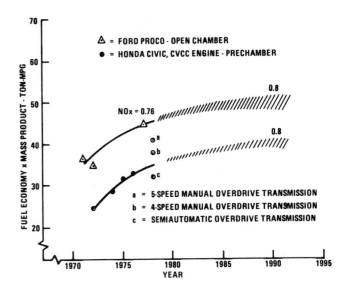

Fig. 6 - Apparent learning curve: fuel economy of automobiles with stratified charge engines

of the curves are based on the authors' judgments of plausible slopes of future improvement. In Figure 6, the slopes are similar to those of the baseline (see Fig. 4).

Possible attainments in NOx control are noted on the curves, in grams per mile for cars of 3000-3500 lb inertia weight. Forecasted entry into production is not necessarily implied by the curves in this paper.

The learning curve for prechamber engines was based entirely on the Honda CVCC engine. Emission control was apparently the main reason for developing this engine. The fuel economy for this engine type has risen appreciably in just a few years, but note that the effect of transmission type may be larger than that of potential engine improvement. We would expect that the rate of improvement, henceforth, will be on a slope similar to the baseline. The prechamber engine may be favored for the smaller cars where the ultimate in thermal efficiency is not required; for the larger cars and light trucks, the choice may be between an open chamber stratified charge engine or the diesel.

It is believed that, at least for small cars, emissions below 0.41/3.4/1.0 can be obtained with good fuel economy.(22) There is some evidence that NOx emissions can be reduced to 0.4 g/mi at least in prototype versions with a fuel economy penalty of about 25%. It is probable that with development, the severity of this penalty could be reduced. It should be noted that an oxidizing catalyst can be used with these engines if needed; however, a catalyst for NOx reduction is not workable in an excess-air exhaust stream.

At the best economy settings, open chamber stratified engines have higher brake thermal efficiencies than either the baseline uniform-charge engine or prechamber stratified charge engine. In some instances, fuel economies similar to the diesel engine have been demonstrated. This is probably due to the open chamber engine's ability to run at high compression ratios without the use of high octane fuel, and to lean, unthrottled operation.

Only a few points were available for the open chamber learning curve, and we have chosen the experimental PROCO engines to represent this type. The curve is subject to revision as more data become available.

Open chamber engines have been tested experimentally with NOx levels reduced to 0.4 g/mi. As with other ICE's, it would appear that these low NOx levels are attainable only with some sacrifice in thermal efficiency.

Ownership costs for open chamber engines are expected to be slightly higher than for baseline engines. We expect that both first costs and maintenance costs will be higher, at least until the benefits of high manufacturing volume and considerable field experience are achieved.

ROTARY COMBUSTION ENGINES - The popularity of the high performance rotary engine suffered a significant setback a few years ago, when public emphasis was suddenly placed upon a need for superior fuel economy. Up until the present time, production versions of rotary combustion engines, although improving, have shown a higher fuel consumption than standard reciprocating baseline engines. It may not be fair to conclude that this will be a permanent disadvantage, however. At least three manufacturers are currently known to be making improvements on rotary engines, including developments in stratified charge versions. References 28-35 portray an interesting account of the recent development of this engine, and give assurances that meaningful improvements in fuel economy will occur.

Figure 7 represents evolution in vehicle fuel economy with rotary engines. The historical portion of this learning curve is from production automobiles built by Toyo Kogyo (Mazda). It is not always possible to separate out influences other than engine improvement in such curves. However, the curve has been drawn through points representing similar transmission and vehicle configurations. A range of projections into the future has been included. This range is bounded by a simple extrapolation from present carbureted rotary engines and by more optimistic expectations from stratified charge rotary engines. A great deal of stratified charge development has been done by Curtiss-Wright; but, to date, these engines have not been tested in automobiles.(30)

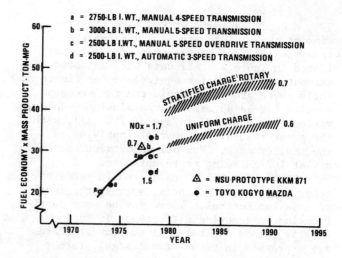

Fig. 7 - Rotary combustion engines: vehicle fuel economy

Unfortunately, it is not easily possible to speculate on the degree to which emissions can ultimately be controlled with rotary engines. The future NOx numbers should, therefore, be regarded as quite uncertain.

DIESEL ENGINES - Rudolph Diesel's invention of the "rational engine" is a remarkable example of an engine which turned out to be much more efficient, not merely different, from contemporary machines. The diesel engine remains unchallenged in the arena of auto engine fuel economy. In addition, there seems to be enough untapped potential remaining, that the diesel can maintain its relative superiority in fuel economy. However, it must be acknowledged that the diesel has its own peculiar problems. Some of these, such as lower specific output and noise, can be alleviated to a degree; but more serious questions can be raised regarding exhaust emissions. Of the presently regulated emissions, NOx will probably be the most troublesome to control. For the smaller automobiles, it is believed that NOx can be brought down to about 1.0 g/mi without destroying the fuel economy advantage. There is some doubt that 0.4 g/mi can be reached in the foreseeable future, at least without significant fuel economy penalties.

Perhaps an even more serious challenge will be the control of presently unregulated emissions, such as particulates and odor.(36-38) Emissions of polynuclear aromatics (PNA) and other particulates are of concern because some are suspected to be carcinogens.

The emissions situation suggests deeper probing into the possible relationships of combustion characteristics to fuel type. City bus operators are well aware of the smoke reductions which accompany the substitution of Number 1 Diesel fuel for Number 2, but most automobile diesel owners still use the Number 2 fuel dispensed at truck stops.

Some interesting aspects of diesel engine development and its emissions problems are found in References 37-47. Figure 8 suggests learning curves for the automotive diesel. One curve starts with historical trends from fairly conventional and conservative engines such as marketed earlier by Mercedes-Benz and Peugeot. New trends seem to be forming with lighter engine weight and turbocharging. It must be remembered that most diesel powered automobiles have lower acceleration performance and lower peak power ratings than their counterparts with gasoline engines. Therefore, a second projected curve is shown adjusted to performance levels of the gasoline engine.

VARIABLE DISPLACEMENT ENGINES - Dual displacement and variable displacement engines are not at all new developments, but there now seems to be a rebirth of interest in engines capable of operating over a range of piston displacements. One historical example was the Enger "Convertible 12", an auto marketed in 1916.(48) This car featured a V-12 engine which could also be operated on six cylinders for better fuel economy. This is the approach taken by the Eaton Corporation and Ford Motor Co. in developing cylinder cut-out devices in which solenoids are used to deactivate the valves of cylinders to be cut out.(49-51) Eaton claims that fuel economy is improved by 15% during highway cruising, 25% during low speed cruise, and by 40% during idle and deceleration.(51) It is the opinion of the authors that the average owner could experience a fuel economy advantage of perhaps 10%-15%.

The other approach is a continuously variable piston displacement concept, exemplified by the Pouliot engine.(52-54) Named for its inventor, this is a five-cylinder engine in which the piston stroke is varied by a mechanical linkage. A range of 43 CID to 190 CID was provided in a test prototype. We estimate that this type of engine, if feasible in a production version, could show a little more gain than the cylinder cut-out device since engine friction can be reduced somewhat in proportion to the displacement. Perhaps a 15%-20% gain in mpg is reasonable to expect.

We do not have learning curves for these engines, but would indicate that the rate of improvement should be similar to that of conventional engine technology. The effect on exhaust emissions is another subject worth investigation. It is possible that NOx emissions would be troublesome due to high thermodynamic loading of the cylinders.

Basically, the action of a variable displacement engine is to change the piston displacement per mile of vehicle travel. Presumably, the same effect could be obtained by the use of gearing or a modified transmission. Economics and operational flexibility may dictate which method is the better.

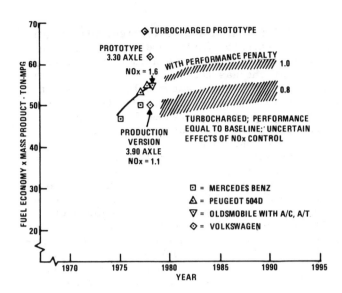

Fig. 8 - Fuel economy with diesel engines. Note adjustment to baseline performance

Transmission gearing, however, cannot by itself, reduce idle fuel consumption.

GAS TURBINES - A gas turbine is a continuous combustion ICE. One cannot very well dispute the technical elegance of these machines. Some advantages are apparent simplicity, the lack of high temperature rubbing parts, and use of air as an available and non-freezing working fluid. With the practical development of regenerators and with the use of high turbine inlet temperatures, it should be possible to raise thermal efficiency to fairly high values under certain load conditions. From Figure 3, it could be inferred that the gas turbine might someday challenge the diesel's fuel economy in some large vehicle applications. Improved efficiency will come from the use of ceramic materials now being developed, allowing the use of higher turbine inlet temperatures. Hydrocarbon and carbon monoxide emissions can be inherently low because of the steady combustion process with excess air.

Figure 3 notwithstanding, the fuel economies of present automobile gas turbines are below the norm, particularly under urban driving conditions. It is very difficult to design a small gas turbine which will give low fuel consumption at part load and idle. Exhaust emissions are low for hydrocarbons and carbon monoxide, but NOx emissions for driveable cars remains at a level of around 2 g/mi. Figure 9 shows different possible views of the future of the automobile gas turbine. A federal research goal for advanced engines is identified in this figure. The goal set by the U. S. Department of Energy and to be demonstrated by 1983 is to achieve fuel economy 30% better than with comparable spark ignition engines.(6) In this field, meaningful research and development will require large investments of time and money. Assuming evolutionary progress, the authors do not believe that the gas turbine will be a competitive engine in the average automobile for at least the next 15 years. These statements are not meant to discourage continued R & D, but only to emphasize the need for research planning toward very long range goals.

Extrapolations for Figure 9 were based on the use of metallic turbine rotors for test autos through the early 1980's. This technology is believed to be limited to about 1925°F turbine inlet temperature, so that improvements will be essentially limited to better component efficiencies and component matching, and, hopefully, to reduced idle fuel consumption. We have assigned a 15% fuel economy improvement for 1983 over the best test point to date.

Ceramics technology may permit turbine operating temperatures up to 2500°F to be exploited. With other improvements, this could lead to a theoretical 50% increase in peak thermal efficiency over that of 1983 turbines. We speculate that by 1995, turbine powered autos could show a 30% improvement in combined fuel economy over 1983. It will take time to assimilate ceramics technology in the most efficient forms. Also, it will be difficult to completely mitigate the problems of idle and part load losses. Variable geometry nozzles and vanes are expected to help.

The trend in downsizing cars and engines will also present difficulties to the turbine designer. It is not easy to maintain the required high component efficiencies as turbomachinery is scaled down in size. Our simple normalization to ton-mpg may not be applicable to turbine cars of widely varying weights.

References 3-6, 11-13, 23, and 55-61 are representative in this field of endeavor.

STIRLING ENGINE - With reference to actual automobile installations, the Stirling engine is in a period of early experience and rapid learning. References 15, 23, and 62-71 pertain specifically to this engine. Much of the development work is being sponsored by the U. S. Department of Energy.(6)

Theoretically, the Stirling cycle has a very high thermal efficiency--mathematically equal to that of the Carnot cycle. This has led to a number of very optimistic forecasts as to the fuel economy potential for this engine. However, developers of new engines usually find that many factors work to discount the efficiencies which can be achieved on the road.

Only a small amount of data have been released concerning the performance and economy of automobiles equipped with Stirling engines. Experimental vehicle installations have been completed by Philips of the Netherlands in conjunction with the Ford Motor Co. and by United Stirling of Sweden. (Ford is no longer active in this work.) United Stirling is now involved in a joint program with Mechanical Technology, Inc., and AM General. Some years ago, General Motors built an experimental hybrid powerplant using a Stirling engine and tested it in an automobile.(70)

Technically, the research results are encouraging; low emissions have been achieved and the fuel economy is almost as good as the spark ignition ICE. Emissions and fuel economy were listed in Table 2. These results are excellent for any engine which has not yet reached maturity.

Our assessment of present and future fuel economies for the Stirling engine is given in Figure 10. As in all the other fuel economy curves in this paper, the estimates are based on visualized testing in automobiles, rather than computer simulations or inferences from the engine laboratory. Promising results from the latter two activities are always available sooner than from tests of vehicles.

Other projections, relating to the United Stirling engine being developed in Sweden, show an increase in peak cycle temperatures to about 820°C (1508°F) in the early 1980's.(68) Present peak temperatures are around 700°C (1292°F). A combined improvement in cycle efficiency and component efficiencies plus better transmission matching could lead to a 15%-20% increase in vehicle fuel economy between 1979 and 1983. The projected estimate is 42 ton-mpg demonstrated by 1983.

Further improvements are forecasted with the use of new materials being developed. Temperatures around 1090°C (2000°F) might be possible in advanced engines.(23) This trend, coupled with other evolutionary refinements, could result in the range 45-50 ton-mpg in 1990. If these predictions are, in fact, borne out, the Stirling engine will be the fastest growing of the alternate engine technologies over the next few years.

Effectiveness of emission control should be relatively unvarying over the life of the engine, and levels of nitrogen oxide emission may drop well below 0.4 g/mi in the future, perhaps to 0.2 g/mi.

In common with the gas turbine, these engines are of high technological density; and thus, they may remain expensive or "premium" engines. Many difficult problems remain to be solved, such as sealing, materials, controls, and configuring for economical manufacture.

RANKINE ENGINES - While engines have been proposed and built which use organic working fluids, steam remains the best working fluid for automotive Rankine engines. A few steam engines have already demonstrated the potential ability to reduce emissions below the federal research goals of 0.4/3.4/0.4.(14, 72-74)

Rankine engines are unlikely ever to compete with some of the other advanced engines on a peak thermal efficiency basis. Neverthe-

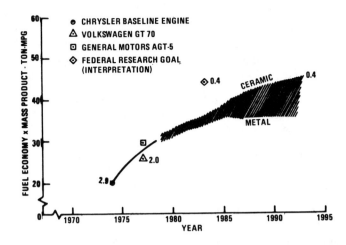

Fig. 9 - Automobile fuel economy with gas turbines: historical, projections, and research goals

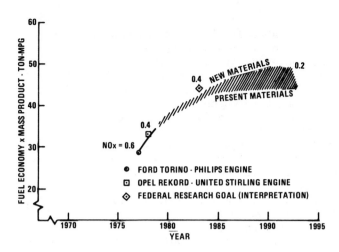

Fig. 10 - Learning curve with projections for Stirling engines. Adequate data are not yet available

less, some hope exists that the part load fuel economy can someday compete with the spark ignition carbureted ICE.(75-77) This indicates a possible, but unlikely, specialized role for vehicles operating in selected metropolitan areas where future emission limits could be forced even lower than 0.4/3.4/0.4. If refined steam cars have a future, perhaps it would be more in competition with the electric vehicle than the ICE.

At the time of this writing, the State of California is sponsoring a low level of R & D in steam engines. The objective is a laboratory test of a steam engine of significantly improved thermal efficiency. The best automobile steam engines of the past had brake thermal efficiencies in the range 15%-20%. The goal of the present project is to demonstrate a thermal efficiency of 27% at some selected part load. A reheat, regenerative compound engine is being built at this time, and test results are expected in late 1979.

COMBINED CYCLE ENGINES - As with many other topics covered in this paper, the combined cycle engine is not a new subject. There is, however, a legitimate renewal of interest in topping and bottoming cycle applications. The most common proposal is to use a Rankine cycle engine to recover some of the waste heat of an ICE. Recent developments, intended mainly for truck application, are described in References 78 and 79. We are not showing learning curves because we do not believe combined cycle engines will be used for passenger cars or light trucks in the foreseeable future. Economic considerations may prompt their use in long-haul trucking. Fuel savings of around 15% are believed possible in such an application. Historically, it has been established that overall thermal efficiencies of 40% or better are possible.(80)

SUMMARY AND CONCLUSIONS

A number of alternative engines which might be used in future automobiles and light trucks were examined, with major emphasis on fuel economy. Fuel economy learning curves were developed, based on historical and current test data. Combined city/highway fuel economy data were obtained from the literature describing tests of actual vehicle installations. The combined federal test cycles, as used by the U. S. EPA, provide a common base for this work. After establishing the historical learning curves, then potential future improvements in fuel economy were projected using plausible extrapolations of the curves.

A baseline of approximately 35 ton-mpg was found representative of 1978 automobiles. This is the product of inertia weight multiplied by the combined city/highway fuel economy. The authors project the baseline for the average uniform charge, carbureted spark ignition engine to be 40 ton-mpg around the year 1990. The value may be less if levels of control for oxides of nitrogen are set at less than one gram per mile.

Of all the engines examined, the authors believe that the diesel engine will remain supreme in the matter of fuel economy over the next 20 years; but this is conditional upon the level of NOx control required. The diesel engine also faces difficult questions with respect to presently unregulated exhaust emissions. Diesel powered autos presently achieve 50-60 ton-mpg; and in the future, values of 50 or more should be attainable with acceleration performance equal to conventional vehicles.

The open chamber stratified charge engine also shows potential for good fuel economy, both in reciprocating and rotary configurations.

Variable displacement engines can provide a fuel economy improvement in the range 10%-20%. However, since the same effect (of varying the piston displacement per mile of travel) can also be achieved with gearing, we raise the question as to which is the better approach.

The authors also examined the gas turbine and external combustion Stirling and Rankine engines. These are engines which hold out a promise of the low exhaust emissions which may be required in the future. All these technologies are relatively immature when compared to the automotive ICE, and the full potential for fuel economy is not well mapped out. Of these three engines, the Stirling is likely to excel in fuel economy, which ultimately (we believe) will fall somewhere between the spark ignition and diesel engines in this respect. With a Stirling engine, this can be done with NOx levels below 0.4 g/mi. Economic barriers may be formidable with these engines because of the apparently high technological density. Quantity production is not foreseen prior to 1990. However, the probable need for clean, efficient engines a quarter century hence may justify priority level R & D now.

Finally, there is every reason to believe that combined cycle engines can substantially reduce fuel consumption; but the investment may not pay off, except for long-haul commercial vehicles.

As to the learning curves themselves, many of the engine candidates are characterized as being in a period of rapid growth. This is not coincidental, in the light of public and government pressures to solve problems of emission control and to improve fuel economy. As the curves level off, there will be much hard work ahead for the developers of these engines.

This examination of an applied use of learning curves has been preliminary but very instructive. The authors believe that such a technique will become even more valuable in the future when more data becomes available. It is certain that characteristics other than fuel economy can be evaluated in a similar manner.

ACKNOWLEDGEMENTS

A number of colleagues provided the authors with valuable insights and criticisms during the course of this study. In particular, we wish to acknowledge the assistance of Jerar Andon, Dr. Alfred G. Cattaneo, Robert Schwarz, and Carl E. Burke. These people and others helped to keep us from straying too far from reality. We also appreciated the guidance and patience of the government's project officers, including Dr. A. C. Malliaris, Mr. Samuel Powel, and Mr. William Basham of the U. S. Department of Transportation.

REFERENCES

1. "Learning," Encyclopedia Americana, Vol. 17, p. 117, 1972 Ed.

2. Dugald Clerk, "The Gas, Petrol, and Oil Engine," (Vol. 1), John Wiley and Sons, Publishers, New York 1909.

3. W. A. Turunen and J. S. Collman, "The General Motors Research GT-309 Gas Turbine Engine," SAE Paper 650714, 1965.

4. R. G. Cadwell, W. I. Chapman, and H. C. Walch, "The Ford Turbine - An Engine Designed to Compete with the Diesel," Paper 720168 presented at SAE Automotive Engineering Congress, Detroit, Mich., Jan. 10-14, 1972.

5. G. Engel and W. S. Anderson, "Compactness of Ground Turbine Depends on Integral Recuperator," SAE Journal of Automotive Engineering, Aug., 1971.

6. G. M. Thur, "Department of Energy Automotive Heat Engine Program," Paper 780698 presented at SAE West Coast Meeting, San Diego, Calif., Aug. 7-10, 1978.

7. EPA Staff, "V.I. Report, 1978 Fuel Economy Program" (for EPA/FEA 1978 Gas Mileage Guide)," U. S. Environmental Protection Agency, Sept. 14, 1977.

8. Private Communication with J. Chao, Haagan-Smit Laboratory, California Air Resources Board, El Monte, Calif., April, 1978.

9. A. J. Scussel, A. O. Simko, and W. R. Wade, "The Ford PROCO Engine Update," Paper 780699 presented at SAE West Coast Meeting, San Diego, Calif., Aug. 7-10, 1978.

10. "Specific Fuel Economy Label Approval List-Federal Vehicles, 1978 Model Year," U. S. Environmental Protection Agency, Dec., 1977.

11. A. E. Barth, "Chrysler Baseline Gas Turbine Vehicle Tests," Report 75-15, U. S. Environmental Protection Agency, Jan., 1975.

12. P. Walzer, "The Automotive Gas Turbine - State of Development and Prospects," Paper presented to 4th International Symposium on Automotive Propulsion Systems (NATO CCMS), April 17-22, 1977.

13. "Public Interest Report, 1977-78," Published by General Motors Corp., May 15, 1978.

14. "Exhaust Emissions Tests of the Carter Steam Car," Report 74-31, Technology Assessment and Evaluation Branch, U. S. Environmental Protection Agency, Ann Arbor, Mich., June, 1974.

15. N. Postma, "Stirling Engine Program," Presentation to U. S. Department of Energy Highway Vehicle Systems Contractors' Coordination Meeting, Dearborn, Mich., Oct. 4-6, 1977.

16. R. Nicholson and G. Niepoth, "Effect of Emission Constraints of Optimum Engine Size and Fuel Economy," Paper 760046 presented at SAE Automotive Engineering Congress, Detroit, Mich., Feb. 23-27, 1976.

17. Private Communication with J. D. Murrell, Emission Control Technology Division, U. S. Environmental Protection Agency, Ann Arbor, Mich., March 7, 1978.

18. T. Date, S. Yagi, A. Ishizuya, and I. Fujii, "Research and Development of the Honda CVCC Engine," Paper 740605 presented at SAE West Coast Meeting, Anaheim, Calif., Aug. 12-16, 1974.

19. P. R. Johnson, S. L. Genslak, R. C. Nicholson, "Vehicle Emission Systems Utilizing a Stratified Charge Engine," Paper 741157 presented at SAE International Stratified Charge Engine Conference, Troy, Mich., Oct. 30-Nov. 1, 1974.

20. W. R. Brandstetter, G. Decker, H. J. Schafer, and D. Steinke, "The Volkswagen PCI Stratified Charge Concept - Results from a 1.6 Liter Air Cooled Engine," Paper 741173 presented at the SAE International Stratified Charge Engine Conference, Troy, Mich., Oct. 30-Nov. 1, 1974.

21. M. C. Turkish, "3-Valve Stratified Charge Engines: Evolvement, Analysis, and Progression," SAE Paper 741163, 1974.

22. H. K. Newhall, "Combustion Process Fundamentals and Combustion Chamber Design for Low Emissions," SAE Paper 751001, 1975.

23. R. R. Stephenson, Principal Investigator, "Should We Have a New Engine?" Vol. II, Jet Propulsion Laboratory, Calif. Inst. of Technology, Aug., 1975.

24. I. N. Bishop and A. Simco, "A New Concept of Stratified Charge Combustion - The Ford Combustion Process (FCP)," SAE Paper 680041, 1968.

25. A. Simko, M. A. Choma, and L. L. Repko, "Exhaust Emission Control by the Ford Programmed Combustion Process - PROCO," Paper 72052, presented at SAE Automotive Engineering Congress, Detroit, Mich., Jan. 10-14, 1972.

26. M. Alperstein, G. H. Schafer, and F. J. Villforth III, "Texaco's Stratified Charge Engine - Multifuel, Efficient, Clean, and Practical," Paper 740563 presented at SAE Southern California Section, May 14, 1974.

27. Charles D. Wood, "Unthrottled Open Chamber Stratified Charge Engines," Paper 780341 presented at SAE Congress, Detroit, Mich., Feb. 27-Mar. 3, 1978.

28. C. Jones, "The Curtiss-Wright Rotating Combustion Engines Today," Paper 886 D presented at SAE West Coast Meeting, Aug., 1964.

29. C. Jones, "New Rotating Combustion Powerplant Development," Paper 650723 presented at SAE Powerplant and Transportation Meeting, Cleveland, Ohio, Oct. 18-21, 1965.

30. C. Jones, H. D. Lamping, D. M. Myers, and R. W. Loyd, "An Update of the Direct Injected Stratified Charge Rotary Combustion Engine Developments at Curtiss-Wright," Paper 770044 presented at SAE Automotive Engineering Congress, Detroit, Mich., Feb. 28-Mar. 4, 1977.

31. Richard van Basshuysen, "Status of the Wankel Engine Development at Audi NSU," Paper presented to Engineering Society of Detroit, Mich., May 19, 1977.

32. K. Yamamoto, T. Muroki and T. Kobayakawa, "Combustion Characteristics of Rotary Engines," SAE Paper 720357, April, 1972.

33. K. Yamamoto and T. Muroki, "Development on Exhaust Emissions and Fuel Economy of the Rotary Engine at Toyo Kogyo," Paper 780417 presented at SAE Congress, Detroit, Mich., Feb. 27-March 3, 1978.

34. R. van Basshuysen and G. Wilmers, "An Update of the Development on the New Audi NSU Rotary Engine Generation," Paper 780418 presented at SAE Congress, Detroit, Mich., Feb. 27-Mar. 3, 1978.

35. H. A. Burley, M. R. Meloeny, and T. L. Stark, "Sources of Hydrocarbon Emissions in Rotary Engines," Paper 780419 presented at SAE Congress, Detroit, Mich., Feb. 27-Mar. 3, 1978.

36. G. P. Gross, "Automotive Emissions of Polynuclear Aromatic Hydrocarbons," Paper 740564 presented at SAE Farm, Construction, & Industrial Machinery and Fuels and Lubricants Meetings, Milwaukee, Wis., Sept. 10-13, 1973.

37. K. J. Springer and R. C. Stahman, "Unregulated Emissions from Diesels Used in Trucks and Buses," Paper 770258 presented at SAE Automotive Engineering Congress, Detroit, Mich., Feb. 28-Mar. 4, 1977.

38. W. H. Lipkea, J. H. Johnson, and C. T. Vuk, "The Physical and Chemical Character

of Diesel Particulate Emissions - Measurement Techniques and Fundamental Considerations," Paper 780108 (SAE SP-430) presented at SAE Congress, Detroit, Mich., Feb. 27-Mar. 3, 1978.

39. K. J. Springer and R. C. Stahman, "Emissions and Economy of Four Diesel Cars," Paper 750332 presented at SAE Automotive Engineering Congress, Detroit, Mich., Feb. 24-28, 1975.

40. R. D. Fleming, "Fuel Economy of Light Duty Diesel Vehicles," Paper 760592 presented at SAE West Coast Meeting, San Francisco, Calif., Aug. 9-12, 1976.

41. J. L. Dooley, "McCulloch is Developing Lightweight Aircraft Diesel," SAE Journal of Automotive Engineering, Sept., 1971.

42. P. Hofbauer and K. Sator, "Advanced Automotive Power Systems, Part 2: A Diesel for a Subcompact Car," Paper 770113 presented at SAE Automotive Engineering Conference, Detroit, Mich., Feb. 28-Mar. 4, 1977.

43. S. H. Hill and J. L. Dodd, "A Low NOx Lightweight Car Diesel Engine," Paper 770430 presented at SAE Automotive Engineering Congress, Detroit, Mich., Feb. 28-Mar. 4, 1977.

44. B. Martin and G. Wright, "High Output Diesel Engine Design Philosophy," Paper 770755 presented at SAE Off Highway Vehicle Meeting, Milwaukee, Wis., Sept. 12-15, 1977.

45. R. W. Talder, J. D. Fleming, D. C. Siegla, and C. A. Amann, "Dynamometer Based Evaluation of Low Oxides of Nitrogen, Advanced Concept Diesel Engine for a Passenger Car," Paper 780343 presented at SAE Congress, Detroit, Mich., Feb. 27-Mar. 3, 1978.

46. K. Oblander, M. Fortnagel, H. J. Feucht, and U. Conrad, "The Turbocharged Five Cylinder Diesel Engine for the Mercedes-Benz 300 SD," Paper 780633 presented at SAE Passenger Car Meeting, Troy, Mich., June 5-9, 1978.

47. B. Wiedemann and P. Hofbauer, "Data Base for Lightweight Automotive Diesel Power Plants," Paper 780634 presented at SAE Passenger Car Meeting, Troy, Mich., June 5-9, 1978.

48. Floyd Clymer, "Historical Motor Scrapbook No. 2," Clymer Motors, Los Angeles, Calif., 1944.

49. B. Bates, J. M. Dosdall, and D. H. Smith, "Variable Displacement by Engine Valve Control," Paper 780145 presented at SAE Congress, Detroit, Mich., Feb. 27-Mar. 3, 1978.

50. R. L. Bechtold, "An Investigative Study of Engine Limiting," U. S. Department of Energy Report BERC/RI-77/13, Feb., 1978.

51. Larry Givens, "A New Approach to Variable Displacement," SAE Journal of Automotive Engineering, May, 1977.

52. H. N. Pouliot, W. R. Delameter, and C. W. Robinson, "A Variable Displacement Spark Ignition Engine," Paper 770114 presented at SAE Automotive Engineering Congress, Detroit, Mich., Feb. 28-Mar. 4, 1977.

53. C. W. Robinson, "Summary of Variable Displacement Engine Project," U. S. Energy Research and Development Administration Contractors Coordination Meeting, Dearborn, Mich., Oct. 4-6, 1977.

54. D. C. Siegla and R. M. Siewert, "The Variable Stroke Engine - Problems and Promises," Paper 780700 presented at SAE West Coast Meeting, San Diego, Calif., Aug. 7-10, 1978.

55. W. E. Goette, "Gas Turbine Project Status," U. S. Department of Energy Highway Vehicle Systems Contractors Coordination Meeting, Troy, Mich., May 9-12, 1978.

56. J. Collman, C. Amann, C. Matthews, R. Stettler, and F. Verkamp, "The GT-225 - An Engine for Passenger Car Gas Turbine Research," Paper 750167 presented at SAE Automotive Engineering Congress, Detroit, Mich., Feb. 24-28, 1975.

57. S. O. Kronogard, "Advanced Three-Shaft Turbine-Transmission Systems," U. S. ERDA Highway Vehicle Systems Contractors Coordination Meeting, Ann Arbor, Mich., Nov. 17-18, 1975.

58. D. L. Hartsock, P. H. Havstad, and C. F. Johnson, "Fabrication and Testing of Silicon Nitride Turbine Rotors," U. S. ERDA Highway Vehicle Systems Contractors Coordination Meeting, Dearborn, Mich., Oct. 4-6, 1977.

59. C. E. Wagner, "Status of Corrective Development Program on Chrysler Upgraded Engine," U. S. Department of Energy Contractors Coordination Meeting, Troy, Mich., May 9-12, 1978.

60. R. A. Johnson, "Single Shaft, Variable Geometry Automotive Gas Turbine Engine Characterization Test," U. S. Department of Energy Contractors Coordination Meeting, Troy, Mich., May 9-12, 1978.

61. C. P. Blankenship, "Automotive Gas Turbine Ceramic Materials Program Overview," U. S. Department of Energy Contractors Coordination Meeting, Troy, Mich., May 9-12, 1978.

62. H. S. J. van Beukering and H. Fokker, "Present State-of-the-Art of the Philips Stirling Engine," Paper 730646 presented at SAE Commercial Vehicle Engineering & Operations and Powerplant Meetings, Chicago, Illinois, June 18-22, 1973.

63. N. K. G. Rosenqvist, Stig Gummesson, and S. Lundholm, "The Development of a 150 KW (200 HP) Stirling Engine for Medium Duty Automotive Application - A Status Report," Paper 770081 presented at SAE Automotive Engineering Congress, Detroit, Mich., Feb. 28-Mar. 4, 1977.

64. P. Kuhlman, "Das Kennfeld des Stirlingmotors," "MTZ Motortechnische Zeitschrift," 34. Jahrg. Nr. 5/1973.

65. N. D. Postma, R. van Giessel, and F. Reinink, "The Stirling Engine for Passenger Car Application," Paper 730648 presented at SAE Commercial Vehicle Engineering & Operations and Powerplant Meetings, Chicago, Illinois, June 18-22, 1973.

66. R. G. Ragsdale and D. G. Beremand, "Stirling Engine Project Status," U. S. Department of Energy Contractors Coordination Meeting, Troy, Mich., May 9-12, 1978.

67. N. Postma, "Automotive Stirling Engine Development Program," U. S. DOE Contractors Coordination Meeting, Troy, Mich., May 9-12, 1978.

68. O. Decker, "MTI Stirling Engine Powertrain Development," U. S. DOE Contractors Coordination Meeting, Troy, Mich., May 9-12, 1978.

69. K. Rosenqvist, "In-Vehicle Stirling Engine Operational Experience," U. S. DOE Contractors Coordination Meeting, Troy, Mich., May 9-12, 1978.

70. P. D. Agarwal, R. J. Mooney, and R. R. Toepel, "Stir-Lec I, A Stirling Electric Hybrid Car," Paper 690074 presented at SAE Automotive Engineering Congress, Detroit, Mich., Jan. 13-17, 1969.

71. "Stirling Demonstration Vehicle Genesis-1," Pamphlet Describing Installation of United Stirling Engine, Published by Mechanical Technology, Inc., Latham, N. Y., 1978.

72. S. Luchter and R. A. Renner, "An Assessment of the Technology of Rankine Engines for Automobiles," Report No. ERDA-77-54, U. S. Energy Research and Development Administration, April, 1977.

73. R. A. Renner and M. Wenstrom, "Experience with Steam Cars in California," Paper 750069 presented at SAE Automotive Engineering Congress, Detroit, Mich., Feb. 24-28, 1975.

74. J. Carter, Jr., "The Carter System - A New Approach for a Steam Powered Automobile," Paper 750071 presented at SAE Automotive Engineering Congress, Detroit, Mich., Feb. 24-28, 1975.

75. R. A. Renner, W. Brobeck, and F. Younger, "Status and Projections of the Development of External Combustion Engines," William M. Brobeck & Assoc., Report 4500-189-7-R1 prepared for State of California Department of Transportation, Jan., 1977.

76. J. Carter, Jr., and W. Wingenbach, "The Carter System - Preliminary Test Results of Second Generation Steam Engine," Paper 760341 presented at SAE Automotive Engineering Congress, Detroit, Mich., Feb. 23-27, 1976.

77. R. L. Demler, "The Application of the Positive Displacement Reciprocation Steam Expander to the Passenger Car," Paper 760342 presented at SAE Automotive Engineering Congress, Detroit, Mich., Feb. 23-27, 1976.

78. P. S. Patel and E. F. Doyle, "Compounding the Truck Diesel Engine with an Organic Rankine Cycle System," Paper 760343 presented at SAE Automotive Engineering Congress, Detroit, Mich., Feb. 23-27, 1976.

79. C. J. Leising, G. P. Purohit, S. P. DeGrey, and J. G. Finegold, "Waste Heat Recovery in Truck Engines," Paper 780686 presented at SAE West Coast Meeting, San Diego, Calif., Aug. 7-10, 1978.

80. "The Still Engine," Marks' Mechanical Engineers' Handbook, Sixth Ed., McGraw-Hill Book Co., New York, 1958.

Light Duty Automotive Fuel Economy ....Trends Through 1979*

J. D. Murrell
U.S. Environmental Protection Agency

VIRTUALLY ALL SEGMENTS of U.S. society are concerned with motor fuel supplies and prices, these issues touching individual consumers as well as those in high levels of industry, finance, and international affairs. Fuel economy figures are perhaps the most widely used quantifiers in analyses related to the usage of automotive fuel.

It is probably inevitable that each interested sector sooner or later seeks out the particular set of fuel economy numbers that best suits its needs and beliefs. The EPA fuel economy figures have come to be a recognized common denominator, a baseline which might be axiomatically accepted, methodically modified, or roundly refuted, but rarely disregarded.

As explained later, three EPA fuel economy values exist: a "city" value, a "highway" value, and a third that combines (and resides between) these two. To accomodate most reader/analysts' needs, this paper employs the combined value in situations where vehicle fuel economy standards seem to be the prime concern, but also gives city values for direct comparison of historical data with the 1979 single-figure EPA Gas Mileage Guide, and highway values for comparison with previous Guides and other publications.

*Paper 790225 presented at the Congress and Exposition, Detroit, 1979.

ABSTRACT

The pre-1975 EPA fuel economy data base has been expanded to over 6,600 cars, and these data on older cars have been adjusted for odometer mileage effects on fuel economy. The data base for model year 1975-1977 certification cars is also updated, reflecting actual sales figures. The resulting trend analyses are thus (for the first time) consistent from year to year with regard to the representation of actual sales weighted new-car fleet fuel economy.

Data on the 1979 fleet is presented in detail, and compared with that of the 1978 fleet, on the basis of projected sales for those two years.

Comparisons with pre-emission control MPG, which serve as measures of technological change accompanying increasingly stringent emission standards, are accordingly revised from previous publications (1)* on this subject.

Passenger car size classes and interior volumes are treated in the trend sense, and also in a study of interactions between vehicle technology, fleet physical attributes, and post-1979 fuel economy standards.

* Numbers in parentheses designate references.

DATA SOURCES

Data on sales distributions of cars of various weight classes were obtained from several sources. For model years 1958 through 1973, the data is from a study performed for EPA by the Aerospace Corporation.(2) The data for model years 1974-1977 are from a study conducted by the EPA and, like the Aerospace data, are based on production figures. For the 1978 and 1979 model years, the data are sales forecasts from the manufacturers' Part I Applications for Certification submitted to EPA each year as part of the emissions certification process. The projected sales figures used are updates as of the summer immediately preceding the start of each model year.

Fuel economy data used in the analysis also came from more than one source. The test procedure used was the same for all model years considered. For the 1975 through 1979 models, the fuel economy data were accumulated during the certification program conducted annually at the EPA's Ann Arbor, Michigan laboratory. For pre-1975 models, the fuel economy data comes from production car testing programs conducted by private laboratories under EPA contract.(3)

All fuel economy data are based on the carbon balance measurement method.

Unless specifically noted otherwise, all fuel economy values reported in this paper are "city/highway combined fuel economy" values. This value is computed from the harmonic, mileage weighted average of city fuel economy (corresponding to the 1975 EPA urban test procedure) and highway economy (corresponding to the EPA highway test procedure). In calculating city/highway combined fuel economy, the city fuel economy value is weighted 55% and the highway fuel economy value is weighted 45% to account for the 55/45 ratio of urban to non-urban mileage accumulation which the U.S. Department of Transportation has stated(4) was typical for passenger vehicle operation in the U.S. during the 1970's. The formula for this combined city/highway fuel economy is thus:

$$MPG_{combined} = \frac{1}{\frac{.55}{MPG_{city}} + \frac{.45}{MPG_{highway}}} \quad (1)$$

This equation gives the ratio of total miles driven to total gallons consumed, for driving that consists of 55% urban mileage and 45% non-urban mileage.

For brevity, this combined fuel economy value will be referred to throughout this report as the "55/45 Fuel Economy", or "55/45 MPG".

For *PRE-1975 MODELS*, detailed sales data for all manufacturer/weight/engine CID/transmission combinations were not available. The fuel economy for a particular model year was determined by computing the harmonic mean fuel economy of all cars tested in each weight class, and sales weighting these values using sales fraction data for each weight class for that model year.

The formula used to calculate sales-weighted fleet fuel economy for a given model year is:

$$SWMPG = \frac{1}{\Sigma f_i (1/MPG_i)} \quad (2)$$

where SWMPG is the sales-weighted fuel economy for the year in consideration; f_i is the total fraction of sales in weight class i; MPG_i is the harmonic mean fuel economy of all available data in weight class i (sample weighted); and i is the running index, i = 1,2,3,...n, over which the sum is taken.

The pre-1975 EPA data base has been updated since the publication of Reference 1, in two ways. First, new data from the EPA Emission Factors program has been incorporated, increasing the total number of test cars in this data base from 4273 to 6685, a 56% increase. The number of cars added varies as to model year represented, such that the populations of the more recent model years have been preferentially expanded.

Secondly, the data were analyzed with regard to the relationship between fuel economy and odometer mileage. This was done by segregating the data into discrete configurations* and regressing MPG against odometer mileage for each configuration. All the data for each configuration were then normalized to the 4000-mile regression MPG, and the resulting dimensionless fuel economy (MPG/MPG_{4K}) regressed against odometer mileage by model year, by model year and manufacturer, and by model year and inertia weight. No statistical justification was found for stratification at any of these sublevels, so the single regression equation** representing all the data,

$$\frac{MPG}{MPG_{4K}} = .846 + .0186(\ln \text{Miles}) \quad (3)$$

was used on the entire data base, to correct <u>each car</u>'s fuel economy to a "4000-mile" value based on its own odometer mileage.

---

* A unique combination of model year, manufacturer, division, standard (FLDV, CLDV, etc.), test weight, engine displacement, no. carburetor barrels, and transmission type (A3, M4, M5, etc.).

** Linear regressions were also performed, but the logarithmic form yielded better fits to the data, both at the configuration level and at the global level.

Table 1 summarizes the magnitudes of the data base expansion, and the mileage characteristics and nominal 4000-mile correction factors for the analysis. The percentage increase in the number of test cars is less significant for the older model years, but the impact of the mileage correction is stronger due to these cars' higher mileages. The opposite is true for the more recent models.

Equation (3) is plotted in Figure 1; for comparison, results of similar mileage analyses performed on 1975 and 1976 Emission Factors and Certification Durability cars are also shown.

Table 1 - Update of Pre-1975 EPA Passenger Car Data Base

	Number of Test Cars:		Odometer Miles:			Average 4000-Mile MPG Correction
	Previous	Present	Minimum	Average	Maximum	
Pre-Control	1017	1218	29,900	83,100	212,100	0.936
1968	411	562	20,500	76,100	148,700	0.941
1969	452	630	12,600	69,900	159,600	0.945
1970	496	739	12,800	64,600	135,900	0.950
1971	576	853	15,200	57,500	120,200	0.956
1972	616	1027	4,400	47,000	133,200	0.964
1973	392	832	6,800	37,800	96,700	0.971
1974	313	824	1,300	26,200	72,000	0.970
TOTAL	4273	6685				

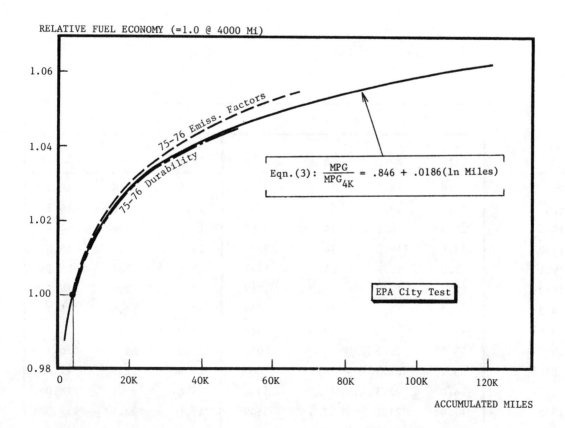

Eqn. (3): $\frac{MPG}{MPG_{4K}} = .846 + .0186(\ln \text{Miles})$

Fig. 1 - *Fuel economy dependence on accumulated (odometer) mileage*

THE 1975-1977 DATA BASES are also updated, reflecting actual production volumes rather than manufacturers' forecasts. The production volume updating was based on data compiled by the R.L. Polk Company. In the case of the domestic manufacturers, volume corrections were applied at the car line (nameplate) level. Within each car line level, the relative proportions of cars and station wagons, automatic and manual transmissions, 49-States and California versions, and 4-cylinder, 6-cylinder and 8-cylinder engines, were corrected to agree with published figures. For the foreign manufacturers, volume corrections were applied at the "submanufacturer" level (e.g. VW, Audi, and Porsche within VWoA). Table 2 compares the "old" and "new" values for fleet average weight, engine CID, and 55/45 MPG for these three model years.

Table 2 - Update of 1975-77 EPA Passenger Car Data Base

	1975	1976	1977
Projected Sales			
Sales (Millions)	11.110	10.338	11.956
Average Weight	4120	3942	3885
Average CID	283	273	272
Avg. 55/45 MPG	15.6	17.7	18.6
Actual Sales			
Sales	8.234	9.722	11.300
Average Weight	4057	4060	3943
Average CID	288	286	279
Avg. 55/45 MPG	15.8	17.5	18.3

FUEL ECONOMY TRENDS, PRE-CONTROL TO 1979

Table 3 and Figure 2 give the sales-weighted 50-states average 4000 mile fuel economy and test weight for the new-car fleets from pre-emission control years through Model Year 1979. More detailed tabulations showing the same data for each weight class can be found in Appendix A.

Among the many factors which can influence year-to-year changes in fleet MPG, those changes associated specifically with powertrain characteristics are directly related to emission control technology used by the industry. These changes can be isolated by recalculating the fuel economy trend using a fixed mix of vehicle weights. This

Table 3 - Trends in Sales-Weighted Fleet Fuel Economy, Passenger Cars

	Each Year's Weight Mix:				1974 Weight Mix:*		
	City	Hwy.	55/45	Avg. Test Wt.	City	Hwy.	55/45
Pre-Control	12.9	18.5	14.9	3812	12.5	17.4	14.3
1968	12.6	18.4	14.7	3863	12.4	17.6	14.3
1969	12.6	18.6	14.7	3942	12.7	18.8	14.9
1970	12.6	19.0	14.8	3877	12.3	18.5	14.5
1971	12.3	18.2	14.4	3887	12.1	17.9	14.1
1972	12.2	18.9	14.5	3942	12.2	18.9	14.5
1973	12.0	18.1	14.2	3969	12.0	18.1	14.1
1974	12.0	18.2	14.2	3968	12.0	18.2	14.2
1975	13.7	19.5	15.8	4057	14.0	19.9	16.1
1976	15.2	21.3	17.5	4060	15.5	21.8	17.8
1977	16.0	22.3	18.3	3943	15.6	22.1	18.0
1978	17.0	24.1	19.6	3649	15.4	22.4	18.0
1979	17.6	24.3	20.1	3508	15.3	21.2	17.5

*Average Test Weight = 3968 lb.

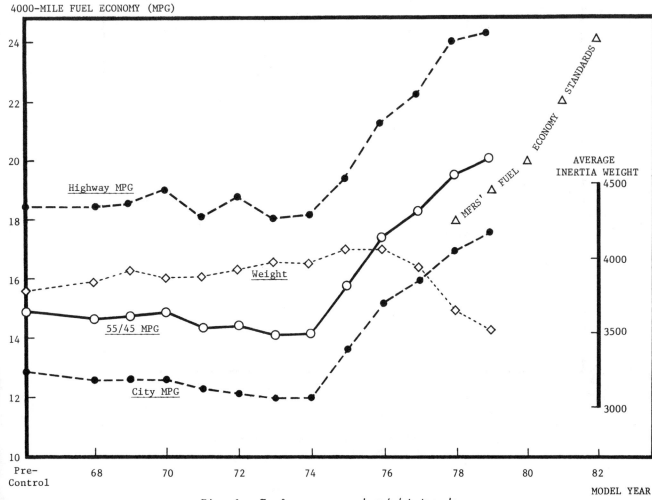

Fig. 2 - *Fuel economy and weight trends, U.S. passenger vehicles*

means that MPG_i data for each year (Appendix A) is used in Equation (2) with one set of f_i values. The analysis (based on the 1974 f_i's) yields the "weight-normalized" trend given in Table 3 and Figure 3.

Comparing the original data with this weight-normal data, we see that the MPG loss from pre-control years to 1974 was caused primarily by adverse shifts in the weight mix, since the actual trend shows a 4.6% drop, while the weight-normal trend shows only an 0.8% drop. This finding of an 0.8% MPG loss due to changes in technology stands in contrast to earlier "emission control penalty" estimates (1) which were derived from the older, smaller data base, with its nonuniformaties in odometer mileage. The 1975 fleet shows a 55/45 fuel economy 11.1% higher than that of 1974; this is the net effect of a 13.4% weight-normal (technology) gain and a 2.0% loss caused by weight mix shifts. Compared to 1975, the 1979 fleet MPG has improved another 27.2%, the combined result of a 17.0% advance due to beneficial weight mix shifts and an 8.7% gain due to improved technology.

The trend in passenger vehicle weight distribution is given in Figure 4. Tracing this trend back to 1960, it may be seen that the presently declining heavy-car era is a recent phenomenon: a peculiarity of the 1969-1977 time period. The 1978-79 relative market shares of the three vehicle weight groupings are not at all unprecedented, being quite similar to what they were in the early 1960's.

DETAILED ANALYSIS OF 1979 MODELS

The maximum, minimum, and average 55/45 fuel economy values for 1979 are shown, by weight class, in Table 4 and Figure 5. Wide variations in the fuel economy of vehicles within the same weight class show, as in prior years, the **significant MPG** influence of factors other than weight. For one weight class (5000 lb), the average MPG is higher than the best fuel economy for vehicles equipped with spark ignition engines, due to a significant sales penetration forecasted for Diesel-powered cars in that weight class.

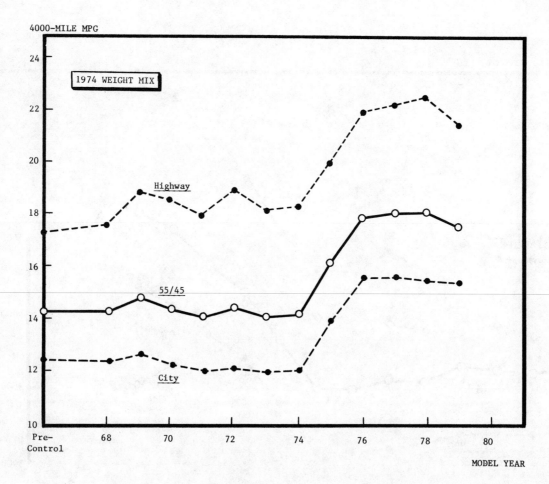

Fig. 3 - Passenger car fuel economy trend with fixed weight mix

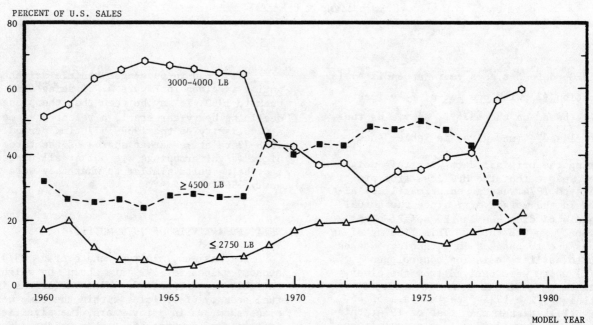

Fig. 4 - Twenty-year trend in passenger car weight distributions

Table 4 - Range of 55/45 Fuel Economies,
1979 50-States Passenger Cars

	WEIGHT CLASS, POUNDS									
	2000	2250	2500	2750	3000	3500	4000	4500	5000	5500
Highest MPG (Diesel)	-	46.0	40.2	-	-	31.6	29.2	23.9	23.1	-
Highest MPG (Spark)	38.9	39.5	33.4	30.4	28.4	24.0	21.5	17.3	13.6	11.4
Sales-Weighted Average (All)	32.7	32.9	27.7	24.0	21.5	20.0	17.9	16.3	14.2	10.7
Lowest MPG	25.3	24.5	23.4	18.5	15.2	13.8	13.9	9.2	12.1	10.0

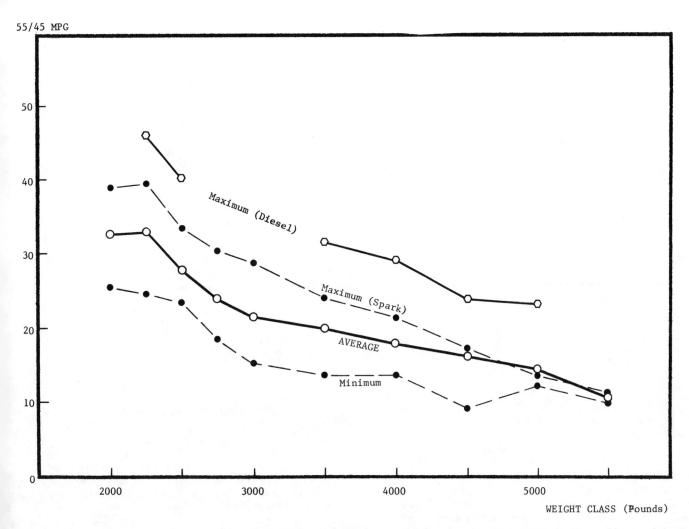

Fig. 5 - 1979 passenger car fuel economy, by test weight

Sales-weighted average fuel economies and engine displacements are shown in Table 5, by weight class, for the fleet and for each of the top-selling fifteen manufacturers sufficiently certified by the data cutoff date. At the fleet level, fuel economy is less than or equal to that of last year in nine of the ten weight classes.

Model eliminations and/or body redesigns account for the disappearance of the highest weight class entries for AMC, Chrysler, and Ford (see Table 3, Ref. 1 for comparable 1978 data). BMW and Honda have added higher weight class entries - opposite to the trend of the above domestic manufacturers. For British Leyland, BMW, Mercedes-Benz and VWoA, 1979 MPG is equal to or better than 1978 in all weight classes. Fuel economy losses in all weight classes were registered by Ford, Toyota, and Fuji.

As illustrated in Figure 6, the 1979 models show a 20 to 30% MPG advantage in most weight classes (along with the 1976-78 cars), when compared to pre-emission control models (1957-1967 test cars treated as a group).

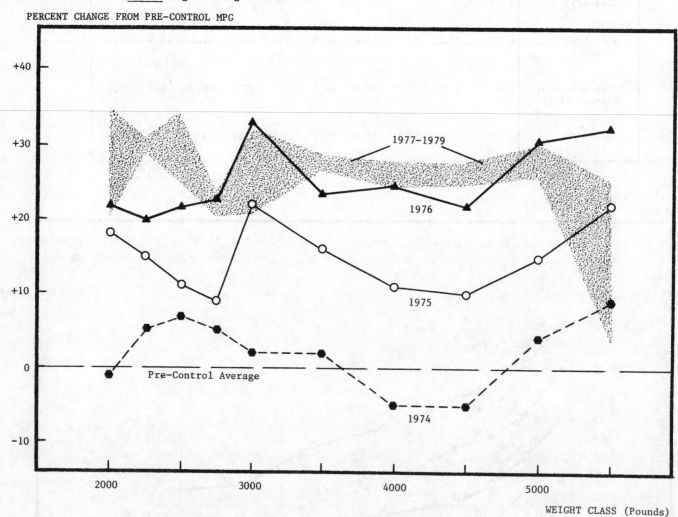

Fig. 6 - Change from pre-emission control MPG, by test weight (4000-mile MPG values)

CAUSE/EFFECT ANALYSIS OF 1979 MPG CHANGES - The allocation of 1978-1979 fuel economy changes to specific causes is summarized in Table 6*. Proceeding from right to left, significant increases in total fleet 55/45 MPG were achieved by Chrysler (mainly due to weight mix shifts) and Peugeot (mainly due to engine mix shifts). Manufacturers who posted significant net losses include Toyota and Fuji - who suffered system optimization penalties - and Honda, who sustained a loss due to a shift in weight mix.

* The methodology used to construct Table 6 is described in Appendix B.

Weight-related fuel economy losses for both Honda and BMW result directly from the appearance of higher weight classes not used in last year's fleets, as mentioned above. Ford's notable MPG gain associated with weight reductions was offset by a system optimization loss, wherein every IW/CID combination is showing lower MPG than last year's models.

Engine mix shifts gave fuel economy gains to Mercedes-Benz, Peugeot, and VW -- all projecting higher Diesel sales fractions in 1979: VW's fleet is approaching 50% Diesel, Mercedes' over 50%, and Peugeot's over 70%. Peugeot's discontinuance of their lowest MPG 1978 engine, the 163 CID, also

Table 5 - Average 55/45 MPG and Engine CID
by Weight Class, 1979 Model Cars

MANUFACTURER	WEIGHT CLASS(POUNDS)									
	2000	2250	2500	2750	3000	3500	4000	4500	5000	5500
AMERICAN MOTORS				24.4 121	22.1 194	18.9 258	16.3 304			
CHRYSLER CORP.	37.8 86	35.9 96	27.9 104	27.0 114	24.6 156	20.3 225	18.3 299	16.8 359		
FORD MOTOR CO.	31.9 98			24.4 140	21.2 170	18.9 257	17.3 312	15.9 325	13.4 400	
GENERAL MOTORS		33.6 98	28.0 99		25.5 157	20.3 261	18.0 314	16.7 372	20.6 360	10.8 425
BRITISH LEYLAND		26.1 91		21.2 116	18.1 215			13.9 281		
BMW				21.6 121		20.0 170	14.9 196			
MERCEDES BENZ						29.1 147	20.0 200	13.9 283		
HONDA	33.0 84	33.8 96	27.9 107							
NISSAN (DATSUN)		31.5 85	26.7 110	25.3 119	20.9 159					
PEUGEOT						26.3 137				
RENAULT	30.6 79			24.4 101						
TOYOTA		34.2 71	26.6 97		21.0 136					
VW-AUDI-PORSCHE		34.9 91	27.9 95		19.6 135	13.8 273				
VOLVO					21.7 130	20.3 143				
FUJI (SUBARU)		31.1 97	28.0 97							
FLEET	32.7 88	32.9 90	27.7 101	24.0 125	21.5 161	20.0 255	17.9 310	16.3 353	14.2 393	10.7 416
1978 FLEET	35.4 91	32.4 89	28.0 101	24.5 124	22.4 168	20.1 264	18.0 304	16.3 359	14.6 404	12.4 445

Table 6 - Allocation of Fuel Economy Changes from 1978 to 1979, Passenger Cars

MANUFACTURER	1978 50-STATES CAR SWMPG*	PERCENT FUEL ECONOMY CHANGE DUE TO :					1979 50-STATES CAR SWMPG*
		SYSTEM OPTIMIZATION	TRANSMISSION MIX SHIFTS	ENGINE MIX SHIFTS	WEIGHT MIX SHIFTS	ALL CHANGES COMBINED	
AMERICAN MOTORS	19.1	2.3	-0.2	-1.2	3.8	4.7	20.0
CHRYSLER CORP.	18.4	1.5	0.1	1.1	9.3	12.1	20.6
FORD MOTOR CO.	18.3	-4.4	0.0	1.6	6.7	3.4	19.0
GENERAL MOTORS	18.8	-0.9	-0.1	1.7	1.2	2.0	19.2
BRITISH LEYLAND	20.8	1.4	-0.2	0.3	-1.5	0.0	20.8
BMW	20.0	0.0	0.4	5.8	-4.6	1.3	20.3
MERCEDES BENZ	19.2	0.6	0.7	3.2	0.4	5.1	20.1
HONDA	33.7	-0.5	-0.1	-0.4	-6.4	-7.4	31.2
NISSAN (DATSUN)	26.3	-0.0	-0.1	0.0	0.7	0.6	26.4
PEUGEOT	21.2	0.3	-0.2	23.8	0.0	23.8	26.3
RENAULT	29.8	-0.8	0.0	0.7	0.8	0.7	30.1
TOYOTA	27.1	-8.6	-0.3	0.2	-2.4	-11.0	24.1
VW-AUDI-PORSCHE	29.0	0.1	-1.1	4.8	1.8	5.6	30.6
VOLVO	21.1	-1.9	0.0	-0.1	0.0	-2.0	20.7
FUJI (SUBARU)	31.6	-5.1	-1.0	0.2	-1.8	-7.5	29.3
FLEET	19.6	-2.3	-0.1	1.5	3.7	2.8	20.1
1978 FLEET		+0.5	-0.5	-0.5	5.4	4.8	

* The manufacturers' fleet average MPG values listed here are not official figures for the purpose of compliance with MPG standards. The analysis uses standard practices for decimalization and handling of captive import sales, rather than computational rules set forth in the fuel economy regulations, and the source data base includes only those vehicle configurations and projected sales in the EPA General Fuel Economy Label files as of the data cutoff date.

figures in their engine mix-based MPG advance. Under the analysis ground rules, BMW's replacement of their 1978 (16 MPG) 182 CID engine with the new 170 CID (at 20 MPG) credits them with an engine mix MPG improvement; however, the incorporation of 3-way catalyst technology on the 170 is responsible for its MPG superiority over the thermal reactor 182 it replaces.

The only mentionable MPG changes associated with transmission mix effects belong to VW and to Fuji, whose fleets are predicted to be 25% automatic in 1979, up from about 20% in 1978.

SYSTEM OPTIMIZATION EFFECTS - As pointed out in Appendix B, a "System Optimization" change reflects the combined MPG impact of several factors:

• Changes in driveline hardware - axle ratios, transmission gear ratios, torque converter;

• Changes in operational characteristics - shift schedules, road load settings; and

• Changes in engine calibrations made for purposes of driveability, performance, fuel economy, and/or emission control improvement

Certain of these effects become obvious when the largest system optimization changes (for Fuji, Toyota, and Ford) are examined more closely.

Fuji's 5.1% fuel economy loss in the system optimization category results from a combination of calibrations and shift schedules. In head-to-head comparisons, their 49-states models show an average loss of some 6%, with one configuration representing 40% of sales down by more than 8%; Fuji's California models, however, have actually improved about 4% over 1978. Looking at Fuji's fleet another way, we find noticeable differences between transmission types, indicating the impact of modified shift schedules on manual-shift cars

Fuji MPG Change from 1978:

	49-states	California
Manuals	> 6% loss	3% gain
Automatics	3% loss	> 7% gain

Changes in road load settings are miniscule for Fuji, as they are also for Toyota. Unlike Fuji, Toyota's system optimization loss (8.6%) occurred in both California and 49-states models, but the effects of revised shift schedules are as evident as they are for Fuji. Toyota's manual transmission models show an average MPG decline of more than 11%, while that for automatics is typically 3%. For both of these manufacturers, the MPG loss for manual transmissions was larger on the urban cycle (where more shifting occurs) than on the highway cycle.

The most significant item that surfaces in a close examination of Ford's data is an increase in road load settings. Ford's average road load increased by 21% over that of its 1978 models. As illustrated in Figure 7, Ford's 1978 models were certified at an average road load considerably below the "cookbook" load(5) for their average vehicle weight, while their 1979's are loaded

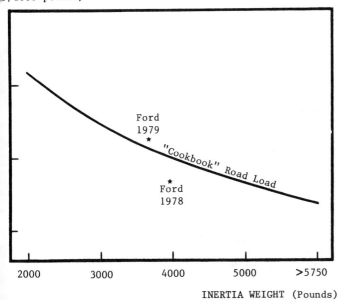

Fig. 7 - Average Ford Motor Co. road loads and test weights, 1978-79

nearer to the cookbook value at 1979 average weight. Using a 55/45 MPG sensitivity to road load of -0.15, this road load increase would account for a fuel economy loss of 3.1%, leaving a system optimization change of -1.3% not associated with road load. With no change in emission standards from 1978 to 1979, this 1.3% loss is assumed to stem from hardware and calibration changes related to driveability and performance.

For a more extensive analysis of road load effects on the 1979 fuel economy of the other domestic manufacturers, for cars and for light duty trucks, see Appendix C.

49-STATES AND CALIFORNIA FUEL ECONOMY - The cause-and-effect analysis comparing the 1979 Federal and California fleets appears in Table 7. By and large, those manufacturers with the largest overall California MPG penalties are the ones with the largest system [de]optimization losses. The bulk of the California system optimization losses are assumed due to engine calibrations used for the more stringent California emission standards. Road load settings and shift schedules are practically the same for Federal and California 1979 cars, for all manufacturers, and are not a significant factor in system optimization changes.

Engine and transmission mix shift effects are relatively small for all manufacturers, and are negligible for the fleet as a whole.

Virtually all of the import manufacturers show a weight-related MPG loss for California as their California fleets are skewed toward higher weights. The same is true for AMC and GM fleets. AMC's cars with inertia weights of 3000 lb or less comprise 25% of their Federal mix, but only 8% of their California fleet (almost exactly the same splits seen in 1978). GM's negative weight mix effect stems from the absence of the 3000 lb class from the California fleet, and from higher sales fractions in California for their three highest weight classes. Nearly 30% of GM's 50-states sales of 5000 and 5500 lb cars is projected for California.

Note in Table 7 that, although 11 of the 15 manufacturers sustained a weight mix-based MPG loss for California, the fleet has an 8.6% gain due to weight mix. This arises due to a shift in the sales mix among manufacturers, described in Table 8. The imports constitute 15% of the Federal fleet, but 36% of the California fleet. The fuel economy and weight advantages of the imports in California are declining, however; from the data in Table 8 and its pre-1979 counterparts (Appendix D), the domestic-to-import differential in weight has decreased in every successive year since 1975, as has the domestic-to-import MPG differential since 1977.

A comparison of 1979 49-States and California fleet weight mixes appears in Figure 8.

Applying a fixed weight mix to the Federal and California fleet data, the variances in weight distributions are eliminated, yielding the weight-independent fuel economy trends shown in Fig. 9. Comparing this data to previous such charts (1), it is clear that the older, non mileage-corrected pre-control data led to a non-trivial underestimation of the recent fuel economy gains that have actually occurred simultaneously with emission reductions.

It is also evident that the 1979 system optimization loss found at the 50-states level (Table 6) was concentrated in the 49-states fleet, since fixed-mix California MPG is higher in 1979 than in 1978. In fact, an inspection of Table 7 of this paper and Table 7 of Ref. 1 shows that seven manufacturers (including three of the Big 4) have higher California fleet MPG than last year, and ten manufacturers show lower Fed-to-Cal system optimization losses than last year. Whatever penalties may have occurred in 1979 due to test/operational factors, they were not of a magnitude sufficient to halt fuel economy progress at current California emission levels.

Table 7 - Allocation of Fuel Economy Differences between 1979 Federal and California Passenger Cars

MANUFACTURER	1979 49-STATES CAR SWMPG*	PERCENT DIFFERENCE IN CALIFORNIA MPG DUE TO:					1979 CALIFORNIA CAR SWMPG*
		SYSTEM OPTIMIZATION	TRANSMISSION MIX SHIFTS	ENGINE MIX SHIFTS	WEIGHT MIX SHIFTS	ALL DIFFERENCES COMBINED	
AMERICAN MOTORS	20.4	-12.1	-0.2	1.5	-2.7	-13.2	17.7
CHRYSLER CORP.	20.7	-11.5	-0.4	1.1	11.2	-0.9	20.5
FORD MOTOR CO.	19.1	-10.2	0.4	-1.2	5.4	-6.1	17.9
GENERAL MOTORS	19.4	-10.9	-0.0	0.7	-1.3	-11.5	17.2
BRITISH LEYLAND	21.1	-2.4	0.0	-0.6	-4.4	-7.2	19.6
BMW	20.6	-4.1	0.0	0.0	-0.4	-4.5	19.6
MERCEDES BENZ	20.4	-0.9	0.0	-1.8	-1.0	-3.7	19.6
HONDA	31.6	-7.4	-0.1	5.3	-2.0	-4.6	30.1
NISSAN (DATSUN)	26.6	-2.3	0.3	0.3	-1.8	-3.6	25.7
PEUGEOT	26.1	0.1	0.5	1.2	0.0	1.8	26.6
RENAULT	30.6	-8.2	0.0	0.0	-0.5	-8.6	28.0
TOYOTA	24.8	-6.8	0.0	0.0	-5.3	-11.8	21.9
VW-AUDI-PORSCHE	31.2	-0.8	-0.5	-2.4	-3.7	-7.2	29.0
VOLVO	21.1	-0.0	-0.9	-2.4	-1.5	-4.7	20.1
FUJI (SUBARU)	29.8	-11.8	0.0	0.0	0.0	-11.8	26.3
FLEET	20.2	-10.1	0.0	-0.1	8.6	-2.4	19.7
1978 FLEET		-11.1	0.0	-1.8	11.7	-2.6	

* The manufacturers' fleet average MPG values listed here are <u>not</u> official figures for the purpose of compliance with MPG standards. The analysis uses standard practices for decimalization and handling of captive import sales, rather than computational rules set forth in the fuel economy regulations, and the source data base includes only those vehicle configurations and projected sales in the EPA General Fuel Economy Label files as of the data cutoff date.

Table 8 - Fuel Economy, Weight, and Sales Characteristics of 1979 Domestic and Import Subfleets, 49-States and California

		49-States	California	50-States
Domestic	55/45 MPG	19.3	17.8	19.2
	Weight, lb.	3707	3660	3703
	% sales	75.8	7.1	82.9
Import	55/45 MPG	26.9	24.9	26.4
	Weight, lb.	2563	2648	2583
	% sales	13.2	4.0	17.1
Total Fleet	55/45 MPG	20.1	19.8	20.1
	Weight, lb.	3538	3298	3508
	% sales	88.9	11.1	100.0

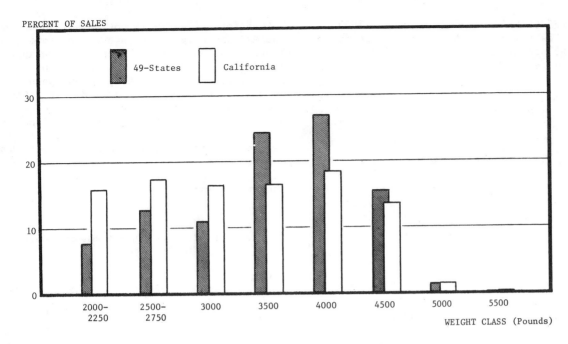

Fig. 8 - 1979 49-states and California weight distributions

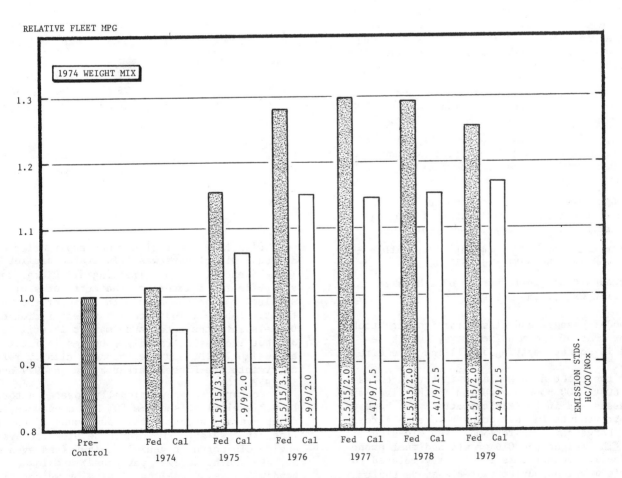

Fig. 9 - Weight-normalized fleet MPG, 49-states and California models

VEHICLE SIZE CLASSIFICATIONS used since 1977 in the EPA/DOE Gas Mileage Guide are described in Table 9; average actual volumes for the 1979 models are also shown.

Table 9 - Vehicle Size Classifications, Passenger Vehicles

Vehicle Size Class	Interior Volume (Cu. Ft.) Specified	1979 Average
Two-Seater	----	----
Minicompact	<85	81.1
Subcompact	85-100	89.3
Compact	100-110	104.5
Midsize	110-120	114.2
Large	>120	129.9
Small Station Wagon	<130	112.7
Midsize Station Wagon	130-160	139.4
Large Station Wagon	≥160	163.0

Table 10 lists the sales-weighted average 55/45 MPG, engine displacement and inertia weight, within each car class, for the fifteen reference manufacturers and the fleet. In all but the Large Car and the Large Station Wagon categories, wide differences in these characteristics exist among the manufacturers. Notable changes from 1978 (see Reference 1, Table 5) are:

- AMC's withdrawal from the Midsize and Large car and Large wagon classes, due to discontinuance of the Matador line;

- Chrysler's disappearance from the Compact car class, which last year included Omni/Horizon (now Sub-compacts) and Aspen/Volare models (now Midsize);

- BLMC's movement into the Compact Car class, due to reclassification of the Jaguar XJ, a Sub-compact last year;

- BMW's entry into the Compact class with two models, the 528I and 733I;

- Mercedes-Benz' introduction of a Midsize Station Wagon, the 300TD.

(Other model changes and class reassignments took place for most of the manufacturers, but these resulted in no net emptying or filling of entire classes).

For the fleet as a whole, weight reductions of some 60 to 200 pounds occurred in the five largest classes and in the Minicompact class; however, clear improvements in fuel economy were only realized for Large cars and for Small and Mid-size Wagons. The weight and CID growth and fuel economy decline in Compact cars is directly related to Chrysler's movement out of that class, as their 1978 Compacts averaged 21.3 MPG, 208 CID, and 3309 pounds.

With THREE YEARS OF CAR CLASS DATA now available, it is useful to recap the behavior of new-car interior volumes in light of the weight and fuel economy changes occurring over this period. In Table 11, we see that the trend in average interior volume for 1977-1979 is not at all as obvious as the MPG and weight trends over the same interval. In terms of what might be called "transport efficiency" (cubic foot miles per gallon) and "specific volume" (cubic feet per 1000 pounds), the overall fleet is improving, a direct reflection of the Domestic cars' improvements in weight and fuel economy without corresponding sacrifices in volume. Japanese autos continue to show not only lighter weight and higher fuel economy than both Domestic and European built cars over these three years, but higher transport efficiency, and noticeably higher specific volume.

Table 11 - Interior Volume Trends and Related Figures of Merit

	Volume (Cu.Ft.)	Weight (Pounds)	55/45 MPG	Cu.Ft.-Mi Gallon	Cu.Ft. 1000 lbs.
Domestic					
1977	116.5	4218	17.2	2004	27.6
1978	117.3	3848	18.7	2193	30.5
1979	116.1	3703	19.2	2223	31.4
European					
1977	91.7	2715	25.1	2303	33.8
1978	89.7	2658	25.1	2254	33.8
1979	89.1	2759	25.5	2265	32.3
Japanese					
1977	99.1	2483	29.3	2907	40.0
1978	92.2	2482	28.2	2596	37.2
1979	96.8	2487	26.8	2588	38.9
Fleet					
1977	113.5	3943	18.3	2078	28.8
1978	113.6	3649	19.6	2223	31.2
1979	112.6	3508	20.1	2264	32.1

DIESELS - The fuel economy superiority of individual Diesel cars was illustrated earlier in Table 4 and Figure 5. Treating all Diesel cars as a "subfleet" of interest, the attributes of this subfleet can be compared with those of the overall fleet. Table 12 indicates that penetration of Diesels into the total auto market is still below 3%, but increasing rapidly. Changes in fleet average weight, CID, and fuel economy clearly reflect entry into the Diesel market by VW in 1977 and GM in 1978.

Compared to conventionally-powered autos, the 1979 Diesel fleet shows a 50% MPG advantage, and a fleet average weight within one pound of the gasoline cars' weight. However, the Diesels' average engine CID per unit vehicle weight (and even more so, average horsepower per pound) are lower. Appendix E gives Diesel cars' data by weight class.

Interior volumes for the last three years' Diesel fleets are given in Table 12, together with transport efficiency and specific volume. Overall average interior room in the Diesel fleet is now comparable to that of the complete fleet; transport efficiency is much higher; and specific volume is typical of European and Domestic made cars.

Table 10 - Average 55/45 MPG, Engine CID, and Test Weight
by Vehicle Size Class, 1979 Model Cars

MANUFACTURER	I--(PASSENGER CARS)--------------------------------I						I--(STATION WAGONS)-----I		
	2-SEATER	MINI COMPACT	SUB COMPACT	COMPACT	MIDSIZE	LARGE	SMALL	MIDSIZE	LARGE
AMERICAN MOTORS			20.8 215 3164	19.1 250 3478			18.6 262 3562		
CHRYSLER CORP.		29.6 108 2518	28.3 107 2511		18.4 297 3958	17.4 334 4254	25.4 146 2956	18.5 268 4015	
FORD MOTOR CO.		24.3 141 2763	22.8 155 2861	18.7 262 3571	17.0 280 3956	16.6 326 4160	23.0 145 3000	19.6 214 3356	16.0 332 4500
GENERAL MOTORS	16.9 350 4000		21.4 220 3282	18.4 285 3933	20.1 271 3555	17.7 341 4208	23.4 178 3240	18.7 279 3918	16.1 337 4516
BRITISH LEYLAND	22.1 114 2630		11.0 326 4500	14.5 274 4500					
BMW			21.3 124 2796	18.3 177 3634					
MERCEDES BENZ	14.2 276 4000			23.5 171 3871	13.9 283 4500			24.6 183 4000	
HONDA		33.0 84 2000	30.1 104 2408				31.8 91 2250		
NISSAN (DATSUN)	20.4 168 3000	25.3 119 2750	28.2 103 2433				26.9 106 2563		
PEUGEOT				26.0 137 3500				28.0 139 3500	
RENAULT		30.1 81 2056							
TOYOTA			24.2 112 2713				23.9 115 2708		
VW-AUDI-PORSCHE	19.9 139 3000	22.5 115 2599	33.6 91 2290	19.3 131 3000			32.2 93 2500		
VOLVO				20.6 140 3278				20.9 136 3500	
FUJI (SUBARU)			30.5 97 2280				28.1 97 2500		
FLEET	20.1 168 3007	27.9 110 2422	24.4 154 2836	19.0 250 3680	18.9 278 3734	17.4 336 4198	24.9 132 2784	19.0 253 3778	16.0 335 4510
1978 FLEET	20.1 172 3002	27.7 119 2551	24.7 163 2864	19.8 242 3613	18.7 293 3800	16.8 358 4391	24.4 141 2843	18.4 273 3869	15.8 364 4679

Table 12 - Diesel Passenger Car Trends

	Sales Fraction	Fuel Economy (MPG) City	Hwy.	55/45	Avg. Wt. (Pounds)	Average CID	Average Volume (Cu. Ft.)	Cu.Ft.-Mi gallon	Cu. Ft. 1000 lb
1975	0.2%	24.7	32.0	27.5	3500	153			
1976	0.3%	23.7	29.8	26.1	3744	162			
1977	0.5%	28.7	36.0	31.6	3137	131	88.6	2800	28.2
1978	1.7%	29.1	39.2	32.9	3152	176	105.3	3464	33.4
1979	2.9%	26.7	35.3	30.0	3509	213	111.5	3345	31.8
1979 Gasoline	97.1%	17.4	24.1	19.9	3508	242			

TODAY'S TECHNOLOGY vs MPG STANDARDS

Numerous options are being considered by the industry and government for achieving the passenger car fuel economy standards established through 1985. This section will investigate just how far current technology can be stretched toward meeting those standards, and what the effects of such a stretching might be upon the mix of types of cars and available interior space. "Current technology" is used here in the broad sense, i.e. with reference to total vehicle technology, embodying not only fuel economy capabilities, but interior volume and weight attributes as well.

The analysis is based on 1978 data. This baseline case is illustrated in Figure 10. Fuel consumption is measured vertically and sales fraction horizontally. The relative consumptions and sales penetrations are shown for five 1978 car classes*. The sales-weighted average fuel consumption is 5.11, which corresponds to the fleet 55/45 average fuel economy, 19.6 MPG. Those car classes whose interior volume exceeds 100 cu.ft. are projected to capture 73% of sales, and the smaller classes (< 100 cu.ft.) the balance, 27%.

Basically, two things can be done to improve fuel economy (decrease consumption) from this baseline: the consumptions for the various classes can be brought down through the use of more fuel efficient technology, or the mix of classes can be modified so that the larger, higher fuel consumption cars become less dominant in sales. The investigation defines four scenarios, i.e. four discrete combinations of 1978 gasoline and Diesel technology, and treats the sales mix of car classes as an independent variable. For any of the four specific blends of technology, various class mixes are possible, with fleet average fuel economy, weight, and interior volume resulting as dependent variables. The scenarios are:

(1) 1978 Base Technology, 2% Diesel: the baseline, with each car class at the MPG and weight values corresponding to the 1978 fleet average for that class; Diesel penetration in each class is at the 1978 level, with a fleetwide sales fraction of 1.7%;

(2) 1978 Base Technology, 33% Diesel: the gasoline component of each car class has MPG and weight characteristics at average 1978 gasoline values; the Diesel components are also at 1978 values, but Diesel sales fractions in each class are increased from their 1978 levels by a factor of 33/1.7, giving a fleet penetration of 33%;

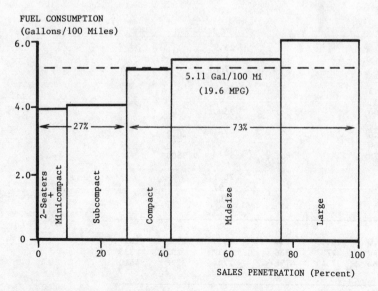

Fig. 10 - Baseline scenario: 1978 models' fuel consumption and sales fractions, by car class groupings

---
* Station Wagons are grouped in the same class as the sedans that they are derivatives of. Two-seaters, which account for less than 3% of 1978 sales, are grouped with Minicompacts. The analysis assumes an interior volume of 50 cu.ft. for the two-seaters.

(3) Best 1978 Technology, 2% Diesel: the gasoline component of each class has the MPG and weight characteristics of that manufacturer with the highest average (gasoline) MPG in that class; the Diesel characteristics and penetrations in each class are the same as case (1);

(4) Best 1978 Technology, 33% Diesel: the gasoline and Diesel components of each class have the same attributes as in case (3), with Diesel sales fractions the same as case (2).

Scenarios (1) and (2) represent situations wherein both gasoline and Diesel powered cars are frozen at their current respective capabilities. In (1), they are proportioned within each car class per 1978 sales figures, and the class mix is thus the only quantity that can change from 1978. In (2), fuel consumption levels are improved by means of increased Dieselization. For Scenario (3) the emphasis is reversed, as Diesel market penetration remains low, and gasoline-powered cars within the respective classes are improved to match the pace-setting manufacturer's average weight and MPG for that class. This scenario is conservative to the extent that "someone has already found a way to put cars that good on the road"; however... those present-day cars are not required to meet the emission standards of the 1980's. Hence our scenario (3) implies adoption of that vehicle/engine technology equivalent to the 1978 pacesetter, with no MPG loss due to tighter emission control. Case (4) of course assumes advances in technology for gasoline cars and advances in sales fraction (but no technology gains) for Diesels.

Although there is no explicit treatment of weight reduction, it is implicit in the analysis, and it definitely shows up in the results. For scenarios 3 and 4, the "best technology" cars are where they are due mainly to their light weight compared to others in their car class. Any move to adopt that vehicle technology would therefore have to include weight reduction. For all scenarios, of course, fleet average weight will vary as classes are remixed.

A few of the consumption/sales plots resulting from the study are displayed in Figure 11 for illustration, and the complete parametric results are given in Figures 12(a)-(e).

Figure 12(a) shows the mix shifts which must occur for each scenario to meet the MPG standards. The 1978 base technology, without increased Diesel penetration, would suffice until 1982, at which point 90% of the fleet must be Two-seaters, Mini-compacts, and Subcompacts, and their derivative station wagons. The 1983-85 MPG standards are achievable with no change in gasoline vehicle technology if the Diesels' market share increases substantially, but the larger car classes would have to recede to 20% market penetration by 1985. For any manufacturer who approaches the vehicle tech-

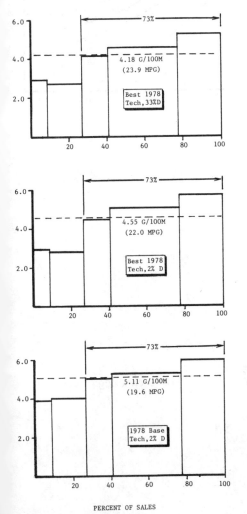
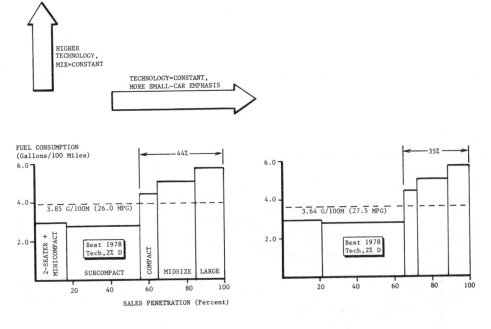

Fig. 11 - Examples of interaction between sales mix, technology utilization, and fleet fuel economy capability

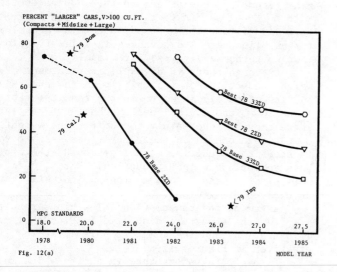
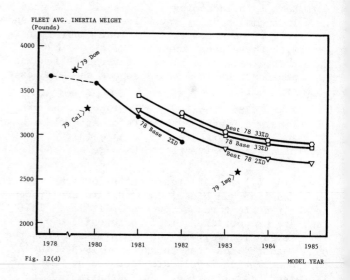
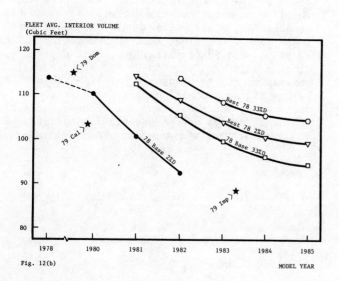
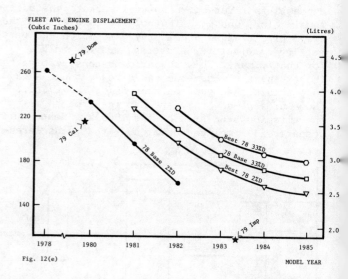
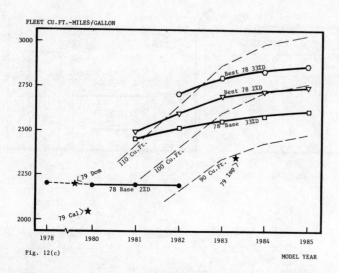

Fig. 12 - Fleet attributes for four combinations of 1978 technology

nology level of the 1978 pacesetter manufacturer in each car class, the required decline in larger car sales would be more gradual, remaining above 33% through 1985, even with no growth in Dieselization. With adoption of the best 1978 gasoline vehicle technology and also more widespread use of Diesels, the break-even mix bottoms at about 47% larger cars for 1985.

The average interior volumes corresponding to the mixes from Figure 12(a) are given in Figure 12(b). By 1985, the most innovative scenario permits a fleet average interior volume comparable to that of 1979 California cars.

Figure 12(c) displays the transport efficiencies resulting from the decreasing volumes and increasing fuel economies. The flatness of the 1978 "base case" curve indicates that, for this scenario, volume must be reduced as fast as fuel economy rises. The three improved technology cases, however, do not lose volume one-for-one in exchange for fuel economy improvement.

In Figure 12(d), we see the effect of the Diesels' much higher MPG capability per unit of weight, as both of the scenarios with high Diesel penetrations are capable of meeting the standards at higher fleet weights. For the low Diesel fraction cases, more weight reduction is necessary, either by remixing base technology for less weight or by moving to "best-technology" reduced weights in less extreme mixes. Even though the two cases with high Diesel fractions give nearly identical weight trends, it must be remembered that they are quite different in terms of mix and average volume, as seen in Figures 12(a) and (b).

Fleet average engine displacement, shown in Figure 12(e), behaves much like average weight, owing to Diesel engines' higher MPG capability per unit engine displacement.

The star symbols spotted on Figures 12(a)-(e) indicate the characteristics of the 1979 models as represented by three subfleets: domestics, imports, and California cars (domestics + imports). The data points are aligned with the year whose MPG standards they "meet" in 1979. It is clear that the imports' high average MPG derives more from vehicle miniaturization than from high technology.

Appendix F is a do-it-yourself kit of numbers from the 1979 fleet, arranged in the same manner as the 1978 data used in this study.

LIGHT DUTY TRUCKS

Data for the 1979 light truck fleet, and historical data for model years 1975-1978, were developed using the same approaches as the passenger car data. The graphs and tables resulting from this analysis are largely self-explanatory and are presented without comment in Appendix G. These data correspond to all non-passenger vehicles certified by EPA with gross vehicle weight ratings (GVWR) up to 6000 pounds.

CONCLUSIONS

1. Updated pre-1975 data now show a total fleet fuel economy loss of 4.6% from pre-emission control years to 1974, of which 3.8% was due to vehicle weight mix changes and 0.8% was due to emission control penalties and other technology changes.

2. Fuel economy of the 1979 fleet is 41.5% better than that of 1974; 23.2% derives from technology advancements (although emission standards have toughened), and the balance results from weight mix improvements.

3. From 1978 to 1979, there was a 2.8% net gain in fleet MPG, with engine mix and weight mix shifts contributing a 5.2% increase and system optimization a 2.3% loss. Roughly half (1.3%) of the loss appears related to 1979 changes in dynamometer road load settings.

4. California cars' fuel economy improved in 1979 at the fleet level, and for three of the four domestic manufacturers and four of the eleven foreign manufacturers analyzed. Fixed weight mix analysis shows that, on the whole, California cars suffered no net MPG setbacks in 1979, road load and all other technological changes notwithstanding.

5. Average vehicle interior volume has not been sacrificed for the recent gains in fleet fuel economy. 1978 fleet MPG improved 4.8% over 1977, with no change in average volume; 1979 MPG improved 2.8% over 1978, with only an 0.9% volume decrease. Japanese cars continue to deliver superior transport efficiency (cu.ft. miles per gallon) and higher specific volume (cu.ft. per 1000 lb) than European or Domestic manufactured autos.

6. Diesels are projected to capture 2.9% of U.S. 50-states sales in 1979; of the four manufacturers offering Diesels, three are substantially Dieselized: their fleets consisting of 45% to over 70% Diesel. The 1979 U.S. Diesel cars have average weight and interior volume figures within 1% of those of the gasoline fleet, a lower average power-to-weight ratio, and a 55/45 MPG that is 50% better. 1979 Diesel fleet MPG is 9% above the 1985 fuel economy standard.

7. Based on the specific technology scenarios studied in this paper, achievement of all MPG standards through 1985 appears feasible with gasoline-powered vehicle technology available today and low sales fractions for Diesel-powered vehicles. However, due to the large influence of Diesels on fleet attributes such as average weight and interior volume, if Diesel penetration remains low, the resulting fleet attributes may differ substantially from those of today's fleet.

REFERENCES

1. J.D. Murrell, "Light Duty Automotive Fuel Economy - Trends through 1978", SAE Paper 780036, SAE Congress and Exposition, Detroit, Michigan, February 1978.

2. The Aerospace Corporation, "Passenger Car Weight Trend Analysis", EPA-460/3-73-006a, El Segundo, California, January 1974.

3. EPA Emission Factors Program Data, Characterization and Analysis Branch, ECTD, EPA, February 1977 and April 1978.

4. U.S. Department of Transportation, Federal Highway Administration, News Release, January 1975.

5. Federal Register, Vol 38, No 124, June 1973.

6. U.S. Environmental Protection Agency, V.I. Report for 1978 Models, February 1978.

7. U.S. Environmental Protection Agency, V.I. Report for 1979 Models, October 1978.

8. Based on unpublished EPA data, and on data from S. Martin and K. Springer, "Influence on Fuel Economy and Exhaust Emissions of Inertia, Road Load, Driving Cycles, and N/V Ratio for Three Gasoline Automobiles", Southwest Research Institute Final Report under Task No. 9, EPA Contract 68-03-2196, June 1977.

9. D. Samples and R. Wiquist, "TFC/IW", SAE Paper 780937, International Fuels and Lubricants Meeting, Toronto, Canada, November 1978.

APPENDIX A

Sales and MPG Figures by Weight Class for Passenger Cars, Pre-Control through 1979

SALES DISTRIBUTION OF WEIGHT CLASSES, 50-STATE CARS

MODEL YEAR	WEIGHT CLASS (POUNDS)										AVERAGE WEIGHT
	2000	2250	2500	2750	3000	3500	4000	4500	5000	5500	
PRE-CONTROL	.03482	.01220	.01101	.03954	.13232	.14683	.35170	.20405	.05630	.01123	3812
1968	.00519	.05429	.02628	.00073	.03840	.26516	.34111	.18726	.07302	.00855	3863
1969	.01333	.05606	.02176	.03069	.03162	.23524	.16229	.33363	.10601	.00938	3942
1970	.01277	.07372	.03444	.05020	.05999	.12609	.23868	.29177	.09148	.02087	3877
1971	.01105	.08822	.07357	.02374	.05484	.11834	.19842	.26572	.12985	.03625	3887
1972	.01162	.04630	.06561	.06801	.04775	.12484	.20328	.24672	.12413	.06173	3942
1973	.01573	.04378	.05724	.09453	.05766	.11807	.12210	.25462	.16819	.06809	3969
1974	.00765	.04929	.04416	.07525	.12786	.09966	.11620	.23141	.16189	.08663	3968
1975	.01272	.04555	.04932	.03410	.10018	.12500	.13222	.21756	.18282	.10052	4057
1976	.01584	.05867	.03589	.02087	.09089	.09602	.20339	.22867	.15373	.09604	4060
1977	.01390	.07535	.04112	.03091	.05172	.06411	.29411	.30942	.09299	.02638	3943
1978	.02342	.07045	.05497	.03262	.07465	.28255	.21191	.19408	.04384	.01151	3649
1979	.02150	.06639	.09497	.03941	.11615	.23474	.25831	.15213	.01623	.00017	3508

DATA THROUGH 1977 FROM PRODUCTION FIGURES

1978-1979 DATA FROM MANUFACTURERS' SALES FORECASTS

EPA 4000-MILE CITY FUEL ECONOMY (1975 MPG)
BY MODEL YEAR AND WEIGHT CLASS, PASSENGER CARS

MODEL YEAR	WEIGHT CLASS (POUNDS)										FLEET AVERAGE
	2000	2250	2500	2750	3000	3500	4000	4500	5000	5500	
PRE-CONTROL	22.48	21.46	18.62	16.98	15.33	13.57	12.32	11.38	10.25	9.35	12.88
1968	18.76	18.87	17.88	15.81	16.06	13.67	12.27	11.00	10.04	10.08	12.59
1969	21.16	20.98	18.85	18.82	15.56	13.44	12.43	11.38	10.08	10.75	12.60
1970	22.80	22.45	20.24	18.91	15.39	13.06	12.01	11.05	10.34	8.30	12.59
1971	21.74	21.28	20.68	17.38	15.85	12.95	11.97	10.66	9.34	9.22	12.27
1972	22.09	20.81	19.58	18.28	15.61	12.73	11.54	10.70	9.95	9.54	12.15
1973	22.37	21.51	19.08	18.69	15.04	12.51	11.56	10.86	9.47	9.16	12.01
1974	21.91	22.34	19.88	17.76	15.14	13.62	11.42	10.54	9.66	9.18	12.03
1975	27.25	23.97	20.92	18.46	18.41	15.91	13.83	12.37	11.11	10.47	13.69
1976	28.39	25.82	22.43	21.13	20.10	17.19	15.58	13.64	12.77	11.57	15.23
1977	31.31	27.80	25.50	21.33	20.16	17.91	15.78	14.46	12.14	10.71	15.99
1978	31.45	28.40	24.66	21.05	19.54	17.53	15.71	13.99	12.47	10.41	16.97
1979	28.97	29.02	24.56	20.62	18.73	17.69	15.69	14.17	12.29	9.66	17.60

EPA 4000-MILE COMBINED CITY/HIGHWAY FUEL ECONOMY (MPG)
BY MODEL YEAR AND WEIGHT CLASS, PASSENGER CARS

MODEL YEAR	WEIGHT CLASS (POUNDS)										FLEET AVERAGE
	2000	2250	2500	2750	3000	3500	4000	4500	5000	5500	
PRE-CONTROL	27.11	25.71	22.15	20.11	17.86	15.81	14.32	13.13	11.29	10.34	14.90
1968	22.11	23.07	21.32	18.77	18.80	16.09	14.44	12.63	11.24	11.32	14.69
1969	24.89	24.69	22.13	22.09	18.35	15.67	14.45	13.38	11.76	12.52	14.74
1970	26.56	26.40	23.23	22.60	18.39	15.53	14.09	13.08	11.99	9.77	14.85
1971	24.74	25.22	24.44	20.59	17.90	15.08	14.01	12.39	11.16	10.90	14.37
1972	26.41	25.14	22.59	22.08	18.41	15.35	13.82	12.69	12.02	11.04	14.48
1973	25.47	25.76	22.66	21.84	17.73	14.54	13.60	12.86	11.23	10.65	14.15
1974	26.14	25.85	23.38	20.89	18.02	15.92	13.41	12.34	11.60	10.88	14.21
1975	31.39	28.32	24.37	21.71	21.44	18.17	15.64	14.27	12.89	12.16	15.79
1976	32.59	29.62	26.25	24.48	23.51	19.42	17.68	15.72	14.63	13.29	17.46
1977	35.99	32.07	29.29	24.79	23.23	20.23	17.89	16.53	14.13	12.62	18.31
1978	35.44	32.37	27.97	24.47	22.37	20.11	18.02	16.32	14.64	12.40	19.57
1979	32.72	32.91	27.66	23.95	21.53	20.04	17.92	16.35	14.24	10.72	20.11

EPA 4000-MILE HIGHWAY FUEL ECONOMY (MPG)
BY MODEL YEAR AND WEIGHT CLASS, PASSENGER CARS

MODEL YEAR	WEIGHT CLASS (POUNDS)										FLEET AVERAGE
	2000	2250	2500	2750	3000	3500	4000	4500	5000	5500	
PRE-CONTROL	36.20	33.92	28.86	25.98	22.37	19.81	17.86	16.17	12.91	11.87	18.46
1968	28.31	31.70	27.89	24.34	23.77	20.51	18.40	15.41	13.16	13.30	18.42
1969	31.73	31.48	28.09	28.05	23.50	19.63	18.03	17.07	14.81	15.69	18.62
1970	33.28	33.67	28.34	29.69	24.17	20.24	17.89	16.90	14.90	12.45	19.01
1971	29.78	32.56	31.43	26.59	21.24	18.90	17.71	15.45	14.66	14.01	18.18
1972	34.69	33.70	27.80	29.62	23.56	20.50	18.24	16.37	16.12	13.65	18.90
1973	30.65	33.99	29.38	27.48	22.70	18.14	17.34	16.61	14.49	13.27	18.07
1974	34.19	31.96	29.82	26.64	23.47	20.03	17.03	15.60	15.36	14.04	18.23
1975	38.54	36.36	30.51	27.66	26.82	21.99	18.60	17.56	16.02	15.14	19.46
1976	39.79	36.12	33.14	30.37	29.67	23.07	21.18	19.33	17.80	16.22	21.27
1977	44.02	39.48	35.79	30.93	28.55	24.05	21.39	20.04	17.68	16.14	22.26
1978	41.94	39.02	33.47	30.55	27.18	24.53	21.97	20.48	18.61	16.19	24.08
1979	38.87	39.37	32.71	29.82	26.35	23.92	21.68	20.15	17.71	12.39	24.34

APPENDIX B

Calculation Methodology for Fuel Economy Change Allocation

The procedure for computing fleet fuel economy changes due to specific factors, such as system optimization and weight mix shifts, involves the construction of matched sets of data from a base fleet (e.g. 1978) and a new fleet (e.g. 1979), and calculation of intermediate sales-weighted fleet fuel economy values for the matched sets. Depending on the degree of matching, the data sets being compared include only certain known changes between the sets, and hence the calculated intermediate fleet MPG values reflect the fuel economy effects of only those specific changes in fleet makeup.

CALCULATION OF DIFFERENCES DUE TO SYSTEM OPTIMIZATION: To determine the differences in fuel economy between the 1978 and 1979 cars due to system optimization, it was necessary to limit the comparison to nominally identical vehicles. For each manufacturer it was established which 1978 and 1979 models were identical in terms of weight, displacement, and transmission type. When this was established a new set of sales fractions was calculated, based on 1978 sales estimates, using only those combinations which were carried over from 1978 to 1979. Two sales-weighted fuel economy values were calculated using equation (2) [see text]: one calculation using 1978 model MPG values and 1978 carryover sales fractions, and one using the 1978 model MPG values, also with 1978 carryover sales fractions. The difference between the two values reflects the change in fuel economy due to what we have called system optimization. Since the weights, displacements, transmissions - and their sales distributions - are matched, any difference in fuel economy is due to other factors. The main factors which could be contributing to such a system optimization change in fuel economy are:

- Emission control system design changes;
- Engine design and/or calibration changes;
- Changes in transmission efficiency, shift scheduling, or gear ratios;
- Axle ratio changes;
- Changes in test procedure which influence fuel economy.

DIFFERENCES DUE TO TRANSMISSION MIX SHIFTS: In the analysis of fuel economy changes due to system optimization, any IW/CID/transmission combination not common to both years was eliminated from consideration, and the sales distribution of those combinations that were carried over was held at the 1978 mix. If the calculation is repeated using only weight/displacement combinations as the determinants for model year carryover, those IW/CID/transmission combinations that are not common to both sets of data are not "sifted out", but remain in their respective data bases; also, each of the data bases retains its own sales split between automatics and manuals within the carryover IW/CID combinations.

Again, two SWMPG values are calculated using equation 2, wherein the first MPG_i is the harmonic mean sales-weighted fuel economy of each manufacturer's 1978 models in IW/CID class i, and the second MPG_i is the fuel economy of his 1979 models in IW/CID class i. Both of these SWMPG values are based on the same mix of the IW/CID classes (the 1978 mix), so the difference between the two is due to system optimization plus all changes in transmission mix.

DIFFERENCES DUE TO ENGINE MIX SHIFTS: Similarly, by sifting for carryover at only the weight class level, all differences in the IW/CID structures of the fleets are allowed to remain. The difference between the two SWMPG values calculated on this basis is thus due to system optimization, transmission mix shifts, and shifts in the mix of engine displacements*.

DIFFERENCES DUE TO WEIGHT MIX SHIFTS: The bottom-line SWMPG values calculated from the full, unperturbed data bases, each with its own sales mix, includes all of the above effects plus the effect of non-carryover weight classes and the 1979 redistribution of sales among carryover weight classes.

Table B-1 summarizes the above calculation methodology, and Figure B-1 shows a diagram of the relationship between the various calculated SWMPG values. Since the methodology is suitable for a comparison between any two vehicle sets (49-states vs. California, cars vs. trucks, manufacturer X vs. Y, etc.), Table B-1 and Figure B-1 are notated for the general case rather than the year-to-year case.

Table B-2 illustrates the equations for separation of individual factors from the combined effects discussed above.

---

* This also includes shifts in the mix of engine standards/systems; Fed vs. Cal. and Spark vs. Diesel.

Table B-1 — Method for Constructing Fuel Economy
Comparisons between Two Vehicle Groups

Configuration Determinants	Vehicle Group "A"			Vehicle Group "B"			A-to-B SWMPG Change Attributed To:
	MPG Base(mpg_i)	Sales Base(f_i)	Fleet SWMPG	MPG Base(mpg_i)	Sales Base(f_i)	Fleet SWMPG	
IW/CID/Transmission Type	A	A	$FE_{AA}ICT$	B	A	$FE_{BA}ICT$	System optimization in carryover I/C/T combinations
IW/CID	A	A	$FE_{AA}IC$	B	A *	$FE_{BA}IC$	Above <u>plus</u> new/discontinued I/C/T combinations <u>plus</u> shifts in transmission mix within carryover I/C combinations
IW	A	A	$FE_{AA}I$	B	A **	$FE_{BA}I$	Above <u>plus</u> new/discontinued I/C combinations <u>plus</u> shifts in engine mix within carryover IW classes
Open	A	A	FE_{AA}	B	B ***	FE_{BB}	Above <u>plus</u> new/discontinued IW classes <u>plus</u> shifts in IW mix among carryover IW classes

* Includes B mix of transmissions within c/o IC classes.
** Includes B mix of CT combinations within c/o weight classes.
*** Includes B mix of all ICT combinations in group B.

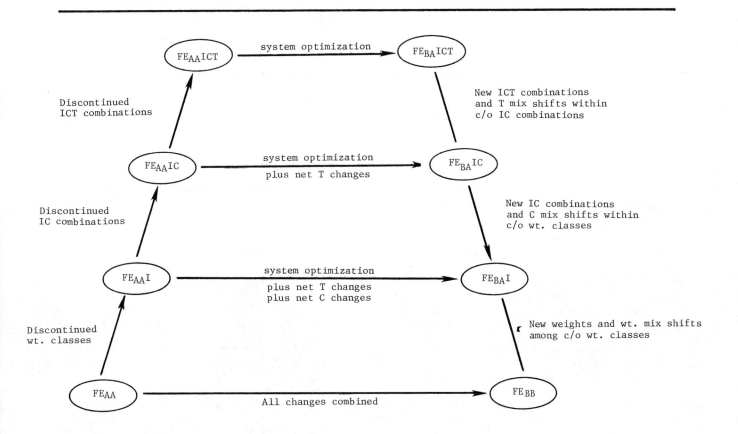

Fig. B-1 — Relationships between SWMPG values from table B-1

Table B-2 - Isolation of Specific Factors
Causing Fuel Economy Change

Percent Change in Fuel Economy Due To:	Calculated By:
Systems Optimization	$\left[\left(\dfrac{FE_{BA}ICT}{FE_{AA}ICT}\right) - 1\right] \times 100$
Transmission Mix Shifts	$\left[\left(\dfrac{FE_{BA}IC}{FE_{AA}IC}\right) \div \dfrac{FE_{BA}ICT}{FE_{AA}ICT} - 1\right] \times 100$
Engine Mix Shifts	$\left[\left(\dfrac{FE_{BA}I}{FE_{AA}I}\right) \div \dfrac{FE_{BA}IC}{FE_{AA}IC} - 1\right] \times 100$
Weight Mix Shifts	$\left[\left(\dfrac{FE_{BB}}{FE_{AA}}\right) \div \dfrac{FE_{BA}I}{FE_{AA}I} - 1\right] \times 100$
All Changes Combined	$\left[\left(\dfrac{FE_{BB}}{FE_{AA}}\right) - 1\right] \times 100$

APPENDIX C

Road Load Changes and Related Fuel Economy Effects

A comparison was made between 1978 and 1979 road load HP settings for AMC, Chrysler, Ford and GM cars and light duty trucks. The HP settings published by EPA in References 6 and 7 were used. cars and light duty trucks. The HP settings published by EPA in References 6 and 7 were used.

Figure C-1 illustrates the comparison for these four manufacturers' passenger cars. For a given manufacturer, each data point denotes a specific engine; the road load and inertia weight coordinates are the average values for all Certification cars using that engine. (Engines not used in both years are omitted for clarity).

As a point of reference, the figure shows so-called "cookbook" road load values from earlier regulations, which were based only on inertia weight. For 1978 models, road loads below the cookbook values were commonplace. For 1979, the average road load values are more in line with the cookbook levels, with loads for lighter cars tending to be below cookbook, and those for the heavier cars usually above.

Since weight reductions occurred between 1978 and 1979 for all these manufacturers, a year-to-year comparison is not valid unless the loads are weight-normalized. (If vehicle weights drop significantly but loads remain unchanged, it would hardly be correct to say there was no road load penalty). The study results in Table C-1 accordingly are expressed in units of HP/1000 lb.

The changes in fuel economy caused by the road load increases can be computed if the sensitivity coefficient* is known. The magnitude of the sensitivity factor used here comes from Ref-

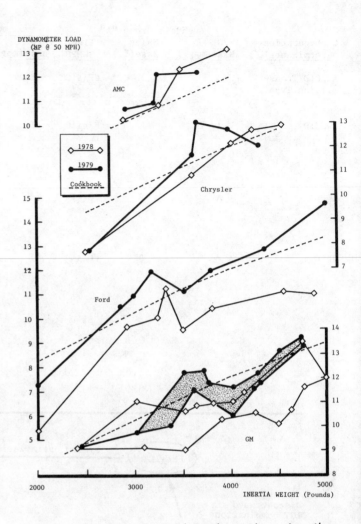

Fig. C-1 - 1978-79 road load settings for the "big 4" manufacturers

erence 8, whose source-specific factors are:

	City	Hwy	55/45
EPA Tests	-.079	-.246	-.155
SwRI Tests	-.112	-.344	-.141
Average	-.096	-.295	-.148

The MPG effects of the road load changes are in data column #4 of the table. A better estimate of the "true" system optimization changes is derived by removing the road load effects from the conventionally-calculated system optimization figures, which include road load effects. These appear in column #5, and the load-independent system changes are listed in the last column.

For both cars and trucks, more than half of the system change is related to road load effects.

---

* The sensitivity coefficient gives the percentage change in MPG per percentage change in road load:

$$S = \frac{\Delta MPG / \overline{MPG}}{\Delta RL / \overline{RL}}$$

Table C-1 - Influence of Dynamometer Load Setting
on "Big-4" Fuel Economy

(Passenger Cars)

	Avg. Road Load, HP/1000 lb* 1978	1979	Change in HP/1000 lb.	F.E. Penalty for ΔRLHP**	Calculated ΔF.E. for System Optimization + ΔRLHP	ΔF.E due to System Optimization Only
AMC	3.41	3.54	+3.8%	-0.6%	+2.3%	+2.9%
Chrysler	3.02	3.19	+5.6%	-0.8%	+1.5%	+2.3%
Ford	2.70	3.26	+20.7%	-3.1%	-4.4%	-1.3%
GM	2.78	3.02	+8.6%	-1.3%	-0.9%	+0.4%
Fleet	2.80	3.04	+8.6%	-1.3%	-2.3%	-1.0%

(Light Duty Trucks)

	1978	1979	Change	F.E. Penalty	Calc. ΔF.E.	ΔF.E Sys Opt Only
AMC	3.15	4.85	+54.0%	-8.0%	-5.0%	+3.3%
Chrysler	2.94	3.94	+34.2%	-5.1%	-3.6%	+1.6%
Ford	2.96	4.04	+36.4%	-5.4%	-13.4%	-8.5%
GM	2.98	3.80	+27.6%	-4.1%	-4.1%	0.0%
Fleet	2.98	3.99	+33.9%	-5.1%	-7.4%	-2.4%

* Based on Actual Dyno HP @ 50 MPH

** Sensitivity = -0.15

APPENDIX D

Fuel Economy, Weight, and Sales Characteristics
of Domestic and Import 49-States and
California Subfleets: 1975-1978

MODEL YEAR 1975

		49-States	California	50-States
Domestic	55/45 MPG	14.9	14.3	14.8
	Weight, lb.	4369	4181	4355
	% sales	76.5	6.2	82.8
Import	55/45 MPG	23.8	21.9	23.3
	Weight, lb.	2609	2711	2633
	% sales	13.2	4.0	17.2
Total Fleet	55/45 MPG	15.7	16.5	15.8
	Weight, lb.	4110	3602	4057
	% sales	89.7	10.3	100.0

MODEL YEAR 1976

		49-States	California	50-States
Domestic	55/45 MPG	16.7	15.3	16.6
	Weight, lb.	4309	4171	4299
	% sales	79.5	6.3	85.8
Import	55/45 MPG	25.7	24.4	25.4
	Weight, lb.	2596	2650	2608
	% sales	10.9	3.3	14.2
Total Fleet	55/45 MPG	17.5	17.5	17.5
	Weight, lb.	4102	3652	4060
	% sales	90.5	9.5	100.0

MODEL YEAR 1977

		49-States	California	50-States
Domestic	55/45 MPG	17.3	15.4	17.2
	Weight, lb.	4225	4131	4218
	% sales	77.7	5.9	83.6
Import	55/45 MPG	28.1	26.3	27.7
	Weight, lb.	2533	2600	2549
	% sales	12.7	3.8	16.4
Total Fleet	55/45 MPG	18.3	18.3	18.3
	Weight, lb.	3987	3533	3943
	% sales	90.4	9.6	100.0

MODEL YEAR 1978

		49-States	California	50-States
Domestic	55/45 MPG	18.9	16.7	18.7
	Weight, lb.	3855	3785	3848
	% sales	76.4	8.2	84.6
Import	55/45 MPG	27.2	25.6	26.8
	Weight, lb.	2539	2607	2555
	% sales	11.8	3.6	15.4
Total Fleet	55/45 MPG	19.7	18.7	19.6
	Weight, lb.	3679	3423	3649
	% sales	88.2	11.8	100.0

APPENDIX E

Sales and MPG Figures by Weight Class for Diesel Cars, 1975 through 1979

SALES DISTRIBUTION OF WEIGHT CLASSES, 50-STATES DIESEL CARS

MODEL YEAR	WEIGHT CLASS (POUNDS)								AVERAGE WEIGHT
	2250	2500	2750	3000	3500	4000	4500	5000	
1975					1.00000				3500
1976					.51190	.48810			3744
1977	.29650	.10584			.37053	.22713			3137
1978	.50794	.05543			.06288	.08871	.25901	.02603	3152
1979	.34908	.03897			.05428	.23246	.23998	.08523	3509

EPA CITY FUEL ECONOMY (1975 MPG), DIESEL CARS

MODEL YEAR	WEIGHT CLASS (POUNDS)								FLEET AVERAGE
	2250	2500	2750	3000	3500	4000	4500	5000	
1975					24.68				24.68
1976					25.27	22.17			23.65
1977	39.10	35.30			25.70	22.98			28.67
1978	39.66	35.30			26.14	22.49	21.00	19.20	29.06
1979	39.92	36.30			27.56	23.41	20.96	20.10	26.73

EPA – COMBINED CITY/HIGHWAY FUEL ECONOMY (MPG), DIESEL CARS

MODEL YEAR	WEIGHT CLASS (POUNDS)								FLEET AVERAGE
	2250	2500	2750	3000	3500	4000	4500	5000	
1975					27.52				27.52
1976					27.92	24.39			26.08
1977	44.08	39.79			28.13	24.89			31.56
1978	44.69	39.79			28.40	24.78	24.25	22.01	32.88
1979	44.11	40.22			29.56	26.17	23.93	23.11	30.01

EPA HIGHWAY FUEL ECONOMY (MPG), DIESEL CARS

MODEL YEAR	WEIGHT CLASS (POUNDS)								FLEET AVERAGE
	2250	2500	2750	3000	3500	4000	4500	5000	
1975					32.05				32.05
1976					32.03	27.79			29.81
1977	52.20	47.10			31.81	27.70			36.01
1978	52.87	47.10			31.76	28.29	29.90	26.80	39.17
1979	50.61	46.34			32.44	30.59	28.94	28.30	35.32

APPENDIX F

Attributes of Car Class Groupings for 1979 Models

Class*	Fraction of Sales	55/45 MPG	IW (lb)	CID	CID per 100lb	MPG per CID	TFC** IW	Cu.Ft.-Mi Gallon	Cu.Ft. 1000lb
AVERAGE VALUES — GASOLINE									
2-Seat/Minicomp	.0586	24.49	2634	131.2	4.98	.187	1.55	1710	26.5
Subcompact	.2993	24.02	2850	153.4	5.38	.157	1.46	2206	32.2
Compact	.0520	18.59	3659	253.7	6.93	.073	1.47	1943	28.6
Midsize	.3297	18.82	3733	274.1	7.34	.069	1.42	2218	31.6
Large	.2315	17.03	4226	295.6	6.99	.058	1.39	2286	31.8
MPG PACESETTER VALUES — GASOLINE									
2-Seat/Minicomp		29.35	2135	89.2	4.18	.329	1.60	2049	32.7
Subcompact		33.42	2264	86.3	3.81	.387	1.32	3069	40.6
Compact		20.68	3347	138.8	4.15	.149	1.44	2161	31.2
Midsize		20.11	3530	250.8	7.11	.080	1.41	2370	33.4
Large		17.25	4228	339.1	8.02	.051	1.37	2316	31.8
AVERAGE VALUES — DIESEL									
2-Seat/Minicomp	0	--	--	--	--	--	--	--	--
Subcompact	.0115	43.68	2275	90.0	3.96	.485	1.01	4011	40.4
Compact	.0041	26.75	3941	207.0	5.25	.127	0.95	2795	26.5
Midsize	.0064	25.95	4100	280.1	6.83	.093	0.94	3058	28.7
Large	.0069	23.64	4584	350.0	7.64	.068	0.92	3174	29.3

* Station Wagons grouped with parent sedans, by make/model
** TFC/IW is a measure of fuel consumption per unit weight (lower values are better); TFC/IW=100,000/(MPG x IW). See Reference 9.

APPENDIX G

Data for Light Duty Trucks (up to 6000 pounds GVWR)

Table G-1 - Trends in Sales-Weighted Fleet Fuel Economy, Light Trucks

Model Year	City	Hwy	55/45	Avg. Wt. (Pounds)	Average Engine CID
1975	12.7	17.9	14.6	4222	315
1976	14.1	19.5	16.1	4146	299
1977	16.9	22.9	19.1	3877	260
1978	16.5	22.4	18.7	3847	256
1979	15.7	20.3	17.5	3796	251

Table G-2 - Range of 55/45 Fuel Economies, 1979 50-States Light Trucks

	WEIGHT CLASS (pounds)						
	2500	2750	3000	3500	4000	4500	≥ 5000
Highest MPG	28.1	28.9	30.5	24.0	19.9	19.2	21.5*
Sales-Weighted Average	27.7	22.7	22.8	16.0	17.2	15.0	18.7
Lowest MPG	25.1	21.0	15.5	14.6	14.4	12.4	10.8

*Diesel

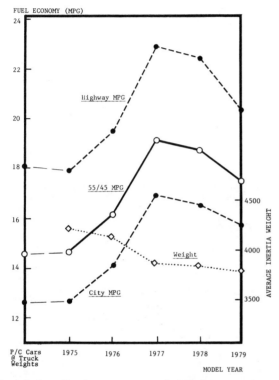

Fig. G-1 - Fuel economy and weight trends, U.S. light duty trucks

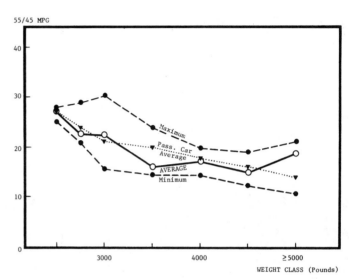

Fig. G-3 - 1979 light truck fuel economy, by test weight

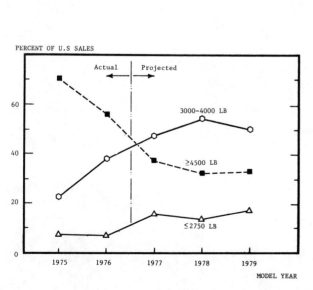

Fig. G-2 - Trend in light truck weight distributions

Table G-3 - Average 55/45 MPG and Engine CID by Weight Class, 1979 Light Trucks

		WEIGHT CLASS (pounds)						
		2500	2750	3000	3500	4000	4500	≥ 5000
American Motors	FE			17.6	15.1			
	CID			258	304			
Chrysler Corp.	FE		23.8			16.8	15.3	
	CID		135			277	302	
Ford Motor Co.	FE			27.2	18.4	17.4	14.7	
	CID			128	140	301	331	
General Motors	FE		27.6	24.5	19.5	18.1	15.0	18.7
	CID		111	111	249	272	309	349
Nissan (Datsun)	FE		23.7		15.5			
	CID		119		119			
Toyo Kogyo (Mazda)	FE			30.2				
	CID			120				
Toyota	FE		20.3		19.8	13.0	11.5	
	CID		134		134	258	258	
Volkswagen	FE				18.7			
	CID				120			
Fuji (Subaru)	FE	27.7						
	CID	97						
FLEET		27.7	22.7	22.8	16.0	17.2	15.0	18.7
		97	126	164	257	287	312	349
1978 FLEET			25.9	25.0	18.7	18.3	15.8	18.2
			121	150	209	288	325	373

Table G-4 - Average 55/45 MPG, Engine CID, and Test Weight by Vehicle Size Class, 1979 Light Trucks

	1978 MODELS:				1979 MODELS:			
	Small Pickups	Standard Pickups	Vans	Special Purpose	Small Pickups	Standard Pickups	Vans	Special Purpose
American Motors				17.0 / 272 / 3290				15.9 / 288 / 3325
Chrysler Corp.		17.9 / 260 / 4040	17.0 / 286 / 4041		23.8 / 135 / 2750	16.8 / 272 / 4000	16.6 / 285 / 4054	16.8 / 279 / 4000
Ford Motor Co.	29.1 / 125 / 3000	18.9 / 315 / 4124	16.8 / 330 / 4500	31.6 / 117 / 3000	27.2 / 128 / 3000	16.6 / 311 / 4115	15.9 / 306 / 4351	18.4 / 140 / 3500
General Motors	26.3 / 111 / 2834	16.6 / 315 / 4344	16.7 / 297 / 4352	19.5 / 185 / 3590	25.7 / 111 / 2901	15.9 / 303 / 4398	15.3 / 298 / 4439	15.7 / 229 / 3938
Nissan (Datsun)	25.9 / 119 / 2750			20.6 / 119 / 3500	23.7 / 119 / 2750			15.5 / 119 / 3500
Toyo Kogyo (Mazda)	31.7 / 110 / 3000				30.2 / 120 / 3000			
Toyota	25.0 / 134 / 2827			13.8 / 230 / 3949	20.3 / 134 / 2750			13.8* / 230 / 3949
Volkswagen			20.0 / 120 / 3500				18.7 / 120 / 3500	
Fuji (Subaru)								27.7 / 97 / 2500
FLEET	26.7 / 122 / 2853	17.3 / 311 / 4255	17.2 / 274 / 4147	16.8 / 252 / 3408	23.8 / 125 / 2832	16.2 / 302 / 4273	16.1 / 277 / 4196	16.3 / 259 / 3324

*1978 Values

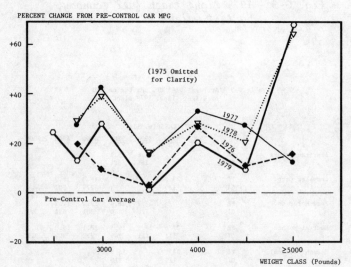

Fig. G-4 - Change from pre-control passenger car fuel economy for light duty trucks

Table G-5 - Allocation of Fuel Economy Changes from 1978 to 1979, Light Trucks

	1978 50-STATES TRUCK SWMPG	PERCENT FUEL ECONOMY CHANGE DUE TO:					1979 50-STATES TRUCK SWMPG
		SYSTEM OPTIMIZATION	TRANSMISSION MIX SHIFTS	ENGINE MIX SHIFTS	WEIGHT MIX SHIFTS	ALL CHANGES COMBINED	
AMERICAN MOTORS	17.0	(-5.0)	(0.0)	-0.4	-1.4	-6.7	15.9
CHRYSLER CORP.	17.3	-3.6	-0.9	0.9	8.2	4.3	18.0
FORD MOTOR CO.	20.6	-13.4	0.0	2.2	-1.9	-13.2	18.0
GENERAL MOTORS	17.3	-4.1	-0.9	2.3	-1.5	-4.4	16.5
NISSAN (DATSUN)	25.6	-10.5	0.0	0.9	-1.3	-10.9	22.8
TOYO KOGYO (MAZDA)	31.7	**	**	-4.6	0.0	-4.6	30.3
TOYOTA	21.3	-12.1	0.0	0.3	0.8	-11.1	19.0
VOLKSWAGEN	20.0	-7.1	-0.1	0.5	0.0	-6.7	18.7
FLEET	18.7	-7.4	-0.8	2.1	-0.9	-6.9	17.5
1978 FLEET		-1.6	0.0	-1.0	0.5	-2.1	

( ) = Estimated
** = Undefined

Table G-6 – Sales and MPG Figures by Weight Class for Light Trucks, 1975 through 1979

SALES DISTRIBUTION OF WEIGHT CLASSES, 50-STATE LT. DUTY TRUCKS

MODEL YEAR	WEIGHT CLASS (POUNDS)						AVERAGE WEIGHT
	2500	2750	3000	3500	4000	4500	≥5000	
1975		.07135	.07132	.01533	.13487	.63592	.07121	4222
1976		.06408	.06022	.02577	.28857	.52509	.03628	4146
1977		.15541	.10444	.04036	.32466	.35866	.01648	3877
1978		.13214	.12926	.05993	.35277	.30932	.01660	3847
1979	.01091	.15859	.11498	.09759	.28767	.31298	.01727	3796

1975-1976 DATA FROM PRODUCTION FIGURES

1977-1979 DATA FROM MANUFACTURERS' SALES FORECASTS

EPA 4000-MILE CITY FUEL ECONOMY (1975 MPG) BY MODEL YEAR AND WEIGHT CLASS, LT. DUTY TRUCKS

MODEL YEAR	WEIGHT CLASS (POUNDS)							FLEET AVERAGE
	2500	2750	3000	3500	4000	4500	≥5000	
1975		19.08	15.83	17.64	13.83	12.01	10.02	12.67
1976		20.61	17.42	14.24	15.86	12.81	11.17	14.12
1977		22.16	22.52	16.14	16.70	14.85	10.73	16.87
1978		23.10	22.13	16.11	16.01	13.91	15.96	16.51
1979	24.15	19.88	20.47	14.52	15.42	13.58	16.85	15.75

EPA 4000-MILE HIGHWAY FUEL ECONOMY (MPG) BY MODEL YEAR AND WEIGHT CLASS, LT. DUTY TRUCKS

MODEL YEAR	WEIGHT CLASS (POUNDS)							FLEET AVERAGE
	2500	2750	3000	3500	4000	4500	≥5000	
1975		28.22	21.10	24.84	19.44	17.07	13.82	17.90
1976		30.42	23.10	19.24	22.21	17.50	15.78	19.49
1977		31.59	30.48	21.66	22.83	19.69	15.67	22.89
1978		30.34	28.77	21.84	22.24	18.85	22.08	22.41
1979	33.72	27.55	26.51	18.23	20.13	17.13	21.53	20.35

EPA 4000-MILE COMBINED CITY/HIGHWAY FUEL ECONOMY (MPG) BY MODEL YEAR AND WEIGHT CLASS, LT. DUTY TRUCKS

MODEL YEAR	WEIGHT CLASS (POUNDS)							FLEET AVERAGE
	2500	2750	3000	3500	4000	4500	≥5000	
1975		22.33	17.84	20.29	15.89	13.86	11.43	14.59
1976		24.11	19.58	16.28	18.20	14.56	12.86	16.12
1977		25.60	25.52	18.23	19.00	16.70	12.50	19.13
1978		25.88	24.69	18.27	18.32	15.77	18.24	18.73
1979	27.69	22.73	22.80	15.98	17.23	14.98	18.67	17.53

THIS BIBLIOGRAPHY is divided into five sections:

    I. Fuel Economy Test Procedures
    II. Vehicle Usage Factors Affecting Fuel Economy
    III. Vehicle Design Factors Affecting Fuel Economy
    IV. Fuel and Lubricant Effects on Fuel Economy
    V. Analysis of Fuel Economy

The literature listed in the bibliography references additional material on automotive fuel economy. Copies of SAE papers are available in original or photocopy format from SAE. For ordering information contact the Publications Division, SAE, 400 Commonwealth Drive, Warrendale, PA 15096.

I. Fuel Economy Test Procedures

C. Marks, A. Fuel Economy Measurement Dilemma - Certification Testing vs. Customer Driving, SAE Paper 780938, 1978.

Many factors can be cited which produce differences between the fuel economy values obtained during the exhaust emission certification process and the economy experienced by car owners. Admittedly, all laboratory tests are compromised by many assumptions, approximations, and practical test limitations. The main value of the EPA test procedure is that it has provided a uniform test method for all manufacturers which produces vast amounts of comparative fuel economy information. Changes to the procedure to make it more "representative" have reduced its usefulness for comparisons to previous years.

The concept of labeling cars with a "representative" fuel economy value is certain to result in some customer misinformation and dissatisfaction. At best, current labeling methods can be expected to indicate real vehicle differences only when label values differ by more than 2 mi/gal (0.85 km/l). Furthermore, wide variations in customer fuel economy (ranging up to 15 mi/gal) for the same EPA label value are bound to make some people regard the label values as misleading.

Changes in new car fuel economy have a significant impact upon future fuel demand projections. A stable fuel economy measurement procedure and an understanding of the factors which relate certification to average customer economy are needed to reduce the uncertainties in such a projection.

N. J. Sheth and T. I. Rice, Identification, Quantification and Reduction of Sources of Variability in Vehicle Emissions and Fuel Economy Measurements, SAE Paper 790232, 1979.

A major problem in Vehicle Exhaust Emissions and Fuel Economy Testing has been the variability in the measurements. An extensive test program was undertaken to identify and quantify the sources of variation. The test program was designed using four test vehicles on five CTE-50 Clayton Dynamometer cells whereby four repeats for each combination were provided for a total of 80 CVS-C/H Emissions and F.E. Tests. During each test a total of 23 test variables were monitored and recorded in real time. Nine other variables were also observed for a total of 32 variables: 5 vehicle related variables, 16 engine related, 4 dynamometer variables, 3 driver related, and 4 environmental variables.

The paper reports various results of this experiment and describes an instrumentation package that was developed for "on-line" data recording. A brief discussion of the software used for "off-line" processing of the voluminous data as well as several statistical analysis techniques which were developed especially for identifying and quantifying the sources of variation, are included. The benefits which result from reduced test variability, namely lower test cost because of improved test efficiency and increased confidence in results, are pointed out.

W. W. Wiers, G. W. Niepoth, and T. D. Hostetter, Emission and Fuel Economy Measurement Improvements, SAE Paper 790233, 1979.

A program was initiated to improve the emission and fuel economy measurement accuracy and test cell to test cell correlation. Improvements were made to the Constant Volume Sampling System, electric dynamometer, and instrument calibration ranges, and system checks were initiated to improve the accuracy of the bag emissions, modal emissions, calculated and measured fuel economy. Unique emission and fuel economy problems associated with gasoline and diesel testing were studied and resolutions effected when possible.

II. Vehicle Usage Factors Affecting Fuel Economy

N. Ostrouchov, Effect of Cold Weather on Motor Vehicle Emissions and Fuel Economy, SAE Paper 780084, 1978.

The effect of soaking temperature on exhaust emissions has been studied using a variety of automobiles representing three different emission control levels and testing them at ambients of 20°C down to -30°C (60°F to -22°F).

It was found that emissions of the three gaseous pollutants demonstrated a mild power relationship with ambient (soaking) temperatures. All regulated pollutants and fuel consumption were higher at -30°C than at 20°C: hydrocarbons (HC) — 3.5 to 9.2 times; carbon monoxide (CO) — 2.4 to 6.4 times; oxides of nitrogen (NO_x) — only 1.1 to 1.4 times; and fuel consumption 1.2 to 1.8 times higher. Analysis of the data has indicated that HC and CO emissions from the cold start phase of the Federal test were the most sensitive to soaking temperature. With NO_x emissions the soaking temperature sensitivity was fairly constant throughout the three phases of the Federal test.

The data also indicate that the temperature sensitivity of both fuel economy and, to a lesser extent emissions, is a function of inertia weight.

R. Herman, R. G. Rule, and M. W. Jackson, Fuel Economy and Exhaust Emissions Under Two Conditions of Traffic Smoothness, SAE Paper 780614, 1978.

The potential fuel economy and exhaust emission benefits that might be obtained by smoothing the flow of traffic have been investigated. Substantial improvements in fuel economy and reductions in exhaust emissions are possible if the flow of traffic is smoothed. Traveling during the smooth flow conditions of the early morning (4 am) as compared to travel on the same urban route during highly congested flow (5 pm rush hour) resulted in a fuel economy improvement of 31% for hot-starts and 35% for cold-starts. Also, traveling during smooth flow conditions resulted in reductions in HC, CO, and NO_x emission levels of 54%, 52%, and 2% respectively for hot-starts and 35%, 52%, and 13% respectively for cold-starts. The reported results were obtained by simulating traffic conditions on a chassis dynamometer.

D. H. Davis, The Effect of Restorative Maintenance on the Relationship Between Short Test and Federal Test Procedure Emission Test Results, SAE Paper 780619, 1978.

The Restorative Maintenance Program conducted by the Environmental Protection Agency in late 1976 and early 1977 provides short test and Federal Test Procedure data on a large fleet of relatively new consumer owned automobiles. The program included testing of 300 vehicles in Chicago, Detroit, and Washington, D.C. The vehicles that indicated a need for maintenance were repaired and retested by the FTP and the five short tests being considered by the agency for the Federal warranty regulations.

These data are examined by conventional regression and correlation methods, contingency table analysis and maintenance effectiveness criteria.

The conclusions of the study indicate that while the mathematical correlation coefficients are quite low for most of the tests, all five tests are effective in identifying vehicles in need of maintenance and significant hydrocarbon and carbon monoxide emissions reductions can be achieved at relative low vehicle failure rates. The transient mode tests are also effective in identifying oxides of nitrogen reduction potentials.

N. E. South and R. Raja, In-Service Fuel Economy, SAE Paper 790227, 1979.

Ford Motor Company surveyed Ford management personnel who drive Ford Motor Company lease vehicles. From the responses, in-use fuel economy data were computed on over 10,600 1978 model year cars. Analyses of the data are presented which include: fuel economy summary statistics; regressions of fuel economy ratings versus in-use fuel economy; measures of the ability of EPA ratings to rank in-use fuel economy; and the influence of car size class or transmission type on fuel economy regression and correlation. Wide ranges of fuel economies were found for each vehicle model type. The analysis shows no significant trend by car size class in the relationship between in-use fuel economy and metro highway fuel economy ratings.

N. Ostrouchov, Effect of Cold Weather on Motor Vehicle Emissions and Fuel Consumption — II, SAE Paper 790229, 1979.

The effect of soaking temperature on exhaust emissions and fuel consumption was investigated using a variety of automobiles representing different emission control levels including diesel engine powered vehicles. Tests were performed at soaking and ambient temperatures of 20°C down to -20°C (68°F to -4°F).

It was found that emissions and fuel consumption are dependent on soaking temperature. Hydrocarbon and carbon monoxide emissions were higher at -20°C than at 20°: hydrocarbon (HC), 1 to 4 times; carbon monoxide (CO), 1 over 3 times. The smallest increase of 1 to 1.04 time belonged to vehicles equipped with diesel engines. Nitrogen oxides (NO_x) emissions were higher or lower at -20°C than at 20°C depending on emission control technologies -0.75 to 1.11 times. Analysis of the data has indicated that HC and CO emissions from the cold start phase of the Federal test were the most sensitive to soaking temperature. With NO_x emissions the soaking temperature sensitivity was fairly constant throughout the three phases of the test.

It appears that temperature sensitivity of fuel consumption in vehicles equipped with diesel engines and lean burn gasoline engines is considerably lower in comparison to the vehicles equipped with other control technologies, and is higher at -20°C than at 20°C: for diesel engines and lean burn -1.15 times; for other vehicles -1.55 times. The data also indicate that the temperature sensitivity of fuel consumption is a function of inertia weight.

III. Vehicle Design Factors Affecting Fuel Economy

A. Morelli, L. Fioravanti, A. Cogotti, The Body Shape of Minimum Drag, SAE Paper 760186, 1976.

After a short review of the work done in the past to reduce the aerodynamic drag of a body moving in the vicinity of the ground, a new theoretical method is developed in order to determine the shape of the body when a certain lift distribution is imposed. Considerations on the induced drag suggest that the total lift is zero as also should be zero the pitching moment for stability reasons. These conditions together with that of gradual variation of the area and shape of the cross sections of the body lead to the determination of the basic shape of the body. A model was realized and tested on the Pininfarina wind tunnel. The results show a good substantiation of the theory and a very low drag coefficient. The dimensions of the model where such as to achieve the actual Reynolds numbers of motor cars. Energy implications of a reduction on the aerodynamic drag are also indicated.

G. W. Carr, Reducing Fuel Consumption By Means of Aerodynamic 'Add-On' Devices, SAE Paper 760187, 1976.

The scope for reduction of aerodynamic drag of passenger cars by 'add-on' devices has been examined by MIRA in a study sponsored by the U.K. Government. On a car of the popular 'hatch-back' style the drag was reduced by up to 31 per cent by these means without detriment to other aerodynamic characteristics. Fuel consumption tests with a similar car showed that with a combination of 'add-on' devices giving a 14 per cent drag reduction an overall fuel saving of 6 per cent could be obtained in mixed road conditions.

A. Ciccarone, C. Antonini, and U. Virgilio, Fuel Consumption in European Passenger Cars Powered by Gasoline, Diesel, and Direct Injection Stratified Charge Engines, SAE Paper 760796, 1976.

A comparison has been made between the fuel consumption of a typical European passenger vehicle powered by gasoline, Diesel and Stratified Charge Engines having performance characteristics allowing equal vehicle acceleration. The study shows that on the average the gasoline and Diesel fuel consumption are similar, while with the S.C.E. a benefit of from 7–20% was observed. Compliance with the European regulations is the main exhaust emission constraint within the study that has been conducted. But, when available, exhaust emissions data for the U.S.E.P.A. "FTP" cycle have also been quoted.

E. A. Rishavy, S. C. Hamilton, J. A. Ayers, and M. A. Keane, Engine Control Optimization for Best Fuel Economy with Emission Constraints, SAE Paper 770075, 1977.

This new approach for improving fuel economy uses computer programs to optimize and tailor an engine's fuel, EGR and spark control in the laboratory. New forms of engine and vehicle test data are used as inputs. This includes a simple simulation of the catalytic converter. The emission engineer is in control of the process via a special interactive program at a computer terminal. He combines his know-how with the computer programs to create a feasible engine control calibration for vehicle evaluation. The programs can also be used to study trade-offs of optimized fuel economy vs. emissions for a vehicle. Although presently limited to warmed up operation, the optimization procedure has proved to be valid and useful.

D. E. Hatch and J. L. Jorstad, Aluminum Structural Castings Result in Automobile Weight Reduction, SAE Paper 780248, 1978.

High integrity aluminum castings are potential replacements for cast iron in current vehicle weight reduction programs. Domestically, several cast aluminum structural-type components are already realities, saving weight and contributing to improved fuel economy; wheels, brake drums, master brake cylinders and power steering housings. In Europe, suspension components, wheel hubs and disc brake calipers are cast in aluminum for some car models, indicating the functional and economic feasibility of such parts.

Alloy and process technology already exist to enable production of reliable, high strength aluminum castings. Domestic automotive product engineers are urged to carefully consider and thoroughly test such aluminum castings along with the many other weight reduction possibilities currently being investigated.

K. Yamamoto and T. Muroki, Development on Exhaust Emissions and Fuel Economy of the Rotary Engine at Toyo Kogyo, SAE Paper 780417, 1978.

The fuel economy was sharply improved from our

1976 model rotary engine cars mainly through modifications to the engine and improvements in the reactivity of the thermal reactor system.

The progress of our development on these items of improvement and their effects are presented in this paper.

Our advance program includes several development items in addition to the above, and the progress of development on these items and their future prospects are also presented.

L. T. Wong and W. J. Clemens, Powertrain Matching for Better Fuel Economy, SAE Paper 790045, 1979.

This paper discusses current powertrain matching methodology and its applications. Modular computer programs, which model each component of the vehicle/powertrain system, simulate the vehicle over specified driving cycles to project fuel economy and performance. Fuel economy opportunities due to better powertrain matching are discussed, including optimum engine sizing, torque converter matching, transmission gear ratio spacing and shift scheduling, axle ratio and vehicle weight effect. An emission projection technique utilizing time weighted engine speed/load points generated either by experiment or by analytical models is used to quantify fuel economy/emissions trade-offs.

IV. Fuel and Lubricant Effects on Fuel Economy

C. E. Goldmann, A Synthesized Engine Oil Providing Fuel Economy Benefits, SAE Paper 760854, 1976.

Utilizing extensive synthesized hydrocarbon fluid (SHF) technology, a superior quality light viscosity automotive engine oil has been developed providing optimized engine performance. This product has been shown to provide significant fuel economy benefits while maintaining excellent performance in the areas of oil economy, engine cleanliness and wear protection. In the past, this level of performance has not been possible using conventionally refined mineral oils.

The superior performance of this product is documented in extensive laboratory engine, chassis dynamometer and field tests. Particular emphasis is placed on extended duration API Sequence tests and EPA fuel economy testing.

B. M. O'Connor, W. S. Romig, and L. F. Schiemann, Energy Conservation Through The Use of Multigraded Gear Oils In Trucks, SAE Paper 770833, 1977.

Studies of selected automotive gear lubricants in heavy truck tandem axles and transmissions have revealed improvements in fuel economy associated with the viscosity of lubricants tested (grades 75W, 75W-90, and 80W-140). The testing included a heavy truck on-highway fleet test and test track operation.

Standard laboratory gear tests on light viscosity monograde (SAE 75W) oils indicate that oils of this type may be deficient in EP protection. Combined observations show that there may be a critical balance between axle lubricant fuel economy benefits and axle durability in field service.

P. A. Willermet and L. T. Dixon, Fuel Economy - Contribution of the Rear Axle Lubricant, SAE Paper 770835, 1977.

Axle dynamometer tests were carried out to evaluate the effects of rear axle lubricant viscosity-temperature behavior and frictional characteristics on vehicle fuel economy. Using a Ford 9 inch 2.75:1.0 ratio axle, a set of input speed and load conditions was selected to permit simulation of the CVS and EPA highway driving cycles. Lubricant temperature was varied from -30°C to 100°C to simulate seasonal climatic effects. Data obtained for three lubricants differing in viscosity-temperature behavior were interpreted assuming a lubrication model including both elastohydrodynamic and mixed lubrication conditions. From these data, fuel economy projections were made using a vehicle simulation computer program.

The results predict that improvements in vehicle fuel economy on the order of a few percent can be made at low temperatures by use of low viscosity synthetic lubricants, but only small effects are projected for the CVS and EPA highway cycles. Insufficient data were obtained to quantify the contribution of friction modifiers to fuel economy. However, the data indicate under what viscosity values and input conditions friction modifiers will become effective and suggest areas for future evaluation.

W. E. Waddey, H. Shaub, J. M. Pecoraro, and R. A. Carley, Improved Fuel Economy Via Engine Oils, SAE Paper 780599, 1978.

A fuel-efficient passenger car engine oil has been developed that showed an average of 4.6% improvement in fuel economy compared with several premium, SAE 10W-40 oils (API Engine Service Classification SE) in a road fleet test. In the combined Environmental Protection Agency city/highway testing, this oil gave 5.5% better fuel economy on average than an SE premium 10W-40 oil. The excellent overall performance of this lubricant was confirmed by ASTM Engine Sequence tests and evaluation in severe taxicab operations.

The new 10W-40 oil, which incorporates friction-reducing properties, is formulated with petroleum base stocks and known additives, many of which are conventionally used. It is expected that advances in friction-related technology developed in this work may lead to oils with even higher levels of fuel economy improvement in the future.

C. A. Passut and R. E. Kollman, Laboratory Techniques for Evaluation of Engine Oil Effects on Fuel Economy, SAE Paper 780601, 1978.

Test methods to evaluate the effects of engine oil formulations on fuel economy were evaluated using a variety of experimental and commercial oils. The oils were tested in a motor driven engine which showed reduced power requirements for low viscosity oils at low temperatures and friction reducing additives at high temperatures. A single cylinder engine generator was able to show trends for improved fuel economy with friction reducing additives. Automotive engine dynamometer tests were run at several operating conditions and showed fuel economy improvements for low viscosity oils and for oils containing friction reducing additives. Data from vehicles tested on an All Weather Chassis Dynamometer provided an excellent simulation of field service. The laboratory test procedures correlated well with results obtained in the chassis dynamometer.

J. L. Bascunana and R. C. Stahman, Impact of Gasoline Characteristics on Fuel Economy, SAE Paper 780628, 1978.

This paper discusses information relevant to the impact of the characteristics of gasoline on fuel economy of motor vehicles. In particular, the paper analyzes the impact of the density, volatility, octane rating, and cleanliness of gasoline, as well as the impact of its additives. It also studies the relationship between fuel characteristics, fuel metering, and fuel economy. Furthermore, the paper serves to help explain some of the differences that may be found between the results of the EPA fuel economy tests and the values observed by vehicle owners in normal service.

The paper is based on the analysis of the available literature on the subject, as well as on specific information submitted to the EPA by companies engaged in petroleum refining, fuel additives production, and automobile manufacturing.

J. E. Riester and W. B. Chamberlin, A Test Track Comparison Of Fuel-Economy Engine Oils, SAE Paper 790213, 1979.

A fuel economy track test comparing two experimental oils to two commercial fuel-efficient engine oils was conducted in 1978 compact sedans. Three of the test lubricants consistently provided a significant decrease in fuel consumption compared to one of the commercial fuel efficient lubricants. All four test lubricants provided the most improvement in fuel economy after the first 2 400 km (1 500 mile).

An experimental formulation from the track test was used as a reference to explore the effects of performance additive, viscosity improver, and viscosity grade on fuel economy. The comparisons were made in different makes of cars using modified EPA fuel economy procedures. Performance additive and viscosity improver selection directionally reduced fuel consumption in the test cars. The response to viscosity grade changes was not consistent among the test vehicles.

V. Analysis of Fuel Economy

A. C. Malliaris, T. Trella, and H. Gould, Engine Cycle Simulations and Comparisons to Real Engine Performance, SAE Paper 760155, 1976.

An analytical framework is assembled, suitable for the quantitative evaluation of engine performance (fuel consumption, WOT, NOx emissions) and the appreciation of engine performance sensitivities and trade-offs. Emphasis is placed on complete, conventional spark-ignition engines in production. The framework is assembled primarily from existing and well documented analyses that deal effectively with various specific aspects of the internal combustion engine. A new and non-conventional approach is followed in the treatment of engine breathing dynamics. The engine-manifold system is treated and documented as a sequence of many small but finite control volumes, each obeying individually the field form of conservation equations. The qualifications of the analytical framework are evaluated by an extensive comparison of analytical results to dynamometer test results of engines with fairly well known design and control parameters, which are also presented. Results of two production (one 8- and one 4-cylinder) engines are used over a wide range of loads and speeds. Extensive comparisons are illustrated of BMEP, BSFC, BSNOX and WOT air flow. Agreement ranging from good to fair is observed depending on operating conditions, with the comparison becoming poorer at very low loads and high EGR rates. Various factors, limiting the resolution of applications are discussed.

W. K. Juneja, W. J. Kelly, and R. W. Valentine, Computer Simulations of Emissions and Fuel Economy, SAE Paper 780287, 1978.

A simulation of the exhaust emission testing system including a vehicle, dynamometer and driver on General Motors' analog-digital hybrid computer is described. The simulation predicts instantaneous and average results for hydrocarbons (HC), carbon monoxide (CO), oxides of nitrogen (NO_x), carbon dioxide (CO_2), and fuel economy over a predefined speed schedule such as the Environmental Protection Agency's (EPA's) city and highway schedules.

The simulation facilitates the study of sensitivity of several parameters which influence emissions and fuel economy. The usefulness of the simulation to study the effect of test parameters, test operating conditions and product changes such as the engine, transmission, axle, etc., on emissions and fuel economy is outlined. Results from some of these sensitivity studies are also presented.

H. Oetting, Impact of Emission Standards on Fuel Economy and Consumer Attributes, SAE Paper 790230, 1979.

Volkswagen had a contract with DOT-TSC to generate a data base which would yield a consistent set of information about the factors affecting fuel economy of engines especially the impact of different emission standards and their influence on unregulated emissions and consumer attributes as well. Main important standards were 0.41/3.4/1.0 and 0.41/3.4/0.4 gpm HC/CO/NO_x. It was found that for setting engineering goals at least a factor of 0.5 was necessary regarding HC and CO, and a factor of 0.25 regarding NOx. The concepts achieving engineering goals mainly consistent with Volkswagen engine families are described. All fuel economy, emissions, noise, startability, driveability, acceleration performance, and gradeability data obtained with these engine families are included.